T0319563

Delay-Adaptive Linear Control

PRINCETON SERIES IN APPLIED MATHEMATICS

Ingrid Daubechies (Duke University); Weinan E (Princeton University); Jan Karel Lenstra (Centrum Wiskunde & Informatica, Amsterdam); Endre Süli (University of Oxford), Series Editors

The Princeton Series in Applied Mathematics features high-quality advanced texts and monographs in all areas of applied mathematics. The series includes books of a theoretical and general nature as well as those that deal with the mathematics of specific applications and real-world scenarios.

For a full list of titles in the series, go to https://press.princeton.edu/catalogs/series/title/princeton-series-in-applied-mathematics.html.

Delay-Adaptive Linear Control

Yang Zhu and Miroslav Krstic

PRINCETON UNIVERSITY PRESS

PRINCETON AND OXFORD

Copyright © 2020 by Princeton University Press

Published by Princeton University Press
41 William Street, Princeton, New Jersey 08540

In the United Kingdom: Princeton University Press
6 Oxford Street, Woodstock, Oxfordshire, OX20 1TR

All Rights Reserved

Library of Congress Cataloging-in-Publication Data

Names: Zhu, Yang, 1988–author. | Krstić, Miroslav, author.
Title: Delay-adaptive linear control / Yang Zhu and Miroslav Krstic.
Description: Princeton : Princeton University Press, [2020] | Includes
    bibliographical references and index.
Identifiers: LCCN 2019029728 (print) | LCCN 2019029729 (ebook) |
    ISBN 9780691202549 (hardback) | ISBN 9780691203317 (ebook)
Subjects: LCSH: Adaptive control systems—Mathematical models. | Time delay
    systems—Mathematical models. | Linear control systems—Mathematical
    models. | Linear time invariant systems—Mathematical models. | Differential
    equations, Linear. | Engineering mathematics.
Classification: LCC TJ217 .Z57 2020 (print) | LCC TJ217 (ebook) |
    DDC 629.8/32—dc23
LC record available at https://lccn.loc.gov/2019029728
LC ebook record available at https://lccn.loc.gov/2019029729

British Library Cataloging-in-Publication Data is available

Editorial: Susannah Shoemaker and Lauren Bucca
Production Editorial: Sara Lerner
Text and Jacket Design: Lorraine Doneker
Production: Jacquie Poirier
Copyeditor: Wendy Washburn

Jacket Credit: Stojan Ćelić, *Sinopsis*, 1986

This book has been composed in LaTeX

Printed on acid-free paper. ∞

press.princeton.edu

Printed in the United States of America

10 9 8 7 6 5 4 3 2 1

Contents

List of Figures and Tables

Figures

Tables

Preface

THIS YEAR IS the sixtieth anniversary of Otto J. Smith's 1959 publication of a control design idea for the compensation of actuator delays, which is commonly referred to as the Smith predictor. As actuator and sensor delays are among the most common dynamic phenomena in engineering practice, and have adverse impact on stability and transient performance under feedback, the Smith predictor finds a widespread application in industry and is perhaps second in popularity only to proportional integral differential (PID) controllers, especially among process control practitioners. Since the Smith predictor is limited to plants that are open-loop stable, improvements have been needed, and generalized predictors of various kinds have emerged over the last six decades. Research on predictors has been occurring in waves, the most notable being in the 1970s for linear systems and over the last ten years for nonlinear systems.

In parallel, the subject of adaptive control has been one of the most important in the field of control theory and engineering. Modeling based on first principles gives rise to model structures that may faithfully capture the system dynamics, but the parameters in such first-principles models are often unknown, or vary broadly over time and with the system's operation and wear. Adaptive control, namely, a simultaneous real-time combination of system identification (the estimation of the model parameters) and feedback control, is one of the most important needs of a control engineer. Adaptive control has occupied control research for about as long as delay systems, with the most intense period of innovation being over the last three decades of the last century, namely, from the 1970s through the 1990s, with the 1970s dedicated to pioneering the methods of adaptive control of linear systems, the 1980s dedicated to the studies of robustification and stability of adaptive systems, and the 1990s dedicated to the development of adaptive controllers for nonlinear systems. Continued, very active research in adaptive control builds on the foundations laid in the 1990s and proceeds in the direction of various generalizations, including those for multi-agent systems.

In the development of adaptive control, the method of integrator backstepping, developed over the 1990s, played the key role, enabling Lyapunov-based and other designs for systems with arbitrary orders and relative degrees. More importantly, backstepping resulted for the first time in adaptive control designs that provide transient performance guarantees, a challenge facing adaptive control for decades and resolved using adaptive backstepping with tuning functions in the early 1990s.

In adaptive control, input or sensor delays have traditionally been viewed as perturbations—effects that may challenge the robustness of the adaptive controllers and whose ignored presence should be studied with the tools for robustness analysis and redesign.

Likewise, or analogously, in traditional predictor feedback the model needs to be known, including its parameters, and the predictor's focus is on compensating for the delay in the known models.

The simultaneous presence of unknown plant parameters and actuator/ sensor delays poses challenges that arise straight out of the control practice, and that are not addressed by conventional predictor and adaptive control methods.

The turning point for the development of the ability to simultaneously tackle delays and unknown parameters was the introduction of the partial differential equation (PDE) *backstepping* framework for parabolic PDEs and the resulting backstepping interpretation of the classical predictor feedback. Similar to the role that finite-dimensional adaptive backstepping had played in the development of adaptive control in the 1990s, PDE backstepping of the 2000s furnished the Lyapunov functionals needed for the study of stability of delay systems under predictor feedback laws. It also enabled the further design of adaptive controllers for systems with delays and unknown parameters from the late 2000s onward.

When plant parameters are unknown and significant delays are present, multiple possible scenarios arise. If the delay is large but known, while classical robust redesigns do not suffice and a sophisticated predictor-based redesign is needed, the problem is not as challenging as when the delay is unknown. One could argue that the situation with a highly uncertain delay, even if the plant parameters are known, is a more difficult problem than all of the delay-free adaptive control of the 1970–1990s. Hence, in order to develop delay-adaptive controllers, as generalizations of the traditional versions of both adaptive control and of predictor feedback, much innovation was needed in both the analysis and the synthesis of adaptive and predictor-based feedback laws.

Backstepping—of both the ordinary differential equation (ODE) and PDE varieties—has played the key role in developing delay-adaptive controllers over the last decade or so. Even when the plant, with an input delay, is linear, the resulting adaptive predictor feedback must be nonlinear and the closed-loop system must consequently be nonlinear and infinite-dimensional. Backstepping has enabled the construction of the needed nonlinear infinite-dimensional analysis and design tools to synthesize and analyze the delay-adaptive controllers.

While adaptive backstepping designs have already resulted in adaptive controllers for PDEs, first of the parabolic kind, in the late 2000s, and most recently of the hyperbolic kind, this book restricts attention to adaptive control of ODEs with large input delays. Since the delay is a special class of a hyperbolic PDE and the plant is an ODE, the focus of this book can be viewed as being on a special cascaded class of hyperbolic PDE systems and linear ODEs with unknown parameters.

What Does the Book Exactly Cover?

While restricting attention to linear ODEs with input delays, this book is comprehensive in the treatment of this problem. A finite-dimensional linear time-invariant (LTI) system with discrete or distributed input delays may come with the following five types of basic uncertainties:

- unknown delay,
- unknown delay kernel,
- unmeasured actuator state,
- unknown parameters in the finite-dimensional part of the plant,
- unmeasured state of the finite-dimensional part of the plant.

We construct control laws for both single-input LTI systems with input delay and multi-input LTI systems with distinct input delays, respectively. A variety of adaptive techniques to deal with different uncertainty combinations are synthesized into a unified framework from which the reader can make his or her own selection to address a vast class of relevant problems.

The adaptive control schemes in this book are based on the PDE framework, in which the systems with input delays are treated as ODE-PDE cascade systems with boundary control. When all information about the plant and the delay is known, the PDE-backstepping design for LTI-ODEs with actuator delays recovers the classical predictor designs (finite spectrum assignment, modified Smith predictor, reduction approach). The PDE-based approach yields backstepping-based Lyapunov-Krasovskii functionals that make the control design constructive and enables stability analysis with quantitative estimates. More importantly, the PDE-based approach enables the extension of predictor feedback design to adaptive control for systems with various uncertainties in delays, parameters, and states.

The book is divided into three parts. Part I is devoted to adaptive control of uncertain single-input LTI systems with discrete input delay, and the results are easily (or even trivially) extended to systems with multiple inputs when the delay is the same in all the input channels. In contrast, part II is devoted to adaptive control of uncertain multi-input LTI systems with distinct discrete input delays. Finally, part III contributes to adaptive and robust control of uncertain LTI systems with distributed input delays. Each of parts I, II, and III starts with a chapter (chapter 2 in part I, chapter 6 in part II, and chapter 10 in part III) that provides a basic PDE-based predictor feedback framework for single-input and multi-input uncertainty-free systems. For readers in the field of control of delay systems who are not familiar with the PDE-based framework, the control schemes in chapters 2, 6, and 10 are presented in three alternative but equivalent notations—standard ODE delay notation, transport PDE notation, and rescaled unity-interval PDE notation.

Chapters 3 to 5 in part I are dedicated to adaptive control for uncertain single-input systems with discrete input delay, whereas chapters 7 to 9 in part II are dedicated to adaptive control for uncertain multi-input systems with distinct

Table 1. Uncertainty Collections of LTI Systems with Input Delays

Organization of the Book	Delay	Delay Kernel	Parameter	ODE State	PDE State
Chapters 2 and 6	known	—	known	known	known
Chapter 3 (3.2)	known	—	known	*unknown*	known
Chapters 3 (3.3) and 7 (7.2)	*unknown*	—	known	known	known
Chapter 7 (7.3)	*unknown*	—	known	*unknown*	known
Chapters 3 (3.4) and 8	*unknown*	—	known	known	*unknown*
Chapters 3 (3.5) and 7 (7.4)	*unknown*	—	*unknown*	known	known
Chapters 3 (3.6) and 9 (9.2)	*unknown*	—	known	*unknown*	*unknown*
Chapters 3 (3.7) and 9 (9.3)	*unknown*	—	*unknown*	known	*unknown*
Chapters 4 and 5 (5.2)	*unknown*	—	*unknown*	*unknown*	known
Chapter 5 (5.3)	*unknown*	—	*unknown*	*unknown*	*unknown*
Chapter 10	known	known	known	known	known
Chapters 11 (11.1) and 12 (12.1)	*unknown*	known	known	known	known
Chapter 11 (11.2)	*unknown*	*unknown*	known	known	known
Chapters 11 (11.3) and 12 (12.2)	*unknown*	*unknown*	*unknown*	known	known
Chapters 11 (11.5) and 12 (12.3)	*unknown*	*unknown*	known	*unknown*	known
Chapters 11 (11.4) and 12 (12.4)	*unknown*	*unknown*	known	known	*unknown*

discrete input delays. Chapter 11 in part III focuses on adaptive and robust control for uncertain single-input systems with distributed input delay, whereas chapter 12 in part III concentrates on adaptive and robust feedback for uncertain multi-input systems with different distributed input delays. Different collections of five aforementioned uncertainties (delay, delay kernel, parameter, ODE state, and PDE state) are tackled by a variety of control techniques. The basic idea of certainty-equivalence-based adaptive control is using an estimator (a parameter estimator or a state estimator) to replace unknown variables in the nominal control scheme and designing update laws according to Lyapunov-based analysis. Different combinations of uncertainties considered in the book result in distinct control designs and they are summarized in table 1, from which the readers can make their own selections to address a vast class of relevant problems.

Whom Is the Book For?

The potential readers of the book are all researchers working on control of delay systems and adaptive control of uncertain systems—engineers, graduate students, and mathematicians.

Mathematicians and specialists in control theory with interest in the broad area of feedback of linear systems, and adaptive control of uncertain systems in particular, may be interested in the book because it tackles a variety of different uncertainty combinations with different techniques. These problems may stimulate further research on the stabilization of ever-expanding classes of uncertain finite-dimensional or infinite-dimensional systems.

All of our designs are given by explicit formulas. As a result, the book should be of interest to any engineer in realistic environments who has faced diverse uncertainties and delay-related challenges and is concerned with actual implementations: chemical engineers and process dynamics researchers, electrical and computer engineers (in delayed telecommunication systems and networks), mechanical and aerospace engineers (working on combustion systems and machining), and civil/structural engineers (in the presence of long and varying delays in traffic flow dynamics). All of them may find the book useful because it provides systematic control synthesis techniques, as well as analysis tools for establishing stability.

Where Does This Book Fit in the Literature on Adaptive Control, PDE Control, and Delay Systems?

Both adaptive control and control of PDEs are challenging fields. In adaptive control, the challenge comes from the need to design feedback for a plant whose dynamics may be highly uncertain (plant parameters are highly unknown) and open-loop unstable, requiring control and learning at the same time. In the control of PDEs, the challenge lies in the infinite-dimensional essence of the system dynamics. As a result, the adaptive control problem for PDEs is not an easy task and its difficulties are even greater than the sum of the difficulties of the two components, namely, (1) adaptive control for ODEs and (2) control of PDEs.

The classical books for adaptive control, [3, 48, 57, 89, 108, 117, 126], mainly focus on uncertain linear or nonlinear systems whose dynamics are represented by ODEs. These conventional adaptive control methods for ODEs cannot be trivially used to address uncertain delay systems whose dynamics are infinite-dimensional. To deal with parametric uncertainty in PDEs (and in their boundary conditions), a systematic summary of adaptive boundary control schemes for unstable parabolic PDEs is presented in [122]. However, the systems with actuator delays belong to a special class of convective/first-order hyperbolic PDEs (also called transport PDEs) rather than parabolic PDEs. Adaptive control of hyperbolic PDEs [1] requires additional tools as compared to those presented in [122]. By modeling the actuator state under input delay as a transport PDE and regarding the propagation speed (the delay's reciprocal) as a parameter in the infinite-dimensional part of the ODE-PDE cascade, we develop adaptive control of delay systems in part III of the book [80]. But the results in [80] are limited to uncertain single-input systems with discrete input delay.

Going far beyond [80], this book not only extends adaptive control of single-input systems to multi-input systems with distinct discrete input delays, but also solves the adaptive and robust control problems of linear systems with distributed input delays.

Sizable opportunities exist for further development of the subject of delay-adaptive control, including systems with simultaneous input and state delays, nonlinear systems with state-dependent delays, and PDE systems with delays.

Nanyang Technological University, Singapore   *Yang Zhu*
University of California, San Diego, USA   *Miroslav Krstic*
June 2019

Acknowledgments

THE AUTHORS WOULD like to thank Delphine Bresch-Pietri for her contribution in chapter 3, Daisuke Tsubakino and Tiago Roux Oliveira for their contributions in section 6.1, and Nikos Bekiaris-Liberis for his pioneering work for chapter 10. Yang Zhu thanks Nikos Bekiaris-Liberis for his generous help in editing the appendix. Yang Zhu is particularly grateful to Delphine Bresch-Pietri and Daisuke Tsubakino for their unselfish exchanges of ideas, and explanations of technical details about their respective work on adaptive control and multi-input delay systems.

List of Abbreviations

LTI	Linear time-invariant
ODE	Ordinary differential equation
PDE	Partial differential equation
PID	Proportional integral differential
NCS	Networked control system
LMI	Linear matrix inequalty
3D	Three-dimensional

Chapter One

Introduction

1.1 Time-Delay Systems

Time-delay systems, also called systems with aftereffect or dead-time, hereditary systems, equations with deviating argument, or differential-difference equations [42], are ubiquitous in practice. Some representative examples are found in

- chemical industry: rolling mills, milling processes, cooling systems, combustion systems, chemical processes,
- electrical and mechanical engineering: networked control systems, teleoperation, robotic manipulators, unmanned aerial vehicles,
- biomedical engineering: 3D printing/additive manufacturing, neuromuscular electrical stimulation,
- management and traffic science: traffic flow, supply chain, population dynamics.

The most common forms of time delay in dynamic phenomena that arise in engineering practice are actuator and sensor delays [133, 134]. Due to the time it takes to receive the information needed for decision-making, to compute control decisions, and to execute these decisions, feedback systems often operate in the presence of delays. As shown in figure 1.1, actuator and sensor delays are involved in feedback loops.

Another big family of time delays are transmission delays in networked control systems (NCS). As the name implies, NCSs are systems where the plant is controlled via a communication network. The typical feature of an NCS, as illustrated in figure 1.2, is that information (plant output, control input, etc.) is exchanged through a network among system components (sensor nodes, controller nodes, actuator nodes, etc.). In comparison with traditional feedback control systems, where the components are usually connected via point-to-point cables, the introduction of communication network media brings great advantages, such as low cost, reduced weight, simple installation/maintenance, and long-distance control. However, the performance of an NCS is heavily affected by the communication delays in both the controller-to-actuator and sensor-to-controller channels.

Although the presence of delays sometimes may have a stabilizing effect, for some special systems, time delay is a source of instability in general. In

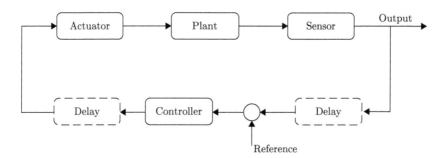

Figure 1.1. Feedback systems with actuator and sensor delay

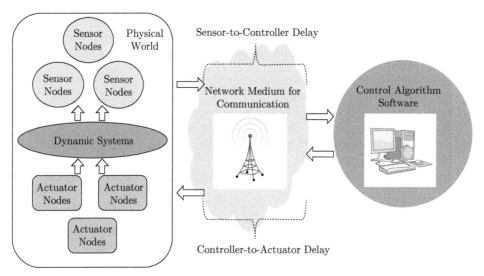

Figure 1.2. Networked control systems with transmission delays in controller-to-actuator and sensor-to-controller channels

accordance with Bode's performance limitation formulas, the existence of delays can severely hamper the performance of physical systems. As a result, the poor performance and catastrophic instability created by delays require the designer to take into account the delays in the control synthesis.

Besides their practical importance, time-delay systems also contribute many mathematical challenges to the analysis and design in control theory. As is well known, the time-delay systems belong to the class of functional differential equations that are infinite-dimensional, as opposed to the finite-dimensional ordinary differential equations (ODE). The state of the delay system is a function (or a vector of functions) rather than a vector. Its characteristic equation is not a polynomial, but involves exponentials. Thus the stability analysis requires Krasovskii functionals rather than Lyapunov functions. For the aforementioned reasons, time-delay systems are both of theoretical and practical importance, and they have drawn the attention of scholars including control engineers, scientists and mathematicians.

Over the past seventy years, delay systems have been an active area of research in control engineering, and major breakthroughs have been reported to deal with delays. The first systematic work of methodological significance may go back to 1946 in the paper by Tsypkin [129]. A couple of breakthroughs in stability analysis were reported in the late 1950s/early 1960s with the papers by Razumikhin in 1956 [115] and by Krasovskii in 1962–1963 [74, 75]. As a counterpart of Lyapunov methods for systems without delay, two main methods are named after their authors: the Krasovskii method of Lyapunov functionals and the Razumikhin method of Lyapunov functions. They are frequently used for stability analysis of time-delay systems. Almost during the same period, Smith introduced the celebrated Smith predictor in 1959 [119], which has seen success in industrial applications, and perhaps been second in popularity to proportional integral differential (PID) controllers, especially among process control practitioners [106].

The next well-known landmark in the control of time-delay systems occurred during the years 1978–1982, when the framework of "finite spectrum assignment" and the "reduction" approach were proposed by Manitius and Olbrot in 1979 [104], Kwon and Pearson in 1980 [94], and Artstein in 1982 [2]. Their basic idea was to make use of the future values of the state, i.e., the so-called "predictor state," in the feedback laws, to compensate for the delay. After the control signal reaches the plant, in a time interval equal to the delay, the state evolves as if there were no delay at all.

Another burst of research activity occurred in the 1990s after the introduction of linear matrix inequalities (LMI). Because of the unavailability of efficient numerical algorithms for the general form of LMI, most of the earlier works on the stability of linear systems via the Lyapunov method were formulated in terms of Lyapunov equations and algebraic Riccati equations. With the realization that LMI is a convex optimization problem, the direct Lyapunov method for linear ODEs leading to stability conditions in terms of LMIs became popular. The development of the efficient interior point method led to the formulation of many control problems and their solutions in the form of LMIs [21], which are capable of providing the desired stability/performance guarantees [41, 42].

In subsequent years, thousands of papers and dozens of books have appeared to deal with the control problem of time-delay systems by adopting various techniques coming from both finite-dimensional and infinite-dimensional systems. The most recent significant books and surveys summarizing the achievements are those by Gu and Niculescu [53], Niculescu [109], Gu, Kharitonov, and Chen [51], Zhong [136], Richard [116], Krstic [80] and Fridman [41, 42]. Still, many basic control problems for time-delay systems remain unsolved. Control of delay systems remains a very active area of research.

1.2 Delay Compensation or Not

Among the basic control problems for time-delay systems, the most challenging ones may arguably be those in which the system has to be controlled through a delayed input, which are non-trivial even when the system is linear and even when the delay is constant. To illustrate the multitude of possible methods in control of time-delay systems, it is useful to consider control of the standard linear state-space model

$$\dot{X}(t) = AX(t) + BU(t - D), \tag{1.1}$$

where $X(t)$ is the state vector, $U(t)$ is the control input with the constant delay $D > 0$, and A and B are system and input matrices with appropriate dimensions. If the delay value is zero, the plant (1.1) is the standard linear time-invariant (LTI) system such that

$$\dot{X}(t) = AX(t) + BU(t) \tag{1.2}$$

and the standard control feedback is such that

$$U(t) = KX(t), \tag{1.3}$$

where K renders the closed-loop system matrix $A + BK$ Hurwitz.

1.2.1 No Compensation and State Delay

When the delay value is larger than zero and no compensation is employed for the input delay, the feedback law becomes

$$U(t - D) = KX(t - D), \tag{1.4}$$

which renders the closed-loop system such that

$$\dot{X}(t) = AX(t) + BKX(t - D), \tag{1.5}$$
$$= AX(t) + A_1X(t - D), \tag{1.6}$$

where $A_1 = BK$. The system (1.6) represents the simplest LTI system with a single discrete constant state delay. As a result, the control problem with input delay is transformed into a problem with state delay. Concentrating on the system (1.6), a large share of the literature investigates the control gain K and the maximum upper bound of delay that still guarantees the stability without compensation [49, 54, 109–111]. Furthermore, the control problem for the single discrete state delay can be nontrivially extended to the case of multiple discrete state delays such that $\dot{X}(t) = AX(t) + \sum_{k=1}^{N} A_kX(t - D_k)$, the case of distributed state delay such that $\dot{X}(t) = AX(t) + \int_{-D}^{0} A_d X(t+\theta)d\theta$, the case of time-varying state delay such that $\dot{X}(t) = AX(t) + A_1 X(t - D(t))$, and so on. To deal with distinct types of state delay, a variety of LMI-based methods are proposed, such as the H_∞ control [45, 47, 125], the descriptor system approach [39], and so on. Different LMI techniques like the Schur complement lemma, Jensen's inequality, Wirtinger's inequality, and Halanay's inequality could result in delicate differences on the maximum upper bound of delay to ensure stability [38–47, 49–54, 118]. And the different Lyapunov-based functionals lead to different delay-dependent or delay-independent stability conditions, which have distinct conservatism. Furthermore, the sampled-data networked control systems, which are widely used in computer implementation nowadays, can also be transformed into time-delay systems and be analyzed using the methods for time-delay systems [95–100].

1.2.2 Delay Compensation by Predictor Feedback

As presented above, a large portion of the available results are finite-dimensional control laws without compensation for delay, whose applicability are limited to systems with "short" delays. In contrast, infinite-dimensional control laws are expected to compensate for the delay so that the stabilization of general systems with long delays can be achieved. Considering the LTI systems with input delay (1.1), instead of the control law (1.4) with no compensation, we aim to compensate for the delay by the control design such that $U(t - D) = KX(t)$ or $U(t) = KX(t + D)$. The effective method to compensate for the delay with arbitrarily finite length is the well-known "predictor feedback," which is also called the "finite spectrum assignment" or "reduction" method. Just as its name implies, the class of predictor feedback laws make use of the future (predicted) values of the state, so that they are able to compensate for the delay. In other words, after the control signal reaches the state of the plant, the delay is compensated for and the state evolves as if there were no delay at all. A challenge in developing predictor feedback laws is the determination of an implementable form for the future values of the state. This is overcome by a distributed integration of the actuator state over the past time interval. Having determined the predictor state, we then obtain the control law by replacing the state in a nominal state feedback law (which stabilizes the delay-free plant) with the state's predictor.

The first mathematically and practically significant development in control design for delay compensation was the introduction of the Smith predictor by Otto Smith in 1959 [119], in which the linear stable systems with constant input delays were addressed. Later on, the control design for the linear stable time-delay systems was extended to the linear unstable time-delay systems by Mayne in 1968 [105], Manitius and Olbrot in 1979 [104], and Kwon and Pearson in 1980 [94], in which the finite spectrum assignment technique for the compensation of input delays was introduced and the open-loop stability restriction was removed. In 1982 Artstein [2] systematized this methodology and named the method a "reduction" to emphasize that the stabilization problem for linear systems with input delays can be reduced to the stabilization problem of a delay-free (reduced) system.

Among the numerous difficulties in the design and analysis of predictor-based control laws, one key challenge is the construction of a Lyapunov-Krasovskii functional for stability analysis of the closed-loop system under predictor feedback. Such a construction is not easy because the overall state of the system consists of the finite-dimensional internal states of the plant and the infinite-dimensional actuator state. With the novel idea of treating the input delay as a first-order hyperbolic transport partial differential equation (PDE), in 2008 Krstic and Smyshlyaev [92, 93] introduced an infinite-dimensional backstepping transformation that enabled the construction of a Lyapunov-Krasovskii functional for linear systems with constant input delays under predictor feedback. Simultaneously, the first predictor feedback design for nonlinear systems was introduced by Krstic in 2008 [78]. After that, based on the PDE framework, the predictor feedback has seen a booming growth in compensating for various kinds of input delays in diverse system classes. For example, with the utilization of PDE and nonlinear control concepts [58, 60, 61, 62, 64, 65, 69–71, 86, 87, 130–131], for linear or nonlinear systems [4–20, 59, 63, 66, 67], delays are handled in forms of single-input delay [22–32], [153, 154, 159], multi-input delays [73, 127–128, 155–157], time-varying delays [33, 84], state-dependent delays [34, 35], distributed delays, and delays on the state, which are summarized in the books [15, 68, 80].

It is worth mentioning that the classic predictor-based controllers are feedback laws of an infinite-dimensional state and require care in their practical implementation [107]. Zhou developed a truncated predictor feedback that involves finite-dimensional state feedback in [137–148]. In [68] and the references therein, concentrating on both linear and nonlinear systems, Karafyllis and Krstic offered three different implementations of the predictor feedback law: the direct implementation, the dynamic implementation, and the hybrid implementation. The three different ways of implementing lead to the corresponding closed-loop systems with important differences.

1.2.3   Examples

For the purpose of more intuition and understanding of delay impact on systems, we offer several examples.

Example 1.1. **Networked Control Systems**

A general class of sampled-data networked control systems with time delay are shown in figure 1.3.

To concentrate on the input delay, we assume that the sensor is able to continuously measure the plant state and there is no transmission delay in the sensor-to-controller channel. Furthermore, we distinguish the actuator and the controller signals by denoting them as $V(t)$ and $U(t)$, respectively. As a result, the sampled-data NCSs in figure 1.3 are reduced to NCSs in figure 1.4.

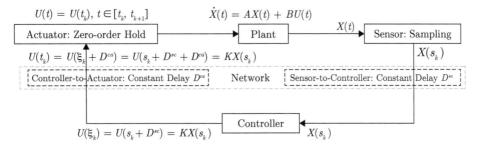

Figure 1.3. Sampled-data NCSs with controller-to-actuator and sensor-to-controller delays

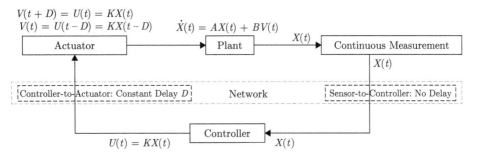

Figure 1.4. NCSs without compensation for input delay under continuous measurements

As shown in figure 1.4, there is no predictor feedback to compensate for the constant controller-to-actuator delay D. Thus we have

$$\begin{aligned}
\dot{X}(t) &= AX(t) + BV(t) \\
&= AX(t) + BU(t - D) \\
&= AX(t) + BKX(t - D) \\
&= AX(t) + A_1 X(t - D),
\end{aligned} \tag{1.7}$$

where the input-delay problem is transformed into a state-delay problem. In this case, the delay D has to be short and the problem was discussed in section 1.2.1.

In contrast, for figure 1.5, the predictor feedback is brought in to compensate for the constant controller-to-actuator delay D. Thus we have, for $t \geq D$,

$$\begin{aligned}
\dot{X}(t) &= AX(t) + BV(t) \\
&= AX(t) + BU(t - D) \\
&= AX(t) + BKX(t) \\
&= (A + BK)X(t).
\end{aligned} \tag{1.8}$$

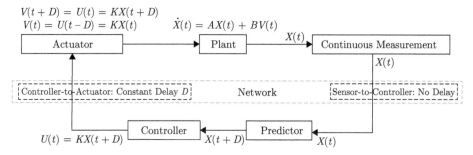

$V(t + D) = U(t) = KX(t + D)$
$V(t) = U(t - D) = KX(t)$ $\dot{X}(t) = AX(t) + BV(t)$

| Actuator | | Plant | | Continuous Measurement |

$X(t)$

$X(t)$

Controller-to-Actuator: Constant Delay D Network   Sensor-to-Controller: No Delay

$U(t) = KX(t + D)$ Controller $X(t + D)$ Predictor $X(t)$

Figure 1.5. NCSs with compensation for input delay under continuous measurements

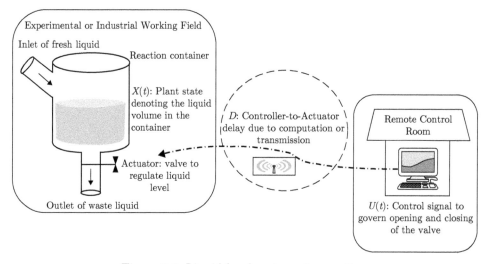

Figure 1.6. Liquid level systems in practice

As a result, the system evolves as if there is no delay at all after $t \geq D$, and the delay value D can be arbitrarily large, which was discussed in section 1.2.2.

Example 1.2. **Liquid Level Systems in Chemical Industry or Bioengineering**

As illustrated in figure 1.6, the liquid level systems exist widely in the chemical industry and bioengineering. For the liquid level system, the fresh liquid continuously enters the reaction container through the inlet, whereas the waste liquid is drained through the outlet, which is regulated by a valve. The plant state $X(t)$ represents the liquid volume in the container, and the control input is employed to govern the valve's opening and closing. The reaction container with the outlet valve is located in the working field, whereas the control command is produced from a remote control room, and they are connected by a

- **Case 1: No Control**
  - $\dot{X}(t) = X(t),\ X(0) = 1,$
  - $\Rightarrow X(t) = e^t$

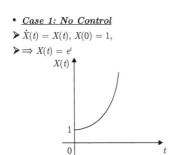

- **Case 2: No Delay**
  - $\dot{X}(t) = X(t) + U(t),\ U(t) = -2X(t) + 0.8$
  - $\Rightarrow \dot{X}(t) = -X(t) + X(t) + 0.8$

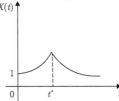

- **Case 3: Control with no compensation for delay**
  - $\dot{X}(t) = X(t) + U(t - D),\ U(t - D) = -2X(t - D) + 0.8$
  - $\Rightarrow \dot{X}(t) = X(t) - 2X(t - D) + 0.8$

- **Case 4: Prediction with compensation for delay**
  - $\dot{X}(t) = X(t) + U(t - D),\ U(t - D) = -2X(t) + 0.8$
  - $\Rightarrow \dot{X}(t) = -X(t) + 0.8$

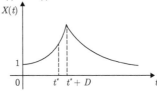

Figure 1.7. Four different cases of control of liquid level systems

long-distance communication network that is subject to a constant controller-to-actuator delay D. For simplicity, we assume that the liquid volume satisfies the following scalar dynamic equation with the initial condition $X(0) = 1$:

$$\dot{X}(t) = X(t) + U(t - D). \tag{1.9}$$

As demonstrated in Case 1 of figure 1.7, the control input is zero such that $U(t - D) \equiv 0$. This means, if the outlet valve is kept closed, the liquid persistently enters the container through the inlet with an exponentially increasing volume.

For Case 2 of figure 1.7, there is no input delay such that $D \equiv 0$. This means, at a certain moment t^*, once the control command is activated, the valve is opened and the control input $U(t) = -2X(t) + X_r$ (where $X_r = 0.8$ is a constant set-point representing the desired reference liquid level in the container) kicks in immediately. After the time instant t^*, the volume of liquid flowing into the container is less than the volume of liquid getting out of the container, and finally the liquid level will be reduced to a fixed position of the reference set-point and arrive at a steady state.

In Case 3 of figure 1.7, the input delay is not zero such that $D > 0$ and there is no compensation for the delay. This means, at a certain moment t^*, we send a command to open the valve to slow down the rising liquid level and let $U(t) = -2X(t) + X_r$ by using the current state of that time. However, because

| Stay in petrol station for fueling | Move on highway with a moderate speed | Move on freeway with a high speed | Stop at intersection to wait for traffic light |

Figure 1.8. Car-following behavior in realistic traffic environment

of the delay, this valve is kept closed and the control $U(t) = -2X(t) + X_r$ does not act on the plant until D unit time later. Thus after that moment $t^* + D$, we have $\dot{X}(t+D) = X(t+D) + U(t) = X(t+D) - 2X(t) + X_r$, which is equivalent to $\dot{X}(t) = X(t) + U(t-D) = X(t) - 2X(t-D) + X_r$. From Case 3 of figure 1.7, when the delay value is large, it is apparent that the closed-loop time-delay system is not stable and the liquid level cannot be regulated to the desired set-point.

In Case 4 of figure 1.7, the input delay is not zero such that $D > 0$, but the predictor feedback is introduced to compensate for the delay. This means, at a certain moment t^*, we send a command to open the valve to slow down the rising liquid level. Due to the existence of a delay, we set $U(t) = -2X(t+D) + X_r$ by predicting the state of D unit time later. Thus at the moment $t^* + D$, this valve opens and the control $U(t) = -2X(t+D) + X_r$ acts on the plant and we have $\dot{X}(t+D) = X(t+D) + U(t) = X(t+D) - 2X(t+D) + X_r = -X(t+D) + X_r$, which is equivalent to $\dot{X}(t) = X(t) + U(t-D) = X(t) - 2X(t) + X_r = -X(t) + X_r$. That is to say, after $t^* + D$, the volume of liquid entering into the container is less than the volume of liquid exiting the container, and thus the liquid level will be reduced to a fixed position to achieve the reference set-point regulation.

Example 1.3. **Platoon Control of Car-Following Vehicles**

As demonstrated in figure 1.8, car-following behaviors for a platoon of vehicles (in which one vehicle follows another) are very common in realistic traffic environments. Generally speaking, the main task of platoon control for car-following models is to regulate a chain of vehicles to an ideal steady state, at which the following vehicles track the leading vehicle with a desirable velocity and simultaneously keep a safe and comfortable inter-vehicle spacing. The widely employed third-order linear vehicle dynamics [102, 113, 135] are taken into account as follows:

$$\dot{p}_i(t) = v_i(t), \tag{1.10}$$

$$\dot{v}_i(t) = a_i(t), \tag{1.11}$$

$$\dot{a}_i(t) = \frac{1}{\tau_i} u_i(t - D_i) - \frac{1}{\tau_i} a_i(t), \tag{1.12}$$

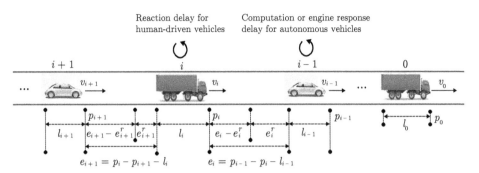

Figure 1.9. Platoon control of car-following vehicles with input delay

where $p_i(t)$, $v_i(t)$, and $a_i(t)$ denote the position, velocity, and acceleration of the ith vehicle; $u_i(t)$ is the ith vehicle's control input, which actually denotes the engine's driving/braking torque governed by the throttle/brake pedal command. τ_i is the engine time constant. D_i is the time delay, which might be either the reaction delay for human-driver vehicles or computation or engine response delay for autonomous vehicles.

As shown in figure 1.9, the inter-vehicle distance is introduced as

$$e_i(t) = p_{i-1}(t) - p_i(t) - l_{i-1}, \tag{1.13}$$

where l_i for $i = 0, 1, \cdots, n$ is the length of the vehicle i, and accordingly the desired/reference spacing is defined by $e_i^r(t)$. The control objective is to regulate the actual inter-vehicle spacing to achieve the desired value such that $\lim_{t \to \infty} \left(e_i(t) - e_i^r(t) \right) = 0$.

To achieve the above control target, the input delay D_i has to be handled carefully, otherwise a big time delay of the actuator could cause a catastrophic car accident.

1.3   Adaptive Control for Time-Delay Systems and PDEs

The stabilization of systems with actuator delay is not an easy task, especially when the system information is incomplete [36, 85, 114]. An effective method to deal with uncertainties in ODE systems is adaptive control [132, 152, 158, 160, 161], which has been extensively studied and summarized in the classic books [3, 48, 57, 108, 117, 126]. Early work on adaptive control of uncertain time-delay systems [37, 110, 112] focused on the uncertainty of parameters rather than the delay. Adaptive backstepping control of uncertain linear systems with input delay had been proposed by [150, 151]; however, by regarding the input delay as unmodelled dynamics, neither identification nor compensation for the delay is considered by the control design.

The standard prediction-based compensation for input delay has the premise that the delay value of every actuator channel is known, as well as that the information is available on the plant and actuator state and plant parameters [2, 18, 20, 72, 73, 94, 104, 119, 127, 128, 142]. However, the fact is that such a prior knowledge may be hard to acquire in practice. Due to the infinite-dimensional nature of time-delay systems, conventional adaptive control methods for finite-dimensional ODE systems cannot be directly applied to addressing uncertain delays. Instead, adaptive control approaches for infinite-dimensional PDE systems provide potential solutions. In [101], a boundary control law and an adaptation law were proposed for Burgers' equation with unknown viscosity. In [90, 120, 121], adaptive boundary control schemes with different update laws (Lyapunov-based, passivity-based, and swapping-based) were developed for unstable parabolic PDEs with unknown parameters in the PDE or boundary condition. A systematic summary of adaptive control methodology for parabolic PDEs was given in [91, 122]. Focusing on another big family of PDEs—anti-stable wave PDEs—[30] and [79, 81], respectively, provided full-state and output feedback adaptive controls under the large uncertainty in anti-damping coefficient. The adaptive-based infinite-dimensional backstepping transformation initially motivated and developed for the boundary control and stabilization of uncertain PDE systems shed new light on addressing uncertain LTI-ODE systems with input delays.

An important step forward on compensation for input delay was made in [92]. With the novel idea of treating the actuator delay as a first-order hyperbolic PDE, LTI-ODE systems with input delays are converted into a class of ODE-PDE cascades. Under the assumption that all system information is available, a prediction-based boundary control and a backstepping transformation were developed. A tutorial introduction of such ODE-PDE cascade representation and PDE-backstepping-based predictor feedback with its extensions was offered in [82]. By regarding the input delay as a parameter in the transport PDE, this kind of PDE-based framework makes it possible for the application of adaptive control of PDEs to LTI-ODEs with input delays [88, 149].

1.4   Results in This Book: Adaptive Control for Uncertain LTI Systems with Input Delays

LTI systems with discrete or distributed input delays usually come with the following five types of basic uncertainties:

- unknown actuator delay,
- unknown delay kernel,
- unknown plant parameter,
- unmeasurable finite-dimensional plant state,
- unmeasurable infinite-dimensional actuator state.

In this book, we design adaptive and robust predictor feedback laws for the compensation of the five above uncertainties for general LTI systems with input delays. When some of the five above variables are unknown or unmeasured, the basic idea of certainty-equivalence-based adaptive control is to use an estimator (a parameter estimator or a state observer) to replace the unknown variables in the nominal backstepping transformation and the predictor feedback of the uncertainty-free systems. The PDE boundary control law is designed to satisfy the stabilization boundary condition of the target closed-loop system, the observer is chosen to estimate the unmeasurable variables with a vanishing estimation error, and the adaptive update law of the unavailable signal is selected to cancel the estimation error term in Lyapunov-based analysis.

In addition, the main content of the book is mainly concerned with the delay-adaptive linear control of linear systems with input delays. Results for delay-adaptive control of nonlinear systems can be found in [29], and the predictor control of various nonlinear systems can be found in [20, 76, 83]. Results for delay-adaptive control of a class of linear feedforward systems with simultaneous state and input delays can be found in [8], whereas delay-robust control of nonlinear or linear systems with time-varying delays can be found in [16, 31, 67, 77].

1.5 Book Organization

The book has three parts.

Part I: In the first part of the book we consider adaptive control of uncertain single-input LTI systems with discrete input delay. The results in this part could be easily (or even trivially) extended to systems with multiple inputs when the delay is the same in all the input channels.

In chapter 2, by assuming all information of interest about the system is known to designers, we provide the introductory idea of predictor feedback and infinite-dimensional backstepping through the nominal single-input LTI systems with discrete input delay. To help readers in the field of control of delay systems who are not accustomed to the PDE notation, we first present an alternative view of the backstepping transformation based purely on standard ODE delay notation. Then the backstepping transformation is described in PDE and rescaled unity-interval transport PDE notation. It is convenient for readers to compare the relationship among the different notational frameworks and the subtle differences in distinct notations.

In chapter 3, a variety of adaptive predictor control techniques are provided to deal with different uncertainty collections from four basic uncertainties (delay, parameter, ODE state, and PDE state). In the presence of a discrete actuator delay that is long and unknown, but when the actuator state is available for measurement, a global adaptive stabilization result is obtainable. In contrast, the problem where the delay value is unknown, and where the actuator state is not measurable at the same time, is not solvable globally, since the problem is not linearly parameterized in the unknown delay. In this case, a local stabilization is feasible, with restrictions on the initial conditions such that not only do the initial values of the ODE and actuator state have to be small, but also the initial value of the delay estimation error has to be small (the delay value is allowed to be large but the initial value of its estimate has to be close to the true value of the delay). Please note that among different uncertainty combinations in chapter 3, we do not consider the case of coexistence of an unknown parameter vector and unmeasurable ODE state of the finite-dimensional part of the plant.

In chapters 4 and 5, output-feedback adaptive control problems involving uncertainties in both an unknown plant parameter and unmeasurable ODE plant state are taken into account. In this case, the relative degree plays a major role in determining the difficulty of a problem. As a result, chapter 4 considers the relatively easier case where the relative degree is equal to the system dimension, whereas chapter 5 considers the more difficult case where there is no limitation on the relative degree. Please note that trajectory tracking requires additional tools, as compared to problems of regulation to zero.

Part II: In the second part of the book we consider adaptive control of uncertain multi-input LTI systems with distinct discrete input delays.

In chapter 6, by assuming all the system information is known to the designer, we provide the elementary PDE-based framework of the exact predictor feedback and the infinite-dimensional backstepping transformation for the nominal multi-input LTI systems with distinct discrete input delays.

In parallel with the single-input systems, chapter 7 provides the adaptive global stabilization results for uncertain multi-input systems via the assumption that the actuator states in distinct control channels are measurable, whereas chapters 8 and 9 provide the adaptive local stabilization results for uncertain multi-input systems under the assumption that the measurements of actuator states are unavailable.

Part III: In the third part of the book we address adaptive and robust control problems of uncertain LTI systems with distributed input delays.

In chapter 10, under the premise of an uncertainty-free system, a fundamental predictor feedback framework to compensate for distributed input delays is developed for both single-input and multi-input systems. By modeling a linear system with distributed delay as an ODE-PDE cascade, we employ a couple of important conversions: the reduction-based change of variable converts a stabilization problem of a delayed system into a stabilization problem of a delay-free (reduced) system, and the forwarding-backstepping transformation contributes to a convenient stability analysis of the target closed-loop system.

When some information of the system (such as delay, delay kernel, parameter, ODE state, and PDE state) is unavailable, chapters 11 and 12 solve the adaptive and robust stabilization problems for single-input and multi-input systems, respectively. When both the finite-dimensional plant state and the infinite-dimensional actuator state are measurable, the global adaptive state feedback is used to handle the uncertainties in delay, delay kernel, and plant parameter. When the delay, delay kernel, and plant state are simultaneously uncertain, the

Table 1.1. Uncertainty Collections of Linear Systems with Actuator Delays

Part I: Predictor Feedback of Single-Input LTI Systems with Discrete Input Delay

Organization of the Book	Input Delay	Parameter	ODE State	PDE State
Chapter 2	known	known	known	known
Section 3.2	known	known	*unknown*	known
Section 3.3	*unknown*	known	known	known
Section 3.4	*unknown*	known	known	*unknown*
Section 3.5	*unknown*	*unknown*	known	known
Section 3.6	*unknown*	known	*unknown*	*unknown*
Section 3.7	*unknown*	*unknown*	known	*unknown*
Chapter 4 and section 5.2	*unknown*	*unknown*	*unknown*	known
Section 5.3	*unknown*	*unknown*	*unknown*	*unknown*

Part II: Predictor Feedback of Multi-Input LTI Systems with Discrete Input Delays

Organization of the Book	Multi-Input Delays	Parameter	ODE State	PDE States
Chapter 6	known	known	known	known
Section 7.2	*unknown*	known	known	known
Section 7.3	*unknown*	known	*unknown*	known
Section 7.4	*unknown*	*unknown*	known	known
Chapter 8	*unknown*	known	known	*unknown*
Section 9.2	*unknown*	known	*unknown*	*unknown*
Section 9.3	*unknown*	*unknown*	known	*unknown*

Part III: Predictor Feedback of LTI Systems with Distributed Input Delays

Organization of the Book	Delay	Delay Kernel	Parameter	ODE State	PDE State
Chapter 10	known	known	known	known	known
Sections 11.1,12.1	*unknown*	known	known	known	known
Section 11.2	*unknown*	*unknown*	known	known	known
Sections 11.3,12.2	*unknown*	*unknown*	*unknown*	known	known
Sections 11.5,12.3	*unknown*	*unknown*	known	*unknown*	known
Sections 11.4,12.4	*unknown*	*unknown*	known	known	*unknown*

adaptive control problem may not be globally solvable in its most general form as the relative degree plays an important role in the output feedback. Thus the robust output feedback is employed. Similarly, when the delay and actuator state are unknown at the same time, the adaptive control can only achieve local stabilization both in the initial state and in the initial parameter error. For this case, instead of the adaptive method, the robust output feedback is considered.

To clearly describe the book is organization, different combinations of uncertainties considered in the book are summarized in table 1.1, from which the readers can make their own selections to address a vast class of relevant problems they face.

1.6 Notation

Throughout the book, the actuator dynamics (of the control input with delays) are modeled by a transport PDE whose state is $u(x,t)$, where t is time and x is a spatial variable that takes values in the interval $[0,1]$ or $[0,D]$. When the plant being controlled (or whose state is estimated) is an ODE, then the overall state of the system is the state of the ODE, $X(t)$, along with the state of the PDE, $u(x,t)$.

For a finite-dimensional ODE vector $X(t)$, its Euclid norm is denoted by $|X(t)|$.

For an infinite-dimensional PDE function $u(x,t)$, $u : [0,1] \times \mathbb{R}^+ \to \mathbb{R}$,

$$\|u(x,t)\| = \left(\int_0^1 u(x,t)^2 dx \right)^{1/2}. \tag{1.14}$$

Part I

Single-Input Discrete Delay

Chapter Two

Basic Predictor Feedback for Single-Input Systems

IN THIS CHAPTER, we introduce the basic idea of a PDE backstepping approach for single-input LTI ordinary differential equation (ODE) systems with discrete input delay. This chapter is the core for all the developments in the rest of the book. The results in this chapter are obtained under the assumption that all information about the considered systems is known to designer, and they easily (or even trivially) extend to systems with multiple inputs when the delay is the same in all the input channels.

By treating the delayed input as a first-order hyperbolic transport PDE, we transform the LTI-ODE systems with input delay into the ODE-PDE cascade systems. The PDE-based design recovers a classical control formula obtained through various other approaches such as a modified Smith predictor, finite spectrum assignment, and the Artstein-Kwon-Pierson "reduction" approach [2, 94, 104, 119]. The key point of the backstepping approach lies in it providing a systematic construction of an infinite-dimensional transformation of the actuator state, which yields a cascade system of transformed stable actuator dynamics and stabilized plant dynamics. The cascade system consisting of such infinite-dimensional stable actuator dynamics and finite-dimensional stabilized plant dynamics is referred to as the closed-loop "target system" through the book. The PDE-based design results in the construction of an extremely simple explicit Lyapunov-Krasovskii functional and an explicit exponential stability estimate for the target system, while being quite complex for the original system, which is the main advantage of this approach.

To assist readers in the control field of time-delay systems who are not familiar with the PDE notation, we provide three alternative representations for the infinite-dimensional actuator state: (1) using standard delay notation of the delayed state, (2) using a representation with a transport PDE for the actuator state, and (3) using a rescaled unity-interval transport PDE for the actuator state. Correspondingly, we provide three alternative representations for the backstepping transformation. The three alternative but mathematically equivalent representations constitute chapter 2 which is structured as follows:

To help readers compare the relationships and subtle differences among the aforementioned three representations of predictor feedback and backstepping transformation, we use a few tables to summarize the three frameworks in different notations.

The infinite-dimensional backstepping transformation was initially inspired by and developed for the boundary control and stabilization of PDE systems. The main results of this chapter are provided through some main theorems and their proofs.

In the next chapter, based on the framework of rescaled unity-interval transport PDE notation, the backstepping transformation and predictor feedback design are extended to adaptive control of uncertain single-input LTI systems with input delay.

2.1 Basic Idea of Predictor Feedback for LTI Systems with Input Delay

Consider a general class of LTI systems with discrete input delay as follows:

$$\dot{X}(t) = AX(t) + BU(t - D), \quad t \geq 0, \tag{2.1}$$

where $X(t) \in \mathbb{R}^n$ is the system state vector, $U(t) \in \mathbb{R}$ is a scalar control input delayed by D units of time, and $D \geq 0$ is a constant. The pair of matrices (A, B) with appropriate dimension is stabilizable. The result of single-input delay could be trivially extended to systems with multiple inputs when the delays of each input channel are equal to each other. When the delay D is equal to zero, (2.1) becomes the standard LTI system without delay in state-space form such that

$$\dot{X}(t) = AX(t) + BU(t), \quad t \geq 0 \tag{2.2}$$

and the feedback law is designed as

$$U(t) = KX(t), \quad t \geq 0, \tag{2.3}$$

where the stabilization gain K is chosen to render $A + BK$ Hurwitz.

As shown in figure 2.1, if there is no prediction for compensation for the delay, the feedback law (2.3) yields

$$U(t - D) = KX(t - D) \tag{2.4}$$

and the resulting closed-loop system is

$$\dot{X}(t) = AX(t) + BKX(t - D) = AX(t) + A_1 X(t - D), \tag{2.5}$$

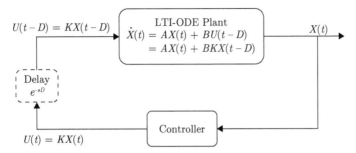

Figure 2.1. Feedback of single-input systems with no compensation for delay

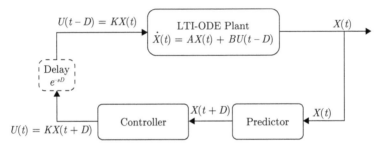

Figure 2.2. Predictor feedback of single-input systems with compensation for delay

where $A_1 = BK$. The LTI systems with the state delay (2.5) can be analyzed by the Lyapunov-Krasovskii functional or Lyapunov-Razumikhin function methods in [41, 42]. By a variety of LMI techniques in [41, 42], the maximum upper bound of the time delay D and the stabilization gain K could be derived to stabilize the system with time delay in state (2.5). The LMI-based control methods do not compensate for the delay in the actuator; thus they can only deal with systems with "short" time delays.

For LTI systems with arbitrarily long finite delays D in the input, as shown in figure 2.2, the predictor feedback law is chosen so that

$$U(t - D) = KX(t), \tag{2.6}$$

which is equivalent to

$$U(t) = KX(t + D). \tag{2.7}$$

Please note that the future state $X(t + D)$ cannot be employed in a literal sense. With the variation-of-constants formula, treating the current state $X(t)$ as the initial condition, (2.7) is equivalent to the implementable control law

$$U(t) = KX(t+D),\tag{2.8}$$

$$= K\left[e^{A(t+D-t)}X(t) + \int_t^{t+D} e^{A(t+D-s)}BU(s-D)ds\right],\tag{2.9}$$

$$= K\left[e^{AD}X(t) + \int_{t-D}^t e^{A(t-\theta)}BU(\theta)d\theta\right].\tag{2.10}$$

The controller (2.10) was first derived in the years 1978 to 1982 in the framework of "finite spectrum assignment" and the "reduction approach" [2, 94, 104]. The idea is to introduce the predictor state:

$$P(t) = X(t+D),\tag{2.11}$$

$$= e^{AD}X(t) + \int_{t-D}^t e^{A(t-\theta)}BU(\theta)d\theta.\tag{2.12}$$

Taking the time derivative on both sides of (2.12) and utilizing dynamics (2.1), we obtain the reduced finite-dimensional system:

$$\dot{P}(t) = AP(t) + BU(t), \quad t \geq 0.\tag{2.13}$$

The resulting control law is simply

$$U(t) = KP(t).\tag{2.14}$$

The predictor transformation (2.12) and the simple, intuitive design based on the reduction approach do not equip the designer with a tool for Lyapunov-Krasovskii stability analysis. The reason for this is that the transformation $P(t)$ is only a transformation of the ODE state $X(t)$, rather than also providing a suitable change of variable for the infinite-dimensional actuator state. As a result, the analysis in the literature does not capture the entire system consisting of the ODE plant and the infinite-dimensional subsystem of the input delay.

2.2   Backstepping Transformation in Standard ODE Delay Notation

The main idea of backstepping is to find an invertible transformation of the original state variables such that the system in the new variables has some desirable properties (such as stability) or it is easier to analyze than the original system. We refer from now on to this transformed system as the "target system."

In (2.10), the function $U(\theta)$ on the entire sliding window $\theta \in [t-D, t]$ is the actuator's state. It is infinite-dimensional, since it is a function, rather than a vector or a scalar. To prove the stability of the closed-loop system consisting of the plant (2.1) and the control law (2.10), one has to find an appropriate backstepping transformation of the state variables (X, U). Towards that end, similar to (2.9)–(2.10), using the variation-of-constants formula with initial state $X(t)$ and current time $\theta + D$, from (2.1) the invertible transformation $(X, U) \leftrightarrow (X, W)$ is considered as follows:

$$W(\theta) = U(\theta) - KX(\theta + D), \tag{2.15}$$

$$= U(\theta) - K \left[e^{A(\theta + D - t)} X(t) + \int_t^{\theta + D} e^{A(\theta + D - s)} BU(s - D) ds \right], \tag{2.16}$$

$$= U(\theta) - K \left[e^{A(\theta + D - t)} X(t) + \int_{t-D}^{\theta} e^{A(\theta - \sigma)} BU(\sigma) d\sigma \right], \tag{2.17}$$

$$U(\theta) = W(\theta) + K \left[e^{(A + BK)(\theta + D - t)} X(t) + \int_{t-D}^{\theta} e^{(A + BK)(\theta - \sigma)} BW(\sigma) d\sigma \right], \tag{2.18}$$

where $\theta \in [t - D, t]$ and $t \in [0, +\infty)$. Setting $\theta = t - D$ in (2.17), solving the resulting equation as $U(t - D) = KX(t) + W(t - D)$ and substituting the expression into (2.1), we arrive at the target system,

$$\dot{X}(t) = (A + BK)X(t) + BW(t - D), \tag{2.19}$$
$$W(t) = 0, \quad \forall t \geq 0. \tag{2.20}$$

Since the matrix $A + BK$ is Hurwitz and the "disturbance" $W(t - D)$ vanishes in finite time (after $t \geq D$), it is evident that the target system (2.19)–(2.20) possesses some desirable properties in terms of stability. It is easy to find the closed-loop system consisting of the plant (2.1), and the control law (2.10) is mapped to the target system (2.19)–(2.20) by the invertible backstepping transformation (2.17)–(2.18).

The above predictor feedback and backstepping transformation in standard ODE notation are summarized in table 2.1.

Table 2.1. Predictor Feedback Framework in Notation of Standard ODE

The closed-loop system:

$$\dot{X}(t) = AX(t) + BU(t - D), \quad t \geq 0 \tag{2.21}$$

$$U(t) = K \left[e^{AD} X(t) + \int_{t-D}^{t} e^{A(t-\theta)} BU(\theta) d\theta \right] \tag{2.22}$$

(*continued*)

Table 2.1. Continued

The invertible backstepping transformation: $\forall \theta \in [t-D, t]$, $\forall t \in [0, +\infty)$:

$$W(\theta) = U(\theta) - K\left[e^{A(\theta+D-t)}X(t) + \int_{t-D}^{\theta} e^{A(\theta-\sigma)}BU(\sigma)d\sigma\right] \tag{2.23}$$

$$U(\theta) = W(\theta) + K\left[e^{(A+BK)(\theta+D-t)}X(t) + \int_{t-D}^{\theta} e^{(A+BK)(\theta-\sigma)}BW(\sigma)d\sigma\right] \tag{2.24}$$

The target system:

$$\dot{X}(t) = (A+BK)X(t) + BW(t-D) \tag{2.25}$$

$$W(t) = 0, \quad \forall t \geq 0 \tag{2.26}$$

The Lyapunov-Krasovskii functional:

$$V(t) = X(t)^T P X(t) + \frac{a}{2}\int_{t-D}^{t}(1+\theta+D-t)W(\theta)^2 d\theta \tag{2.27}$$

Theorem 2.1. *The closed-loop system consisting of the plant (2.21) with the controller (2.22) is exponentially stable at the origin in the sense of the norm*

$$|X(t)| + \left(\int_{t-D}^{t} U(\theta)^2 d\theta\right)^{\frac{1}{2}} \tag{2.28}$$

for all $t \geq 0$.

Proof. A Lyapunov-Krasovskii functional of the closed-loop target system (2.19)–(2.20) is given as

$$V(t) = X(t)^T P X(t) + \frac{a}{2}\int_{t-D}^{t}(1+\theta+D-t)W(\theta)^2 d\theta, \tag{2.29}$$

where

$$P = P^T > 0, \quad Q = Q^T > 0 \tag{2.30}$$

such that

$$(A+BK)^T P + P(A+BK) = -Q \tag{2.31}$$

and

$$a = \frac{4\lambda_{\max}(PBB^T P)}{\lambda_{\min}(Q)}. \tag{2.32}$$

Taking the time derivative of (2.29) along (2.19)–(2.20), we have

$$\dot{V}(t) = X^T(t)\left((A+BK)^T P + P(A+BK)\right)X(t) + 2X^T(t)PBW(t-D)$$
$$+\frac{a}{2}(1+D)W(t)^2 - \frac{a}{2}W(t-D)^2$$
$$-\frac{a}{2}\int_{t-D}^{t} W(\theta)^2 d\theta. \tag{2.33}$$

Bearing in mind that $W(t) = 0$ from (2.20), substituting (2.31) into (2.33), we have

$$\dot{V}(t) = -X^T(t)QX(t) + 2X^T(t)PBW(t-D) - \frac{a}{2}W(t-D)^2$$
$$-\frac{a}{2}\int_{t-D}^{t} W^2(\theta)d\theta, \tag{2.34}$$
$$\leq -X^T(t)QX(t) + 2|X^T(t)PB||W(t-D)| - \frac{a}{2}W(t-D)^2$$
$$-\frac{a}{2}\int_{t-D}^{t} W^2(\theta)d\theta, \tag{2.35}$$
$$= -X^T(t)QX(t)$$
$$-\frac{2}{a}\left(|X^T(t)PB| - \frac{a}{2}W(t-D)\right)^2$$
$$+\frac{2}{a}|X^T(t)PB|^2$$
$$-\frac{a}{2}\int_{t-D}^{t} W^2(\theta)d\theta, \tag{2.36}$$
$$\leq -X^T(t)QX(t) + \frac{2}{a}|X^T(t)PB|^2 - \frac{a}{2}\int_{t-D}^{t} W^2(\theta)d\theta, \tag{2.37}$$
$$= -X^T(t)QX(t) + \frac{2}{a}X^T(t)PBB^T PX(t) - \frac{a}{2}\int_{t-D}^{t} W^2(\theta)d\theta, \tag{2.38}$$
$$\leq -\lambda_{\min}(Q)|X(t)|^2 + \frac{2}{a}\lambda_{\max}(PBB^T P)|X(t)|^2 - \frac{a}{2}\int_{t-D}^{t} W^2(\theta)d\theta$$
$$= -\frac{\lambda_{\min}(Q)}{2}|X(t)|^2 - \frac{2\lambda_{\max}(PBB^T P)}{\lambda_{\min}(Q)}\int_{t-D}^{t} W^2(\theta)d\theta, \tag{2.39}$$

$$\leq -\frac{\lambda_{\min}(Q)}{2}|X(t)|^2 - \frac{2\lambda_{\max}(PBB^T P)}{\lambda_{\min}(Q)(1+D)}\int_{t-D}^t W^2(\theta)d\theta, \qquad (2.40)$$

$$= -\frac{\lambda_{\min}(Q)}{2}|X(t)|^2 - \frac{a}{2(1+D)}\int_{t-D}^t W^2(\theta)d\theta, \qquad (2.41)$$

$$\leq -\mu V(t), \qquad (2.42)$$

where $\mu = \min\left\{\frac{\lambda_{\min}(Q)}{2\lambda_{\max}(P)}, \frac{1}{1+D}\right\} > 0$ is a constant. Using the comparison principle [55] (also see Lemma B.7 in appendix B), we can obtain the exponential stability in the (X, W). Then by the invertible backstepping transformation, the exponential stability in the (X, U) can be implied. Thus the proof is completed.   □

2.3   Backstepping Transformation in Transport PDE Notation

As shown in figure 2.3, representing the actuator state $U(\theta)$ for $\theta \in [t-D, t]$ with a transport PDE,

$$u(x,t) = U(t+x-D), \quad x \in [0, D], \qquad (2.43)$$

the LTI system (2.1) is written as the ODE-PDE cascade as follows:

$$\dot{X}(t) = AX(t) + Bu(0,t), \quad t \geq 0, \qquad (2.44)$$

$$u_t(x,t) = u_x(x,t), \quad x \in [0, D], \qquad (2.45)$$

$$u(D,t) = U(t). \qquad (2.46)$$

Through the invertible backstepping transformation,

$$w(x,t) = u(x,t) - K\left[e^{Ax}X(t) + \int_0^x e^{A(x-y)}Bu(y,t)dy\right], \qquad (2.47)$$

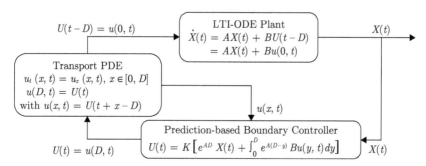

Figure 2.3. Predictor feedback of single-input systems in PDE notation

$$u(x,t) = w(x,t) + K \left[e^{(A+BK)x} X(t) + \int_0^x e^{(A+BK)(x-y)} Bw(y,t)dy \right], \quad (2.48)$$

the ODE-PDE cascade (2.44)–(2.46) is converted into the target system

$$\dot{X}(t) = (A+BK)X(t) + Bw(0,t), \quad t \geq 0, \quad (2.49)$$
$$w_t(x,t) = w_x(x,t), \quad x \in [0,D], \quad (2.50)$$
$$w(D,t) = 0. \quad (2.51)$$

If we substitute $x = D$ into (2.47), the control law is selected to meet the stable boundary condition (2.51) of the target system such that

$$U(t) = u(D,t) = K \left[e^{AD} X(t) + \int_0^D e^{A(D-y)} Bu(y,t)dy \right]. \quad (2.52)$$

Remark 2.2. It is not difficult to find that backstepping transformations in standard ODE time-delay notation (2.17)–(2.18) and transport PDE notation (2.47)–(2.48) are mathematically equivalent by the variable changes

$$u(x,t) = U(t+x-D) = U(\theta), \quad (2.53)$$
$$u(y,t) = U(t+y-D) = U(\sigma), \quad (2.54)$$
$$w(x,t) = W(\theta), \quad (2.55)$$
$$w(y,t) = W(\sigma). \quad (2.56)$$

We apologize to the reader that the transformations (2.47)–(2.48) may appear a bit like "magic." Their choice is guided by a general backstepping design procedure in section 2.2 of [80] and section 2.1.2 of [15], which employs a Volterra operator.

The above predictor feedback and backstepping transformation in transport PDE notation are summarized in table 2.2.

Table 2.2. Predictor Feedback Framework in Notation of Transport PDE

The transport PDE:

$$u(x,t) = U(t+x-D), \quad x \in [0,D] \quad (2.57)$$

The closed-loop system:

$$\dot{X}(t) = AX(t) + Bu(0,t), \quad t \geq 0 \quad (2.58)$$

(*continued*)

Table 2.2. Continued

$$u_t(x,t) = u_x(x,t), \quad x \in [0, D] \tag{2.59}$$

$$u(D, t) = U(t) \tag{2.60}$$

$$U(t) = u(D, t) = K \left[e^{AD} X(t) + \int_0^D e^{A(D-y)} Bu(y, t) dy \right] \tag{2.61}$$

The invertible backstepping transformation:

$$w(x,t) = u(x,t) - K \left[e^{Ax} X(t) + \int_0^x e^{A(x-y)} Bu(y, t) dy \right] \tag{2.62}$$

$$u(x,t) = w(x,t) + K \left[e^{(A+BK)x} X(t) + \int_0^x e^{(A+BK)(x-y)} Bw(y, t) dy \right] \tag{2.63}$$

The target system:

$$\dot{X}(t) = (A + BK)X(t) + Bw(0, t), \quad t \geq 0 \tag{2.64}$$

$$w_t(x,t) = w_x(x,t), \quad x \in [0, D] \tag{2.65}$$

$$w(D, t) = 0 \tag{2.66}$$

The Lyapunov-Krasovskii functional:

$$V(t) = X(t)^T PX(t) + \frac{a}{2} \int_0^D (1+x)w(x,t)^2 dx \tag{2.67}$$

Theorem 2.3. *The closed-loop system consisting of the plant (2.58)–(2.60) and the controller (2.61) is exponentially stable at the origin in the sense of the norm*

$$|X(t)| + \left(\int_0^D u(x,t)^2 dx \right)^{\frac{1}{2}} \tag{2.68}$$

for all $t \geq 0$.

Proof. A Lyapunov-Krasovskii functional of the closed-loop system (2.49)–(2.51) is given as

$$V(t) = X(t)^T PX(t) + \frac{a}{2} \int_0^D (1+x)w(x,t)^2 dx, \tag{2.69}$$

where

$$a = \frac{4\lambda_{\max}(PBB^T P)}{\lambda_{\min}(Q)}. \tag{2.70}$$

Taking the time derivative of (2.69) along (2.49)–(2.51), we have

$$\dot{V}(t) = X^T(t)\left((A+BK)^T P + P(A+BK)\right)X(t) + 2X^T(t)PBw(0,t)$$

$$+ a\int_0^D (1+x)w(x,t)w_t(x,t)dx, \tag{2.71}$$

$$= -X^T(t)QX(t) + 2X^T(t)PBw(0,t)$$

$$+ a\int_0^D (1+x)w(x,t)w_x(x,t)dx. \tag{2.72}$$

Consider the third term on the right side of (2.72), and use integration by parts:

$$a\int_0^D (1+x)w(x,t)w_x(x,t)dx = a(1+D)w(D,t)^2 - aw(0,t)^2$$

$$- a\int_0^D w^2(x,t)dx - a\int_0^D (1+x)w_x(x,t)w(x,t)dx. \tag{2.73}$$

Bearing in mind that $w(D,t)=0$ from (2.51), we have

$$a\int_0^D (1+x)w(x,t)w_x(x,t)dx = -\frac{a}{2}w(0,t)^2 - \frac{a}{2}\int_0^D w^2(x,t)dx. \tag{2.74}$$

Substituting (2.74) into (2.72), we have

$$\dot{V}(t) = -X^T(t)QX(t) + 2X^T(t)PBw(0,t) - \frac{a}{2}w(0,t)^2$$

$$- \frac{a}{2}\int_0^D w^2(x,t)dx, \tag{2.75}$$

$$\leq -X^T(t)QX(t) + 2|X^T(t)PB||w(0,t)| - \frac{a}{2}w(0,t)^2$$

$$- \frac{a}{2}\int_0^D w^2(x,t)dx, \tag{2.76}$$

$$= -X^T(t)QX(t)$$

$$- \frac{2}{a}\left(|X^T(t)PB| - \frac{a}{2}w(0,t)\right)^2$$

$$+ \frac{2}{a}|X^T(t)PB|^2$$

$$-\frac{a}{2}\int_0^D w^2(x,t)dx, \tag{2.77}$$

$$\leq -X^T(t)QX(t) + \frac{2}{a}|X^T(t)PB|^2 - \frac{a}{2}\int_0^D w^2(x,t)dx, \tag{2.78}$$

$$= -X^T(t)QX(t) + \frac{2}{a}X^T(t)PBB^TPX(t) - \frac{a}{2}\int_0^D w^2(x,t)dx, \tag{2.79}$$

$$\leq -\lambda_{\min}(Q)|X(t)|^2 + \frac{2}{a}\lambda_{\max}(PBB^TP)|X(t)|^2 - \frac{a}{2}\int_0^D w^2(x,t)dx$$

$$= -\frac{\lambda_{\min}(Q)}{2}|X(t)|^2 - \frac{2\lambda_{\max}(PBB^TP)}{\lambda_{\min}(Q)}\int_0^D w^2(x,t)dx, \tag{2.80}$$

$$\leq -\frac{\lambda_{\min}(Q)}{2}|X(t)|^2 - \frac{2\lambda_{\max}(PBB^TP)}{\lambda_{\min}(Q)(1+D)}\int_0^D w^2(x,t)dx, \tag{2.81}$$

$$= -\frac{\lambda_{\min}(Q)}{2}|X(t)|^2 - \frac{a}{2(1+D)}\int_0^D w^2(x,t)dx, \tag{2.82}$$

$$\leq -\mu V(t), \tag{2.83}$$

where $\mu = \min\left\{\frac{\lambda_{\min}(Q)}{2\lambda_{\max}(P)}, \frac{1}{1+D}\right\} > 0$ is a constant. Using the comparison principle [55], we can obtain the exponential stability in the (X, w). Then by the invertible backstepping transformation, the exponential stability in the (X, u) can be implied. Thus the proof is completed. □

2.4 Backstepping Transformation in Rescaled Unity-Interval Transport PDE Notation

It is evident that in the last chapter the transport PDE representing the actuator state $u(x,t) = U(t+x-D)$ for $x \in [0, D]$ in (2.43) is not linearly parameterized in time delay D, and this will be less convenient for adaptive control in later chapters. As illustrated in figure 2.4, in order to overcome this shortcoming and produce a nominal design framework for the rest of the book, we consider the rescaled unity-interval transport PDE as follows:

$$u(x,t) = U(t+D(x-1)), \quad x \in [0,1]. \tag{2.84}$$

Thus the LTI system (2.1) is transformed into the ODE-PDE cascade as follows:

$$\dot{X}(t) = AX(t) + Bu(0,t), \quad t \geq 0, \tag{2.85}$$

$$Du_t(x,t) = u_x(x,t), \quad x \in [0,1], \tag{2.86}$$

$$u(1,t) = U(t). \tag{2.87}$$

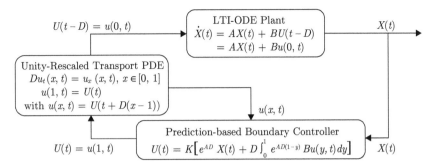

Figure 2.4. Predictor feedback of single-input systems in rescaled unity-interval PDE notation

Here we assume the full information of (2.85)–(2.87) is available, namely, both the finite-dimensional ODE plant state vector $X(t)$ and the infinite-dimensional PDE actuator state $u(x,t)$ representing delayed input $U(\eta)$ over the time window $\eta \in [t-D, t]$ are available for measurement, and the ODE system and input matrices A and B are fully known (there is no unknown element in both of the matrices), and the delay length D is a known constant and can have an arbitrarily large value.

By the invertible backstepping transformation,

$$w(x,t) = u(x,t) - K\left[e^{ADx}X(t) + D\int_0^x e^{AD(x-y)}Bu(y,t)dy\right], \quad (2.88)$$

$$u(x,t) = w(x,t) + K\left[e^{(A+BK)Dx}X(t),\right.$$
$$\left. +D\int_0^x e^{(A+BK)D(x-y)}Bw(y,t)dy\right], \quad (2.89)$$

and the control law

$$U(t) = u(1,t) = K\left[e^{AD}X(t) + D\int_0^1 e^{AD(1-y)}Bu(y,t)dy\right], \quad (2.90)$$

the system (2.85)–(2.87) can be converted into the target system

$$\dot{X}(t) = (A+BK)X(t) + Bw(0,t), \quad t \geq 0, \quad (2.91)$$

$$Dw_t(x,t) = w_x(x,t), \quad x \in [0,1], \quad (2.92)$$

$$w(1,t) = 0. \quad (2.93)$$

Remark 2.4. If we compare the framework in rescaled unity-interval PDE notation (2.84)–(2.93) with the one in traditional PDE notation (2.43)–(2.52), it is not hard to find that they are mathematically equivalent to each other with trivial variable changes. Thus both of the frameworks are able to compensate for the delay to achieve stabilization and possess the same properties.

The above predictor feedback and backstepping transformation in rescaled unity-interval transport PDE notation are summarized in table 2.3.

Table 2.3. Predictor Feedback Framework in Notation of Rescaled Unity-Interval Transport PDE

The transport PDE in rescaled unity-interval:

$$u(x,t) = U(t + D(x-1)), \quad x \in [0,1] \tag{2.94}$$

The closed-loop system:

$$\dot{X}(t) = AX(t) + Bu(0,t), \quad t \geq 0 \tag{2.95}$$

$$Du_t(x,t) = u_x(x,t), \quad x \in [0,1] \tag{2.96}$$

$$u(1,t) = U(t) \tag{2.97}$$

$$U(t) = u(1,t) = K\left[e^{AD}X(t) + D\int_0^1 e^{AD(1-y)}Bu(y,t)dy\right] \tag{2.98}$$

The invertible backstepping transformation:

$$w(x,t) = u(x,t) - K\left[e^{ADx}X(t) + D\int_0^x e^{AD(x-y)}Bu(y,t)dy\right] \tag{2.99}$$

$$u(x,t) = w(x,t) + K\left[e^{(A+BK)Dx}X(t)\right.$$
$$\left. + D\int_0^x e^{(A+BK)D(x-y)}Bw(y,t)dy\right] \tag{2.100}$$

The target system:

$$\dot{X}(t) = (A+BK)X(t) + Bw(0,t), \quad t \geq 0 \tag{2.101}$$

$$Dw_t(x,t) = w_x(x,t), \quad x \in [0,1] \tag{2.102}$$

$$w(1,t) = 0 \tag{2.103}$$

The Lyapunov-Krasovskii functional:

$$V(t) = X(t)^T PX(t) + \frac{a}{2}D\int_0^1 (1+x)w(x,t)^2 dx \tag{2.104}$$

Theorem 2.5. *The closed-loop system consisting of the plant (2.95)–(2.97) and the controller (2.98) is exponentially stable at the origin in the sense of the norm*

$$|X(t)| + \left(\int_0^1 u(x,t)^2 dx \right)^{\frac{1}{2}} \tag{2.105}$$

for all $t \geq 0$.

Proof. A Lyapunov-Krasovskii functional of the closed-loop system (2.91)–(2.93) is given as

$$V(t) = X(t)^T P X(t) + gD \int_0^1 (1+x)w(x,t)^2 dx, \tag{2.106}$$

where $P = P^T > 0$ and $g > 0$ is an analysis coefficient. Taking the time derivative of (2.106) along (2.91)–(2.93), we have

$$\dot{V}(t) = X^T(t) \left((A+BK)^T P + P(A+BK) \right) X(t) + 2X^T(t)PBw(0,t)$$

$$+ 2gD \int_0^1 (1+x)w(x,t)w_t(x,t)dx, \tag{2.107}$$

$$= -X^T(t)QX(t) + 2X^T(t)PBw(0,t)$$

$$+ 2g \int_0^1 (1+x)w(x,t)w_x(x,t)dx, \tag{2.108}$$

where $-Q = (A+BK)^T P + P(A+BK)$. Consider the third term on the right side of (2.108), and use integration by parts:

$$2g \int_0^1 (1+x)w(x,t)w_x(x,t)dx = 4gw(1,t)^2 - 2gw(0,t)^2$$

$$- 2g \int_0^1 w^2(x,t)dx - 2g \int_0^1 (1+x)w_x(x,t)w(x,t)dx. \tag{2.109}$$

Bearing in mind that $w(1,t) = 0$ from (2.93), we have

$$2g \int_0^1 (1+x)w(x,t)w_x(x,t)dx = -gw(0,t)^2 - g \int_0^1 w^2(x,t)dx. \tag{2.110}$$

Substituting (2.110) into (2.108), we have

$$\dot{V}(t) = -X^T(t)QX(t) + 2X^T(t)PBw(0,t) - gw(0,t)^2 - g \int_0^1 w^2(x,t)dx$$

$$\leq -\lambda_{\min}(Q)|X(t)|^2 + 2|X(t)||PB||w(0,t)|$$

$$- gw(0,t)^2 - g \int_0^1 w^2(x,t)dx, \tag{2.111}$$

$$\leq -\frac{\lambda_{\min}(Q)}{2}|X(t)|^2$$

$$-\frac{\lambda_{\min}(Q)}{2}\left(|X(t)| - \frac{2|PB|}{\lambda_{\min}(Q)}w(0,t)\right)^2$$

$$-\left(g - \frac{2|PB|^2}{\lambda_{\min}(Q)}\right)w(0,t)^2$$

$$-g\int_0^1 w(x,t)^2 dx, \tag{2.112}$$

$$\leq -\frac{\lambda_{\min}(Q)}{2}|X(t)|^2 - g\|w(x,t)\|^2, \tag{2.113}$$

$$\leq -\mu V(t), \tag{2.114}$$

where $\mu > 0$ is a constant and $g \geq \frac{2|PB|^2}{\lambda_{\min}(Q)}$. Using the comparison principle, we can obtain the exponential stability in the (X, w). Then by the invertible backstepping transformation, the exponential stability in the (X, u) can be implied. Thus the proof is completed. $\qquad\square$

Chapter Three

Basic Idea of Adaptive Control for Single-Input Systems

IN CHAPTER 2, the general class of single-input LTI systems with discrete input delay was taken into account, which is recapped as follows to avoid reading back and forth:

$$\dot{X}(t) = AX(t) + BU(t - D). \tag{3.1}$$

Through the first-order hyperbolic transport PDE state,

$$u(x, t) = U(t + D(x - 1)), \quad x \in [0, 1], \tag{3.2}$$

the LTI-ODE systems (3.1) could be represented by the following ODE-PDE cascades:

$$\dot{X}(t) = AX(t) + Bu(0, t), \quad t \geq 0, \tag{3.3}$$
$$Du_t(x, t) = u_x(x, t), \quad x \in [0, 1], \tag{3.4}$$
$$u(1, t) = U(t). \tag{3.5}$$

In the control scheme of chapter 2, the full information of the plants (3.3)–(3.5) is assumed to be available. In other words,

- the finite-dimensional ODE plant state vector $X(t)$ is available for measurement,
- the infinite-dimensional PDE actuator state $u(x, t)$ that represents the delayed input $U(\eta)$ over the time window $\eta \in [t - D, t]$ is available for measurement,
- the ODE system and input matrices A and B are fully known (every element in both of the matrices is certain and known),
- the length value of the arbitrarily large input delay D is certain and known.

Under the assumption of a perfect prior plant knowledge, the following predictor-based boundary control law is derived:

$$U(t) = u(1, t) = K \left[e^{AD} X(t) + D \int_0^1 e^{AD(1-y)} Bu(y, t) dy \right]. \tag{3.6}$$

However, a perfect prior knowledge of the systems might be hard to obtain in practice. In that situation, the exact predictor feedback (3.6) cannot be applied due to incomplete system information. An effective approach to overcome this shortcoming is adaptive control for uncertain systems, which is a challenging problem in the presence of actuator delays. The existing theoretical results of adaptive control [37, 110, 112] mainly focus on the uncertainty in the plant para- meter but not on the actuator delay. An adaptive backstepping control with input delay has been proposed by [150, 151]; nevertheless, the input delay is mod- eled as some dynamic perturbation and neither identified nor compensated for by the controller.

In order to fill the gap in control theory for systems with delay uncertainty, chapters 3 to 5 are devoted to the adaptive control of uncertain single-input LTI systems with actuator delay. The certainty-equivalence-version adaptive control scheme is based on the PDE framework (3.1)–(3.6), which was investigated in detail in chapter 2. In parallel with the certain ODE-PDE cascades (3.3)–(3.5), the uncertain LTI systems with discrete actuator delay may come with the foll- owing four types of basic uncertainties:

- the actuator delay D has an arbitrarily large length of which the value is un- known,
- the transport PDE state $u(x,t)$, which represents the infinite-dimensional actuator state $U(\eta)$ over the time-interval $\eta \in [t - D, t]$, is unmeasurable,
- the parameters θ in the ODE-plant matrices $A(\theta)$ and $B(\theta)$ are unknown, which might represent uncertain elements in both of the system and input matrices,
- the ODE state $X(t)$ in the finite-dimensional part of the plant is unmeasurable.

As categorized in [88] and chapter 7 of [80], a total of 14 combinations arise from the four basic uncertainties above. Different uncertainty combinations result in different design difficulties. For instance, when the full-state measure- ment of the transport PDE state $u(x,t)$ is available, the global stabilization is obtained. On the other hand, when the transport PDE state is unmeasurable and a PDE adaptive observer is required, only the local stabilization result is achievable because the solution $u(x,t) = U(t + D(x-1))$ is not linearly param- eterizable (statically or dynamically) in the delay D. In addition, the problem of time-varying trajectory tracking requires additional tools, as compared to problems of zero-equilibrium stabilization and constant set-point regulation.

To clearly describe the organization of part I, different combinations of uncer- tainties considered in part I are summarized in table 3.1, from which interested readers and practitioners can simply make their own selections to address a vast class of relevant problems.

3.1   Model Depiction and Basic Idea

Consider a general class of single-input LTI systems with discrete input delay as follows:

Table 3.1. Uncertainty Combinations of Single-Input LTI Systems with Discrete Input Delay

Organization of Part I	Input Delay D	Parameter θ	ODE State $X(t)$	PDE State $u(x,t)$
Chapter 2	known	known	known	known
Section 3.2	known	known	*unknown*	known
Section 3.3	*unknown*	known	known	known
Section 3.4	*unknown*	known	known	*unknown*
Section 3.5	*unknown*	*unknown*	known	known
Section 3.6	*unknown*	known	*unknown*	*unknown*
Section 3.7	*unknown*	*unknown*	known	*unknown*
Chapter 4 and section 5.2	*unknown*	*unknown*	*unknown*	known
Section 5.3	*unknown*	*unknown*	*unknown*	*unknown*

$$\dot{X}(t) = A(\theta)X(t) + B(\theta)U(t-D), \tag{3.7}$$

$$Y(t) = CX(t), \tag{3.8}$$

where $X(t) \in \mathbb{R}^n$ is the plant state vector of the finite-dimensional ODE, $Y(t) \in \mathbb{R}^q$ is the output vector, and $U(t) \in \mathbb{R}$ is the scalar control input with a constant delay $D \in \mathbb{R}^+$. The matrices $A(\theta) \in \mathbb{R}^{n \times n}$, $B(\theta) \in \mathbb{R}^n$, and $C \in \mathbb{R}^{q \times n}$ are the system matrix, the input matrix, and the output matrix, respectively. The constant parameter vector $\theta \in \mathbb{R}^p$ includes plant parameters and control coefficients, and can be linearly parameterized such that

$$A(\theta) = A_0 + \sum_{i=1}^{p} \theta_i A_i, \quad B(\theta) = B_0 + \sum_{i=1}^{p} \theta_i B_i, \tag{3.9}$$

where θ_i is the ith element of the parameter vector θ and A_0, A_i, B_0, and B_i for $i = 1, \cdots, p$ are known matrices. To stabilize the potentially unstable systems (3.7)–(3.8), we have the following assumptions:

Assumption 3.1. *The constant actuator delay satisfies*

$$0 < \underline{D} \leq D \leq \overline{D}, \tag{3.10}$$

where $\underline{D}$ and $\overline{D}$ are known constants representing lower and upper bounds of D, respectively. The plant parameter vector belongs to

$$\Theta = \{\theta \in \mathbb{R}^p | \mathscr{P}(\theta) \leq 0\}, \tag{3.11}$$

where $\mathscr{P}(\cdot):\mathbb{R}^p \to \mathbb{R}$ is a known, convex, and smooth function, and Θ is a convex set with a smooth boundary $\partial\Theta$.

Assumption 3.2. *The pair $(A(\theta), B(\theta))$ is stabilizable. In other words, there exists a matrix $K(\theta)$ such that $A(\theta)+B(\theta)K(\theta)$ is Hurwitz, namely,*

$$(A+BK)^T(\theta)P(\theta)+P(\theta)(A+BK)(\theta)=-Q(\theta), \qquad (3.12)$$

with $P(\theta)=P(\theta)^T>0$ and $Q(\theta)=Q(\theta)^T>0$.

Referring to section 2.4, we introduce the rescaled unity-interval transport PDE to represent the actuator state as follows:

$$u(x,t)=U(t+D(x-1)), \quad x\in[0,1]. \qquad (3.13)$$

Then the LTI system (3.7)–(3.8) is transformed into the ODE-PDE cascade as follows:

$$\dot{X}(t)=A(\theta)X(t)+B(\theta)u(0,t), \quad t\geq 0, \qquad (3.14)$$
$$Y(t)=CX(t), \qquad (3.15)$$
$$Du_t(x,t)=u_x(x,t), \quad x\in[0,1], \qquad (3.16)$$
$$u(1,t)=U(t). \qquad (3.17)$$

The ODE-PDE cascade (3.14)–(3.17) representing the single-input LTI system with discrete actuator delay may come with the following four types of basic uncertainties:

- unknown actuator delay D,
- unknown plant parameter vector θ,
- unmeasurable finite-dimensional plant state $X(t)$,
- unmeasurable infinite-dimensional actuator state $u(x,t)$.

When the four variables above are fully known, a fundamental PDE-based predictor feedback framework is summarized in table 2.3 of section 2.4. When some of the four variables above are unknown or unmeasured, the basic idea of certainty-equivalence-based adaptive control is to use an estimator (a parameter estimator or a state estimator) to replace the unknown variables in the backstepping transformation and the control law, as shown in figure 3.1.

3.2 Global Stabilization under Uncertain ODE State

From this section, different combinations of four uncertainties above are taken into account. Accordingly, for each kind of uncertainty combination, unique control scheme is proposed. First of all, as shown in the second row of table 3.1,

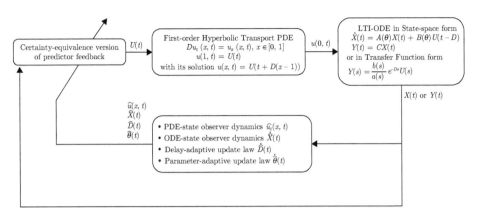

Figure 3.1. Adaptive control framework of uncertain single-input LTI systems with input delay

we consider the ODE-observer-based stabilization when the finite-dimensional plant state $X(t)$ is unmeasurable. Since the parameter vector θ is assumed to be known in this section, for the sake of brevity, we denote

$$A = A(\theta), \quad B = B(\theta). \tag{3.18}$$

Thus the ODE-PDE cascade (3.14)–(3.17) is briefly written as follows:

$$\dot{X}(t) = AX(t) + Bu(0,t), \quad t \geq 0, \tag{3.19}$$
$$Y(t) = CX(t), \tag{3.20}$$
$$Du_t(x,t) = u_x(x,t), \quad x \in [0,1], \tag{3.21}$$
$$u(1,t) = U(t). \tag{3.22}$$

To stabilize the above systems with unmeasured ODE state $X(t)$, we have the following assumption:

Assumption 3.3. *The pair (A,C) is detectable, namely, there exists a matrix $L \in \mathbb{R}^{n \times q}$ to make $A - LC$ Hurwitz such that*

$$(A - LC)^T P_L + P_L(A - LC) = -Q_L, \tag{3.23}$$

with $P_L^T = P_L > 0$ and $Q_L^T = Q_L > 0$.

Then the control scheme is summarized in table 3.2.

Remark 3.4. If we comparing table 3.2 with table 2.3 in section 2.4, it is evident that the main difference is that when the finite-dimensional ODE state $X(t)$ is unmeasurable, the ODE-state $X(t)$ in the control law and the backstepping

Table 3.2. Global Stabilization under Uncertain ODE Plant State

The transport PDE to represent actuator state:

$$u(x,t) = U(t + D(x-1)), \quad x \in [0,1] \tag{3.24}$$

The ODE-PDE cascade:

$$\dot{X}(t) = AX(t) + Bu(0,t), \quad t \geq 0 \tag{3.25}$$
$$Y(t) = CX(t) \tag{3.26}$$
$$Du_t(x,t) = u_x(x,t), \quad x \in [0,1] \tag{3.27}$$
$$u(1,t) = U(t) \tag{3.28}$$

The ODE-state observer:

$$\dot{\hat{X}}(t) = A\hat{X}(t) + Bu(0,t) + L\big(Y(t) - C\hat{X}(t)\big) \tag{3.29}$$

The predictor feedback law:

$$U(t) = u(1,t) = K\left[e^{AD}\hat{X}(t) + D\int_0^1 e^{AD(1-y)}Bu(y,t)dy\right] \tag{3.30}$$

The invertible backstepping transformation:

$$w(x,t) = u(x,t) - K\left[e^{ADx}\hat{X}(t) + D\int_0^x e^{AD(x-y)}Bu(y,t)dy\right] \tag{3.31}$$

$$u(x,t) = w(x,t) + K\left[e^{(A+BK)Dx}\hat{X}(t) + D\int_0^x e^{(A+BK)D(x-y)}Bw(y,t)dy\right]$$
$$\tag{3.32}$$

The closed-loop target system:

$$\dot{\hat{X}}(t) = (A+BK)\hat{X}(t) + Bw(0,t) + LC\tilde{X}(t) \tag{3.33}$$
$$\dot{\tilde{X}}(t) = (A-LC)\tilde{X}(t) \tag{3.34}$$
$$Dw_t(x,t) = w_x(x,t) - DKe^{ADx}LC\tilde{X}(t), \quad x \in [0,1] \tag{3.35}$$
$$w(1,t) = 0 \tag{3.36}$$

where $\tilde{X}(t) = X(t) - \hat{X}(t)$

(continued)

Table 3.2. Continued

The Lyapunov-Krasovskii functional:

$$V(t) = \hat{X}(t)^T P \hat{X}(t) + \tilde{X}(t)^T P_L \tilde{X}(t) + gD \int_0^1 (1+x)w(x,t)^2 dx \qquad (3.37)$$

---

transformation is replaced by its observer $\hat{X}(t)$ and this replacement results in additional terms of the estimation error $\tilde{X}(t) = X(t) - \hat{X}(t)$ in the target system.

Theorem 3.5. *The closed-loop system consisting of the ODE-PDE cascade (3.25)–(3.28), the ODE-state observer (3.29), and the predictor feedback law (3.30) is exponentially stable at the origin in the sense of the norm*

$$\left(|X(t)|^2 + |\hat{X}(t)|^2 + \int_0^1 u(x,t)^2 dt \right)^{\frac{1}{2}}. \qquad (3.38)$$

Proof. If we take the time derivative of $V(t)$ of (3.37) along (3.33)–(3.36), by a variety of techniques of inequalities, it is not hard to show the nonpositive property of $\dot{V}(t)$, and consequently the global stability of the closed-loop system is obtained.   □

3.3   Global Stabilization under Uncertain Delay

In this section, as shown in the third row of table 3.1, we consider the adaptive stabilization when the input delay D is unknown. The control scheme is summarized in table 3.3.

Remark 3.6. If we compare table 3.3 with table 2.3 in section 2.4, it is evident that the main difference is that when the actuator delay D is unknown, the input delay D in the control law and the backstepping transformation is replaced by its estimate $\hat{D}(t)$ and this replacement results in additional terms of the estimation error $\tilde{D}(t) = D - \hat{D}(t)$ and the estimate derivative $\dot{\hat{D}}(t)$ in the target system. Furthermore, the numerator of the delay update law (3.47) is a product of the regulation error $\int_0^1 (1+x)w(x,t)Ke^{A\hat{D}(t)x}dx$ and the regressor $AX(t) + Bu(0,t)$, whereas the denominator of the delay update law is employed to bound cubic and biquadratic terms in the Lyapunov-based analysis by inequalities $\frac{|x|}{1+x^2} \leq 1$ and $\frac{x^2}{1+x^2} \leq 1$. Please note that the numerator of the delay update law (3.47) corresponds to the term $p(x,t)$ of (3.51) multiplying the estimation error $\tilde{D}(t)$

Table 3.3. Global Stabilization under Uncertain Input Delay

The transport PDE to represent actuator state:

$$u(x,t) = U(t + D(x-1)), \quad x \in [0,1] \tag{3.39}$$

The ODE-PDE cascade:

$$\dot{X}(t) = AX(t) + Bu(0,t), \quad t \geq 0 \tag{3.40}$$
$$Du_t(x,t) = u_x(x,t), \quad x \in [0,1] \tag{3.41}$$
$$u(1,t) = U(t) \tag{3.42}$$

The predictor feedback law:

$$U(t) = u(1,t) = K\left[e^{A\hat{D}(t)}X(t) + \hat{D}(t)\int_0^1 e^{A\hat{D}(t)(1-y)}Bu(y,t)dy\right] \tag{3.43}$$

The invertible backstepping transformation:

$$w(x,t) = u(x,t) - K\left[e^{A\hat{D}(t)x}X(t) + \hat{D}(t)\int_0^x e^{A\hat{D}(t)(x-y)}Bu(y,t)dy\right] \tag{3.44}$$

$$u(x,t) = w(x,t) + K\left[e^{(A+BK)\hat{D}(t)x}X(t) + \hat{D}(t)\int_0^x e^{(A+BK)\hat{D}(t)(x-y)}Bw(y,t)dy\right] \tag{3.45}$$

The delay update law:

$$\dot{\hat{D}}(t) = \gamma_D \text{Proj}_{[\underline{D},\overline{D}]}\tau_D(t), \quad \hat{D}(0) \in [\underline{D},\overline{D}] \tag{3.46}$$

$$\tau_D(t) = -\frac{\int_0^1(1+x)w(x,t)Ke^{A\hat{D}(t)x}dx(AX(t)+Bu(0,t))}{1+X(t)^TPX(t)+g\int_0^1(1+x)w(x,t)^2dx} \tag{3.47}$$

The closed-loop target system:

$$\dot{X}(t) = (A+BK)X(t) + Bw(0,t) \tag{3.48}$$
$$Dw_t(x,t) = w_x(x,t) - \tilde{D}(t)p(x,t) - D\dot{\hat{D}}(t)q(x,t), \quad x \in [0,1] \tag{3.49}$$
$$w(1,t) = 0 \tag{3.50}$$

where $\tilde{D}(t) = D - \hat{D}(t)$,

(*continued*)

Table 3.3. Continued

$$p(x,t) = Ke^{A\hat{D}(t)x}\left(AX(t) + Bu(0,t)\right) = Ke^{A\hat{D}(t)x}\left((A+BK)X(t) + Bw(0,t)\right)$$
$$(3.51)$$

$$q(x,t) = KAxe^{A\hat{D}(t)x}X(t) + \int_0^x K(I + A\hat{D}(t)(x-y))e^{A\hat{D}(t)(x-y)}Bu(y,t)dy$$
$$(3.52)$$

The Lyapunov-Krasovskii functional:

$$V(t) = D\log\left(1 + X(t)^T PX(t) + g\int_0^1 (1+x)w(x,t)^2 dx\right) + \frac{g}{\gamma_D}\tilde{D}(t)^2 \quad (3.53)$$

---

in the target system (3.49), whereas the denominator of the delay update law (3.47) corresponds to the term in the log function of the Lyapunov function (3.53). The whole delay update law is used to remove the estimation error term $\tilde{D}(t)$ in the Lyapunov-based stability analysis.

Theorem 3.7. *The closed-loop system consisting of the ODE-PDE cascade (3.40)–(3.42), the predictor feedback law (3.43), and the delay update law (3.46)–(3.47) is stable at the zero solution in the sense of the norm*

$$\left(|X(t)|^2 + \int_0^1 u(x,t)^2 dt + |D - \hat{D}(t)|^2\right)^{\frac{1}{2}} \qquad (3.54)$$

and the convergence $\lim_{t\to\infty} X(t) = 0$ *and* $\lim_{t\to\infty} U(t) = 0$ *is achieved.*

Proof. If we take the time derivative of $V(t)$ of (3.53) along (3.48)–(3.50), by a variety of techniques of inequalities, it is not hard to show the nonpositive property of $\dot{V}(t)$, and consequently the global stability of the closed-loop system is obtained. □

3.4 Local Stabilization under Uncertain Delay and Actuator State

In this section, as shown in the fourth row of table 3.1, we consider the adaptive stabilization when the input delay D is unknown and the actuator state $u(x,t)$ is unmeasurable.

3.4.1   Robustness to Delay Mismatch under Uncertain Delay and Actuator State

To help readers better understand the adaptive control scheme under coexistent uncertainties in both delay and actuator state, we first give a robust control design and evaluate its handling of the delay mismatch. The control scheme is summarized in table 3.4.

Table 3.4. Robustness to Delay Mismatch under Uncertain Input Delay and Actuator State

---

The transport PDE to represent actuator state:

$$u(x,t) = U\big(t + D(x-1)\big), \quad x \in [0,1] \tag{3.55}$$

The PDE-state observer:

$$\hat{D}\hat{u}_t(x,t) = \hat{u}_x(x,t), \quad x \in [0,1] \tag{3.56}$$
$$\hat{u}(1,t) = U(t) \tag{3.57}$$

with its solution being $\hat{u}(x,t) = U\big(t + \hat{D}(x-1)\big)$, $x \in [0,1]$, where $\hat{D}$ is a constant estimate of unknown delay D with the update law $\dot{\hat{D}} = 0$

The ODE-PDE cascade:

$$\dot{X}(t) = AX(t) + Bu(0,t), \quad t \geq 0 \tag{3.58}$$
$$Du_t(x,t) = u_x(x,t), \quad x \in [0,1] \tag{3.59}$$
$$u(1,t) = U(t) \tag{3.60}$$

The predictor feedback law:

$$U(t) = \hat{u}(1,t) = K\left[e^{A\hat{D}} X(t) + \hat{D} \int_0^1 e^{A\hat{D}(1-y)} B\hat{u}(y,t)dy \right] \tag{3.61}$$

The invertible backstepping transformation:

$$\hat{w}(x,t) = \hat{u}(x,t) - K\left[e^{A\hat{D}x} X(t) + \hat{D} \int_0^x e^{A\hat{D}(x-y)} B\hat{u}(y,t)dy \right] \tag{3.62}$$

$$\hat{u}(x,t) = \hat{w}(x,t) + K\left[e^{(A+BK)\hat{D}x} X(t) + \hat{D} \int_0^x e^{(A+BK)\hat{D}(x-y)} B\hat{w}(y,t)dy \right] \tag{3.63}$$

(*continued*)

Table 3.4. Continued

The closed-loop target system:

$$\dot{X}(t) = (A + BK)X(t) + B\hat{w}(0,t) + B\tilde{u}(0,t) \qquad (3.64)$$

$$D\tilde{u}_t(x,t) = \tilde{u}_x(x,t) - \tilde{D}r(x,t) \qquad (3.65)$$

$$\tilde{u}(1,t) = 0 \qquad (3.66)$$

$$\hat{D}\hat{w}_t(x,t) = \hat{w}_x(x,t) - \hat{D}Ke^{A\hat{D}x}B\tilde{u}(0,t) \qquad (3.67)$$

$$\hat{w}(1,t) = 0 \qquad (3.68)$$

where $\tilde{D} = D - \hat{D}$,

$$\tilde{u}(x,t) = u(x,t) - \hat{u}(x,t) \qquad (3.69)$$

$$r(x,t) = \frac{\hat{w}_x(x,t)}{\hat{D}} + K(A + BK)e^{(A+BK)\hat{D}x}X(t)$$

$$+ KB\hat{w}(x,t) + \hat{D}\int_0^x K(A+BK)e^{(A+BK)\hat{D}(x-y)}B\hat{w}(y,t)\,dy \qquad (3.70)$$

The Lyapunov-Krasovskii functional:

$$V(t) = X(t)^T P X(t) + gD\int_0^1 (1+x)\tilde{u}(x,t)^2 dx$$

$$+ g\hat{D}\left(\int_0^1 (1+x)\hat{w}(x,t)^2 dx + \int_0^1 (1+x)\hat{w}_x(x,t)^2 dx \right) \qquad (3.71)$$

Remark 3.8. If we compare table 3.4 with table 2.3 in section 2.4, it is obvious that the main difference is that the unknown actuator delay D and the unmeasured actuator state $u(x,t)$ in the control law and the backstepping transformation are replaced by the constant delay estimate $\hat{D}$ and PDE-state observer $\hat{u}(x,t)$. This replacement adds extra terms of the estimation error $\tilde{D} = D - \hat{D}$ and observer error $\tilde{u}(x,t) = u(x,t) - \hat{u}(x,t)$ to the target system. Furthermore, the reader may view (3.56)–(3.57) as an open-loop certainty equivalence observer of the transport equation (3.59)–(3.60). By "certainty equivalence" we are referring to the fact that the parameter D is replaced by the constant estimate $\hat{D}$ and there exists an estimation error $\tilde{D} = D - \hat{D}$, which could be treated as the delay mismatch that is either positive or negative, whereas by "open-loop" we are referring to the fact that no output injection is used in the observer since the transport equation is open-loop exponentially stable.

Theorem 3.9. *The closed-loop system consisting of the ODE-PDE cascade (3.58)–(3.60), the PDE-state observer (3.56)–(3.57), and the predictor feedback* *law (3.61) is exponentially stable at the zero solution in the sense of the norm*

$$\left(|X(t)|^2 + \int_0^1 u(x,t)^2 dt + \int_0^1 \hat{u}(x,t)^2 dt + \int_0^1 \hat{u}_x(x,t)^2 dt \right)^{\frac{1}{2}} \qquad (3.72)$$

if there exists $M > 0$ such that the estimation error of the constant estimate satisfies

$$|\tilde{D}| = |D - \hat{D}| \le M. \qquad (3.73)$$

Proof. If we take the time derivative of $V(t)$ of (3.71) along (3.64)–(3.68), by a variety of techniques of inequalities, it is not hard to show the nonpositive property of $\dot{V}(t)$, and consequently the stability of the closed-loop system is obtained.   □

3.4.2   Adaptive Stabilization under Uncertain Delay and Actuator State

In this section, as shown in the fourth row of table 3.1, we consider the adaptive stabilization when the input delay D is unknown and the actuator state $u(x,t)$ is unmeasurable. The delay-adaptive problem without the measurement of $u(x,t)$ is unsolvable globally because it cannot be formulated as linearly parameterized in the unknown delay D. That is to say, when the controller uses an estimate of $u(x,t)$, not only do the initial values of the ODE state and the actuator state have to be small, but the initial value of the delay estimation error also has to be small (the delay value is allowed to be large but the initial value of its estimate has to be close to the true value of the delay). The control scheme is summarized in table 3.5.

Remark 3.10. If we compare table 3.5 with table 2.3 in section 2.4, it is obvious that the main difference is that the unknown actuator delay D and the unmeasured actuator state $u(x,t)$ in the control law the delay update law and the backstepping transformation are replaced by the delay estimate $\hat{D}(t)$ and PDE-state observer $\hat{u}(x,t)$. This replacement adds extra terms of the estimation error $\tilde{D}(t) = D - \hat{D}(t)$ and observer error $\tilde{u}(x,t) = u(x,t) - \hat{u}(x,t)$ to the target system. Furthermore, please note that the denominator of the delay update law in (3.47) is absent from (3.84). This is because the global result is not obtainable, so the normalization by the denominator is not needed.

Theorem 3.11. *The closed-loop system consisting of the ODE-PDE cascade (3.77)–(3.79), the PDE-state observer (3.75)–(3.76), the predictor feedback law (3.80), and the delay update law (3.83)–(3.84) is stable at the zero solution in the sense of the norm*

Table 3.5. Local Adaptive Stabilization under Uncertain Input Delay and Actuator State

---

The transport PDE to represent actuator state:

$$u(x,t) = U\big(t + D(x-1)\big), \quad x \in [0,1] \tag{3.74}$$

The PDE-state observer:

$$\hat{D}(t)\hat{u}_t(x,t) = \hat{u}_x(x,t) + \dot{\hat{D}}(t)(x-1)\hat{u}_x(x,t), \quad x \in [0,1] \tag{3.75}$$

$$\hat{u}(1,t) = U(t) \tag{3.76}$$

with its solution being $\hat{u}(x,t) = U\big(t + \hat{D}(t)(x-1)\big)$, $x \in [0,1]$

The ODE-PDE cascade:

$$\dot{X}(t) = AX(t) + Bu(0,t), \quad t \geq 0 \tag{3.77}$$

$$Du_t(x,t) = u_x(x,t), \quad x \in [0,1] \tag{3.78}$$

$$u(1,t) = U(t) \tag{3.79}$$

The predictor feedback law:

$$U(t) = \hat{u}(1,t) = K\left[e^{A\hat{D}(t)} X(t) + \hat{D}(t) \int_0^1 e^{A\hat{D}(t)(1-y)} B\hat{u}(y,t)dy \right] \tag{3.80}$$

The invertible backstepping transformation:

$$\hat{w}(x,t) = \hat{u}(x,t) - K\left[e^{A\hat{D}(t)x} X(t) + \hat{D}(t) \int_0^x e^{A\hat{D}(t)(x-y)} B\hat{u}(y,t)dy \right] \tag{3.81}$$

$$\hat{u}(x,t) = \hat{w}(x,t) + K\left[e^{(A+BK)\hat{D}(t)x} X(t) + \hat{D}(t) \int_0^x e^{(A+BK)\hat{D}(t)(x-y)} B\hat{w}(y,t)dy \right] \tag{3.82}$$

The delay update law:

$$\dot{\hat{D}}(t) = \gamma_D \mathrm{Proj}_{[\underline{D},\overline{D}]} \tau_D(t), \quad \hat{D}(0) \in [\underline{D}, \overline{D}] \tag{3.83}$$

$$\tau_D(t) = -\int_0^1 (1+x)\hat{w}(x,t) K e^{A\hat{D}(t)x} dx \big(AX(t) + B\hat{u}(0,t)\big) \tag{3.84}$$

(continued)

Table 3.5. Continued

The closed-loop target system:

$$\dot{X}(t) = (A + BK)X(t) + B\hat{w}(0,t) + B\tilde{u}(0,t) \tag{3.85}$$

$$D\tilde{u}_t(x,t) = \tilde{u}_x(x,t) - \tilde{D}(t)r(x,t) - D\dot{\hat{D}}(t)(x-1)r(x,t) \tag{3.86}$$

$$\tilde{u}(1,t) = 0 \tag{3.87}$$

$$\hat{D}(t)\hat{w}_t(x,t) = \hat{w}_x(x,t) - \hat{D}(t)\dot{\hat{D}}(t)s(x,t) - \hat{D}(t)Ke^{A\hat{D}(t)x}B\tilde{u}(0,t) \tag{3.88}$$

$$\hat{w}(1,t) = 0 \tag{3.89}$$

where $\tilde{D}(t) = D - \hat{D}(t)$, $\tilde{u}(x,t) = u(x,t) - \hat{u}(x,t)$, $r(x,t) = \frac{\hat{w}_x(x,t)}{\hat{D}(t)} + K(A + BK) \times$
$e^{(A+BK)\hat{D}(t)x}X(t) + KB\hat{w}(x,t) + \hat{D}(t)\int_0^x K(A + BK)e^{(A+BK)\hat{D}(t)(x-y)}B\hat{w}(y,t)$
dy, $s(x,t) = (1 - x)\frac{\hat{w}_x(x,t)}{\hat{D}(t)} + KAe^{A\hat{D}(t)x}X(t) + Ke^{A\hat{D}(t)x}B(\hat{w}(0,t) + KX(t))$

The Lyapunov-Krasovskii functional:

$$V(t) = X(t)^T P X(t) + gD \int_0^1 (1+x)\tilde{u}(x,t)^2 dx + \frac{g}{\gamma_D}\tilde{D}(t)^2$$

$$+ g\hat{D}\left(\int_0^1 (1+x)\big(\hat{w}(x,t) + \hat{w}_x(x,t)^2\big)dx\right) \tag{3.90}$$

$$\left(|X(t)|^2 + \int_0^1 u(x,t)^2 dt + \int_0^1 \hat{u}(x,t)^2 dt + \int_0^1 \hat{u}_x(x,t)^2 dt + |D - \hat{D}(t)|^2\right)^{\frac{1}{2}} \tag{3.91}$$

and the convergence $\lim_{t\to\infty} X(t) = 0$ and $\lim_{t\to\infty} U(t) = 0$ is achieved, if there exists $M > 0$ such that the initial condition

$$\left(|X(0)|^2 + \int_0^1 u(x,0)^2 dt + \int_0^1 \hat{u}(x,0)^2 dt + \int_0^1 \hat{u}_x(x,0)^2 dt + |D - \hat{D}(0)|^2\right)^{\frac{1}{2}} \le M \tag{3.92}$$

is satisfied.

Proof. If we take the time derivative of $V(t)$ of (3.90) along (3.85)–(3.89), by a variety of techniques of inequalities and the initial condition, it is not hard to show the non positive property of $\dot{V}(t)$, and consequently the local stability of the closed-loop system is obtained. □

3.5 Global Trajectory Tracking under Uncertain Delay and Parameters

In this section, as shown in the fifth row of table 3.1, we consider the trajectory tracking when both the input delay D and the parameter vector θ are unknown. Instead of the equilibrium stabilization, the trajectory tracking is pursued. The following assumption holds for the reference trajectory.

Assumption 3.12. *For a given smooth time-varying output trajectory $Y^r(t)$, there exist known reference state and reference input functions $X^r(t, \theta)$ and $U^r(t, \theta)$, which are bounded in t and continuously differentiable in the parameter argument θ, and satisfy*

$$\dot{X}^r(t, \theta) = A(\theta)X^r(t, \theta) + B(\theta)U^r(t, \theta), \tag{3.93}$$

$$Y^r(t) = CX^r(t, \theta). \tag{3.94}$$

Remark 3.13. To apply the predictor feedback to solve the tracking problem of linear systems with discrete input delays, the model of the reference system is required to match with the model of the original system. That is the reason why Assumption 3.12 is made. It is evident that the structure of the reference system (3.93)–(3.94) is matched with the structure of the original system (3.7)–(3.8). This "matching" requirement is reasonable. The control objective is to make the system output $Y(t)$ track a desired reference trajectory $Y^r(t)$ that is prespecified. If the parameter θ is known, under Assumption 3.12, the given reference output $Y^r(t)$ could be produced by a pair of reference state and input $(X^r(t, \theta), U^r(t, \theta))$ through the dynamics (3.93)–(3.94), which has a matched structure with (3.7)–(3.8). Thus, the tracking problem $Y(t) \to Y^r(t)$ is equivalent to the tracking problem $X(t) \to X^r(t, \theta), U(t - D) \to U^r(t, \theta)$ and the predictor feedback can be applied to the tracking error system which is derived from (3.7) minus (3.93). If the parameter θ is unknown, replacing θ with its estimate $\hat{\theta}(t)$, the reference system (3.99) is derived, which could be regarded as a certainty-equivalence version of (3.93)–(3.94).

The control scheme is summarized in table 3.6.

Remark 3.14. If we compare table 3.6 with table 2.3 in section 2.4, the unknown actuator delay D and the unknown plant parameter θ in the control law, the update law, and the backstepping transformation are replaced by the delay estimate $\hat{D}(t)$ and parameter estimate $\hat{\theta}(t)$. This replacement adds extra terms of the estimation errors $\tilde{D}(t) = D - \hat{D}(t)$ and $\tilde{\theta}(t) = \theta - \hat{\theta}(t)$ and the estimate derivatives $\dot{D}(t)$ and $\dot{\theta}(t)$ to the target system. Furthermore, please note that the denominator of the update laws is used for normalization again. This is because the global result is obtainable when the measurement of actuator state $u(x,t)$ is assumed.

Table 3.6. Global Trajectory Tracking under Uncertain Input Delay and Plant Parameters

---

The transport PDE to represent actuator state:

$$u(x,t) = U\big(t + D(x-1)\big), \quad x \in [0,1] \tag{3.95}$$

The ODE-PDE cascade:

$$\dot{X}(t) = A(\theta)X(t) + B(\theta)u(0,t), \quad t \geq 0 \tag{3.96}$$
$$Du_t(x,t) = u_x(x,t), \quad x \in [0,1] \tag{3.97}$$
$$u(1,t) = U(t) \tag{3.98}$$

and its reference trajectory:

$$\dot{X}^r(t,\hat{\theta}) = A(\hat{\theta})X^r(t,\hat{\theta}) + B(\hat{\theta})U^r(t,\hat{\theta}) + \frac{\partial X^r}{\partial \hat{\theta}}(t,\hat{\theta})\dot{\hat{\theta}}(t) \tag{3.99}$$

The predictor feedback law:

$$U(t) = U^r(t + \hat{D},\hat{\theta}) - K(\hat{\theta})X^r(t + \hat{D},\hat{\theta})$$

$$+ K(\hat{\theta})\left[e^{A(\hat{\theta})\hat{D}}X(t) + \hat{D}\int_0^1 e^{A(\hat{\theta})\hat{D}(1-y)}B(\hat{\theta})u(y,t)dy\right]$$

$$\tag{3.100}$$

The invertible backstepping transformation:

$$w(x,t) = e(x,t) - K(\hat{\theta})\left[e^{A(\hat{\theta})\hat{D}x}\tilde{X}(t) + \hat{D}\int_0^x e^{A(\hat{\theta})\hat{D}(x-y)}B(\hat{\theta})e(y,t)dy\right] \tag{3.101}$$

$$e(x,t) = w(x,t) + K(\hat{\theta})\left[e^{(A+BK)(\hat{\theta})\hat{D}x}\tilde{X}(t)\hat{D}\int_0^x e^{(A+BK)(\hat{\theta})\hat{D}(x-y)}B(\hat{\theta})w(y,t)dy\right]$$

$$\tag{3.102}$$

where $\tilde{X}(t) = X(t) - X^r(t,\hat{\theta})$ and $e(x,t) = u(x,t) - u^r(x,t,\hat{\theta})$

(continued)

Table 3.6. Continued

The delay and parameter update law:

$$\dot{\hat{D}}(t) = \gamma_D \text{Proj}_{[\underline{D}, \overline{D}]} \tau_D(t), \quad \hat{D}(0) \in [\underline{D}, \overline{D}] \tag{3.103}$$

$$\tau_D(t) = -\frac{\int_0^1 (1+x)w(x,t)K(\hat{\theta})e^{A(\hat{\theta})\hat{D}x}dx\left((A+BK)(\hat{\theta})\tilde{X}(t)+B(\hat{\theta})w(0,t)\right)}{1+\tilde{X}^T(t)P(\hat{\theta})\tilde{X}(t)+g\int_0^1(1+x)w(x,t)^2dx} \tag{3.104}$$

$$\dot{\hat{\theta}}_i(t) = \gamma_\theta \text{Proj}_\Theta \tau_{\theta_i}(t), \quad \hat{\theta}(0) \in \Theta \tag{3.105}$$

$$\tau_{\theta_i}(t) = \frac{\left(\frac{2}{g}\tilde{X}^T(t)P(\hat{\theta}) - \int_0^1(1+x)w(x,t)K(\hat{\theta})e^{A(\hat{\theta})\hat{D}x}dx\right)\left(A_iX(t)+B_iu(0,t)\right)}{1+\tilde{X}^T(t)P(\hat{\theta})\tilde{X}(t)+g\int_0^1(1+x)w(x,t)^2dx} \tag{3.106}$$

The closed-loop target system:

$$\dot{\tilde{X}}(t) = (A+BK)(\hat{\theta})\tilde{X}(t)+B(\hat{\theta})w(0,t)+A(\tilde{\theta})X(t)+B(\tilde{\theta})u(0,t) - \frac{\partial X^r}{\partial\hat{\theta}}(t,\hat{\theta})\dot{\hat{\theta}} \tag{3.107}$$

$$Dw_t(x,t) = w_x(x,t) - \tilde{D}(t)p_0(x,t) - D\dot{\hat{D}}(t)q_0(x,t) - D\tilde{\theta}^T(t)p(x,t) - D\dot{\hat{\theta}}^T(t)q(x,t) \tag{3.108}$$

$$w(1,t) = 0 \tag{3.109}$$

where $\tilde{D}(t) = D - \hat{D}(t)$, $\tilde{\theta}(t) = \theta - \hat{\theta}(t)$, $A(\tilde{\theta}) = \sum_{i=1}^p \tilde{\theta}_i A_i$, $B(\tilde{\theta}) = \sum_{i=1}^p \tilde{\theta}_i B_i$

The Lyapunov-Krasovskii functional:

$$V(t) = D\log\left(1+\tilde{X}(t)^T P(\hat{\theta})\tilde{X}(t)+g\int_0^1(1+x)w(x,t)^2dx\right)$$

$$+\frac{g}{\gamma_D}\tilde{D}(t)^2 + \frac{gD}{\gamma_\theta}\tilde{\theta}^T(t)\tilde{\theta}(t) \tag{3.110}$$

Theorem 3.15. *Under Assumptions 3.1, 3.2, and 3.12, the closed-loop system consisting of the ODE-PDE cascade (3.96)–(3.98), the predictor feedback law (3.100), and the delay and parameter update laws (3.103)–(3.106) is stable at the zero solution in the sense of the norm*

$$\left(|\tilde{X}(t)|^2 + \int_0^1 e(x,t)^2 dx + |\tilde{D}(t)|^2 + |\tilde{\theta}(t)|^2 \right)^{\frac{1}{2}}, \tag{3.111}$$

and the convergence $\lim_{t \to \infty} \tilde{X}(t) = 0$ *is achieved.*

Proof. If we take the time derivative of $V(t)$ of (3.110) along (3.107)–(3.109), by a variety of techniques of inequalities, it is not hard to show the nonpositive property of $\dot{V}(t)$, and consequently the global trajectory tracking of the closed-loop system is obtained. □

3.6 Local Set-Point Regulation under Uncertain Delay and ODE and PDE States

In this section, as shown in the sixth row of table 3.1, we consider the adaptive set-point regulation when the input delay D is unknown and the ODE plant state $X(t)$ and the PDE actuator state $u(x,t)$ are unmeasurable. As mentioned before, the delay-adaptive problem without the measurement of $u(x,t)$ is unsolvable globally. Instead, the local result is acquirable dependent upon initial conditions. To achieve the constant set-point regulation, we have the following assumption:

Assumption 3.16. *For a given set-point Y^r, there exist known state and input variables $X^r(\theta)$ and $U^r(\theta)$ continuously differentiable in the parameter argument θ, and satisfying*

$$0 = A(\theta)X^r(\theta) + B(\theta)U^r(\theta), \tag{3.112}$$

$$Y^r = CX^r(\theta). \tag{3.113}$$

The control scheme proposed by section 6 of [25] is summarized in table 3.7.

Remark 3.17. If we compare table 3.7 with table 2.3 in section 2.4, the unknown actuator delay D and the unmeasured ODE plant state $X(t)$ and the PDE actuator state $u(x,t)$ in the control law, the delay update law, and the backstepping transformation are replaced by the delay estimate $\hat{D}(t)$, the ODE-state observer $\hat{X}(t)$, and PDE-state observer $\hat{u}(x,t)$. This replacement adds extra terms of the estimation error and observer error to the target system.

Theorem 3.18. *Under Assumption 3.1, 3.2, and 3.16, the closed-loop system consisting of the ODE-PDE cascade (3.117)–(3.119), the PDE-state observer (3.115)–(3.116), the ODE-state observer (3.120), and the predictor feedback law (3.121) is stable at the zero solution in the sense of the norm*

Table 3.7. Local Set-Point Regulation under Uncertain Delay, ODE, and PDE State

---

The transport PDE to represent actuator state:

$$u(x,t) = U\big(t + D(x-1)\big), \quad x \in [0,1] \tag{3.114}$$

The PDE-state observer:

$$\hat{D}(t)\hat{u}_t(x,t) = \hat{u}_x(x,t) + \dot{\hat{D}}(t)(x-1)\hat{u}_x(x,t), \quad x \in [0,1] \tag{3.115}$$

$$\hat{u}(1,t) = U(t) \tag{3.116}$$

with its solution being $\hat{u}(x,t) = U\big(t + \hat{D}(t)(x-1)\big), x \in [0,1]$

The ODE-PDE cascade:

$$\dot{X}(t) = AX(t) + Bu(0,t), \quad t \geq 0 \tag{3.117}$$

$$Du_t(x,t) = u_x(x,t), \quad x \in [0,1] \tag{3.118}$$

$$u(1,t) = U(t) \tag{3.119}$$

The ODE-state observer:

$$\dot{\hat{X}}(t) = A\hat{X}(t) + B\hat{u}(0,t) + L\big(Y(t) - C\hat{X}(t)\big) \tag{3.120}$$

The predictor feedback law:

$$U(t) = U^r - KX^r + K\left[e^{A\hat{D}(t)}\hat{X}(t) + \hat{D}(t)\int_0^1 e^{A\hat{D}(t)(1-y)}B\hat{u}(y,t)dy\right] \tag{3.121}$$

The invertible backstepping transformation:

$$\hat{w}(x,t) = \hat{e}(x,t) - K\left[e^{A\hat{D}(t)x}\Delta\hat{X}(t) + \hat{D}(t)\int_0^x e^{A\hat{D}(t)(x-y)}B\hat{e}(y,t)dy\right] \tag{3.122}$$

$$\hat{e}(x,t) = \hat{w}(x,t) + K\left[e^{(A+BK)\hat{D}(t)x}\Delta\hat{X}(t)\hat{D}(t)\int_0^x e^{(A+BK)\hat{D}(t)(x-y)}B\hat{w}(y,t)dy\right] \tag{3.123}$$

where $\hat{e}(x,t) = \hat{u}(x,t) - u^r$, $\Delta\hat{X}(t) = \hat{X}(t) - X^r$

(*continued*)

Table 3.7. Continued

The closed-loop target system:

$$\dot{\tilde{X}}(t) = (A - LC)\tilde{X}(t) + B\tilde{e}(0, t) \qquad (3.124)$$

$$\frac{d\Delta\hat{X}}{dt}(t) = (A + BK)\Delta\hat{X}(t) + B\hat{w}(0, t) + LC\tilde{X}(t) \qquad (3.125)$$

$$D\tilde{e}_t(x, t) = \tilde{e}_x(x, t) - \tilde{D}(t)f(x, t) \qquad (3.126)$$

$$\tilde{e}(1, t) = 0 \qquad (3.127)$$

$$\hat{D}\hat{w}_t(x, t) = \hat{w}_x(x, t) - \hat{D}Ke^{A\hat{D}x}LC\tilde{X}(t) \qquad (3.128)$$

$$\hat{w}(1, t) = 0 \qquad (3.129)$$

where $\tilde{D}(t) = D - \hat{D}(t)$, $\tilde{X}(t) = X(t) - \hat{X}(t)$, $\tilde{e}(x, t) = u(x, t) - \hat{u}(x, t)$. For the sake of simplicity, the delay update law is assumed to be $\dot{\hat{D}}(t) = 0$.

The Lyapunov-Krasovskii functional:

$$V(t) = \Delta\hat{X}^T(t)P\Delta\hat{X}(t) + \tilde{X}^T(t)P_L\tilde{X}(t) + gD\int_0^1 (1 + x)\tilde{e}(x, t)^2 dx + \frac{g}{\gamma_D}\tilde{D}(t)^2$$

$$+ g\hat{D}\left(\int_0^1 (1 + x)\hat{w}(x, t)^2 dx + \int_0^1 (1 + x)\hat{w}_x(x, t)^2 dx\right) \qquad (3.130)$$

$$\left(|\Delta X(t)|^2 + |\Delta\hat{X}(t)|^2 + \|e(x, t)\|^2 + \|\hat{e}(x, t)\|^2 + \|\hat{e}_x(x, t)\|^2 + |D - \hat{D}(t)|^2\right)^{\frac{1}{2}}, \qquad (3.131)$$

and the convergence $\lim_{t\to\infty}(X(t) - X^r) = 0$ is achieved, where $\Delta X(t) = X(t) - X^r$, $\Delta\hat{X}(t) = \hat{X}(t) - X^r$, $e(x, t) = u(x, t) - u^r$, $\hat{e}(x, t) = \hat{u}(x, t) - u^r$, if there exists $M > 0$ such that the initial condition

$$\left(|\Delta X(0)|^2 + |\Delta\hat{X}(0)|^2 + \|e(x, 0)\|^2 + \|\hat{e}(x, 0)\|^2\right.$$

$$\left. + \|\hat{e}_x(x, 0)\|^2 + |D - \hat{D}(0)|^2\right)^{\frac{1}{2}} \le M \qquad (3.132)$$

is satisfied.

Proof. If we take the time derivative of $V(t)$ of (3.130) along (3.124)–(3.129), by a variety of techniques of inequalities and the initial condition, it is not hard to show the nonpositive property of $\dot{V}(t)$, and consequently the local stability of the closed-loop system is obtained.   □

3.7   Local Set-Point Regulation under Uncertain Delay, Parameters, and PDE State

In this section, as shown in the seventh row of table 3.1, we consider the adaptive set-point regulation when the input delay D and plant parameter θ are unknown and the PDE actuator state $u(x,t)$ is unmeasurable. As mentioned before, the delay-adaptive problem without the measurement of $u(x,t)$ is unsolvable globally. Instead, the local result is acquirable with restriction on initial conditions. The control scheme proposed by section 7 of [25] is summarized in table 3.8.

Remark 3.19. If we compare table 3.8 with table 2.3 in section 2.4, the unknown actuator delay D, the unknown parameter vector θ and the PDE actuator state $u(x,t)$ in the control law, the update laws, and the backstepping transformation are replaced by the delay estimate $\hat{D}(t)$, the parameter estimate $\hat{\theta}(t)$, and PDE-state observer $\hat{u}(x,t)$. This replacement adds extra terms of the estimation error and observer error to the target system.

Table 3.8. Local Set-Point Regulation under Uncertain Delay, Parameters, and PDE State

---

The transport PDE to represent actuator state:

$$u(x,t) = U\big(t + D(x-1)\big), \quad x \in [0,1] \tag{3.133}$$

The PDE-state observer:

$$\hat{D}(t)\hat{u}_t(x,t) = \hat{u}_x(x,t) + \dot{\hat{D}}(t)(x-1)\hat{u}_x(x,t), \quad x \in [0,1] \tag{3.134}$$

$$\hat{u}(1,t) = U(t) \tag{3.135}$$

with its solution being $\hat{u}(x,t) = U\big(t + \hat{D}(t)(x-1)\big), x \in [0,1]$

The ODE-PDE cascade:

$$\dot{X}(t) = AX(t) + Bu(0,t), \quad t \geq 0 \tag{3.136}$$

$$Du_t(x,t) = u_x(x,t), \quad x \in [0,1] \tag{3.137}$$

$$u(1,t) = U(t) \tag{3.138}$$

(*continued*)

Table 3.8. Continued

The predictor feedback law:

$$U(t) = U^r(\hat{\theta}) - K(\hat{\theta})X^r(\hat{\theta}) + K(\hat{\theta})\left[e^{A(\hat{\theta})\hat{D}}X(t) + \hat{D}\int_0^1 e^{A(\hat{\theta})\hat{D}(1-y)}B(\hat{\theta})\hat{u}(y,t)dy\right]$$

$$(3.139)$$

The invertible backstepping transformation:

$$\hat{w}(x,t) = \hat{e}(x,t) - K(\hat{\theta})\left[e^{A(\hat{\theta})\hat{D}x}\tilde{X}(t) + \hat{D}\int_0^x e^{A(\hat{\theta})\hat{D}(x-y)}B(\hat{\theta})\hat{e}(y,t)dy\right] \quad (3.140)$$

$$\hat{e}(x,t) = \hat{w}(x,t) + K(\hat{\theta})\left[e^{(A+BK)(\hat{\theta})\hat{D}x}\tilde{X}(t) + \hat{D}\int_0^x e^{(A+BK)(\hat{\theta})\hat{D}(x-y)}B\hat{w}(y,t)dy\right]$$

$$(3.141)$$

where $\hat{e}(x,t) = \hat{u}(x,t) - u^r(\hat{\theta})$, $\tilde{X}(t) = X(t) - X^r(\hat{\theta})$

The parameter update law:

$$\dot{\hat{\theta}}_i(t) = \gamma_\theta \text{Proj}_\Theta \tau_{\theta_i}(t), \quad \hat{\theta}(0) \in \Theta,$$

$$(3.142)$$

$$\tau_{\theta_i}(t) = \left(\frac{\tilde{X}^T(t)P(\hat{\theta})}{g} - \hat{D}K(\hat{\theta})\int_0^1 (1+x)\left(\hat{w}(x,t) + A(\theta)\hat{D}\hat{w}_x(x,t)\right)e^{A(\hat{\theta})\hat{D}x}dx\right)$$

$$\times \left(A_i X(t) + B_i u^r(\hat{\theta})\right)$$

$$(3.143)$$

The closed-loop target system:

$$\dot{\tilde{X}}(t) = (A+BK)(\hat{\theta})\tilde{X}(t) + B(\hat{\theta})\hat{w}(0,t) + B(\hat{\theta})\tilde{e}(0,t)$$

$$+\tilde{A}X(t) + \tilde{B}u(0,t) - \frac{\partial X^r}{\partial \hat{\theta}}\dot{\hat{\theta}}$$

$$(3.144)$$

$$D\tilde{e}_t(x,t) = \tilde{e}_x(x,t) - \tilde{D}(t)f(x,t)$$

$$(3.145)$$

$$\tilde{e}(1,t) = 0$$

$$(3.146)$$

$$\hat{D}\hat{w}_t(x,t) = \hat{w}_x(x,t) - \hat{D}\dot{\hat{\theta}}(t)^T g(x,t) - \hat{D}\dot{\tilde{\theta}}(t)^T g_0(x,t)$$

$$-\hat{D}K(\hat{\theta})e^{A(\hat{\theta})\hat{D}x}B(\hat{\theta})\tilde{e}(x,t)$$

$$(3.147)$$

$$\hat{w}(1,t) = 0$$

$$(3.148)$$

(continued)

Table 3.8. Continued

---

where $\tilde{D}(t) = D - \hat{D}(t)$, $\tilde{\theta}(t) = \theta - \hat{\theta}(t)$, $\tilde{e}(x,t) = u(x,t) - \hat{u}(x,t)$, $\tilde{A} = \sum_{i=1}^{p} \tilde{\theta}_i A_i$, $\tilde{B} = \sum_{i=1}^{p} \tilde{\theta}_i B_i$. For the sake of simplicity, the delay update law is assumed to be $\hat{\dot{D}}(t) = 0$.

The Lyapunov-Krasovskii functional:

$$V(t) = \tilde{X}^T(t) P(\hat{\theta}) \tilde{X}(t) + gD \int_0^1 (1+x)\tilde{e}(x,t)^2 dx + \frac{g}{\gamma_D} \tilde{D}(t)^2 + \frac{g}{\gamma_\theta}|\tilde{\theta}(t)|^2$$

$$+ g\hat{D}\left(\int_0^1 (1+x)\big(\hat{w}(x,t)^2 + \hat{w}_x(x,t)^2\big) dx\right) \tag{3.149}$$

Theorem 3.20. *Under Assumptions 3.1, 3.2, and 3.16, the closed-loop system consisting of the ODE-PDE cascade (3.136)–(3.138), the PDE-state observer (3.134)–(3.135), the predictor feedback law (3.139), and the parameter update law (3.142)–(3.143) is stable at the zero solution in the sense of the norm*

$$\left(|\tilde{X}(t)|^2 + \|e(x,t)\|^2 + \|\hat{e}(x,t)\|^2 + \|\hat{e}_x(x,t)\|^2 + |D - \hat{D}(t)|^2 + |\theta - \hat{\theta}(t)|^2\right)^{\frac{1}{2}},$$

$$\tag{3.150}$$

and the convergence $\lim_{t\to\infty}(X(t) - X^r(\hat{\theta})) = 0$ is achieved, where $\tilde{X}(t) = X(t) - X^r(\hat{\theta})$, $e(x,t) = u(x,t) - u^r(\hat{\theta})$, $\hat{e}(x,t) = \hat{u}(x,t) - u^r(\hat{\theta})$, if there exists $M > 0$ such that the initial condition

$$\left(|\tilde{X}(0)|^2 + \|e(x,0)\|^2 + \|\hat{e}(x,0)\|^2 + \|\hat{e}_x(x,0)\|^2\right.$$

$$\left. + |D - \hat{D}(0)|^2 + |\theta - \hat{\theta}(0)|^2\right)^{\frac{1}{2}} \le M \tag{3.151}$$

is satisfied.

Proof. If we take the time derivative of $V(t)$ of (3.149) along (3.144)–(3.148), by a variety of techniques of inequalities and the initial condition, it is not hard to show the nonpositive property of $\dot{V}(t)$, and consequently the local stability of the closed-loop system is obtained. □

Chapter Four

Single-Input Systems with Full Relative Degree

AT THE BEGINNING OF CHAPTER 3, we introduced four types of basic uncertainties that usually come with single-input LTI systems with discrete actuator delay:

- unknown actuator delay D, which is allowed to be arbitrarily large,
- unknown plant parameter vector θ in system matrix $A(\theta)$ and input matrix $B(\theta)$,
- unmeasurable finite-dimensional ODE plant state $X(t)$,
- unmeasurable infinite-dimensional PDE actuator state $u(x,t)$, which represents delayed input $U(\eta)$ over the time window $\eta \in [t-D,t]$.

In subsequent sections of chapter 3, we successively dealt with adaptive control problems under uncertainty solely in either $X(t)$ or D, and coexistent uncertainties in $(D, u(x,t))$, (D, θ), $(D, u(x,t), X(t))$, and $(D, u(x,t), \theta)$.

It is evident that in chapter 3, the case of uncertainty combination in both ODE plant state $X(t)$ and parameter vector θ has not been taken into account yet. In such a case, the traditional ODE observer

$$\dot{\hat{X}}(t) = A\hat{X}(t) + BU(t-D) + L\big(Y(t) - C\hat{X}(t)\big) \tag{4.1}$$

cannot be employed, since both matrices A and B contain uncertain parameters. To deal with coexistent uncertainty in $(X(t), \theta)$, the basic predictor feedback framework in table 2.3 of section 2.4 is required to have a nontrivial modification.

From this chapter onwards, the uncertainty collections of $(D, X(t), \theta)$ and the most challenging case of $(D, X(t), \theta, u(x,t))$ are investigated. The primary approach is based on the adaptive backstepping control with Kreisselmeier-filters in chapter 10 of [89]. In output-feedback adaptive problems, namely problems involving both θ and $X(t)$, the relative degree plays a major role in determining the difficulty of a problem. In this chapter, we focus on a special class of LTI systems with its relative degree being equal to its system dimension, whereas in the next chapter this limitation is removed. Furthermore, in this chapter the actuator state $u(x,t)$ is assumed to be measured, whereas in the next chapter, the uncertainty in $u(x,t)$ is considered.

4.1   Problem Formulation

We consider a class of linear single-input single-output systems with discrete input delay

$$Y(s) = \frac{B(s)}{A(s)}e^{-Ds}U(s) = \frac{b_0}{s^n + a_{n-1}s^{n-1} + \cdots + a_1 s + a_0}e^{-Ds}U(s), \qquad (4.2)$$

which can be represented as the following observer canonical form:

$$\dot{X}(t) = AX(t) - aY(t) + bU(t - D)$$
$$Y(t) = X_1(t), \qquad (4.3)$$

where

$$A = \begin{bmatrix} 0 & & & \\ \vdots & & I_{n-1} & \\ 0 & \cdots & & 0 \end{bmatrix}, \quad a = \begin{bmatrix} a_{n-1} \\ \vdots \\ a_1 \\ a_0 \end{bmatrix}, \quad b = \begin{bmatrix} 0 \\ \vdots \\ 0 \\ b_0 \end{bmatrix} \qquad (4.4)$$

and $X(t) = [X_1(t), X_2(t), \cdots, X_n(t)]^T \in \mathbb{R}^n$ is the unmeasured state vector, $Y(t) \in \mathbb{R}$ is the measurable output, and $U(t - D) \in \mathbb{R}$ is the input with an unknown constant time delay $D > 0$, while $a_{n-1}, \cdots, a_1, a_0$ and b_0 are unknown constant plant parameters and control coefficient, respectively.

Remark 4.1. In chapter 4, the plant state $X(t)$ and plant parameter θ are uncertain at the same time, whereas the coexistent uncertainty in $(X(t), \theta)$ did not appear in chapter 3. As a result, the observer canonical form (4.2)–(4.4) is employed in chapter 4, which could be treated as a special case of the parametrization of (3.7)–(3.9) in chapter 3. Take a three-dimensional system of (4.2)–(4.4) as an example.

$$Y(s) = \frac{B(s)}{A(s)}e^{-Ds}U(s) = \frac{b_0}{s^3 + a_2 s^2 + a_1 s + a_0}e^{-Ds}U(s)$$

Denote $X_1(t) = Y(t)$, $X_2(t) = \dot{Y}(t) + a_2 Y(t)$, $X_3(t) = \ddot{Y}(t) + a_2 \dot{Y}(t) + a_1 Y(t)$. Then the above transfer function is transformed into the state-space model

$$\begin{bmatrix} \dot{X}_1(t) \\ \dot{X}_2(t) \\ \dot{X}_3(t) \end{bmatrix} = \begin{bmatrix} 0 & 1 & 0 \\ 0 & 0 & 1 \\ 0 & 0 & 0 \end{bmatrix} \begin{bmatrix} X_1(t) \\ X_2(t) \\ X_3(t) \end{bmatrix} - \begin{bmatrix} a_2 \\ a_1 \\ a_0 \end{bmatrix} Y(t) + \begin{bmatrix} 0 \\ 0 \\ b_0 \end{bmatrix} U(t - D)$$
$$Y(t) = X_1(t),$$

where a_2, a_1, a_0, b_0 are unknown parameters. Substituting $Y(t) = X_1(t)$ into the dynamics $\dot{X}(t)$, we get

$$
\begin{bmatrix} \dot{X}_1(t) \\ \dot{X}_2(t) \\ \dot{X}_3(t) \end{bmatrix} = \begin{bmatrix} -a_2 & 1 & 0 \\ -a_1 & 0 & 1 \\ -a_0 & 0 & 0 \end{bmatrix} \begin{bmatrix} X_1(t) \\ X_2(t) \\ X_3(t) \end{bmatrix} + \begin{bmatrix} 0 \\ 0 \\ b_0 \end{bmatrix} U(t - D),
$$

where

$$
A = \begin{bmatrix} -a_2 & 1 & 0 \\ -a_1 & 0 & 1 \\ -a_0 & 0 & 0 \end{bmatrix} = \begin{bmatrix} 0 & 1 & 0 \\ 0 & 0 & 1 \\ 0 & 0 & 0 \end{bmatrix} + a_2 \begin{bmatrix} -1 & 0 & 0 \\ 0 & 0 & 0 \\ 0 & 0 & 0 \end{bmatrix} + a_1 \begin{bmatrix} 0 & 0 & 0 \\ -1 & 0 & 0 \\ 0 & 0 & 0 \end{bmatrix}
$$

$$
+ a_0 \begin{bmatrix} 0 & 0 & 0 \\ 0 & 0 & 0 \\ -1 & 0 & 0 \end{bmatrix}
$$

$$
B = \begin{bmatrix} 0 \\ 0 \\ b_0 \end{bmatrix} = b_0 \begin{bmatrix} 0 \\ 0 \\ 1 \end{bmatrix}.
$$

It is evident that A and B satisfy (3.9) in chapter 3.

Remark 4.2. Comparing (4.3) with (10.3) appearing in chapter 10 of [89], one has no trouble finding that (4.3) is a special case of (10.3) with $b_m = b_{m-1} = \cdots = b_1 = 0$, which means the relative degree is identical to the system dimension. This kind of system is restrictive, but there are many linear delayed plants in practice satisfying this kind of structure; please refer to examples and simulations in [25] and [26]. When the relative degree is less than the system dimension, to achieve the global stabilization of the closed-loop system, the remaining m-dimensional inverse dynamics ζ that are used just for the analysis in (10.133) of [89] should be included in the denominator of the normalized update law (4.98) of this chapter. However, including ζ in the update law is obviously impossible since ζ is unmeasured and $b_m, b_{m-1}, \cdots, b_0$ are unknown.

For convenience of description, we rewrite plant (4.3) compactly as

$$
\dot{X}(t) = AX(t) + F\big(U(t-D), Y(t)\big)^T \theta
$$
$$
Y(t) = e_1^T X(t), \tag{4.5}
$$

where $p = 1 + n$ dimensional parameter vector θ is defined by $\theta = [b_0, a_{n-1}, \cdots, a_1, a_0]^T$ and e_i for $i = 1, 2, \cdots$ is the ith coordinate vector in corresponding space

$$
F\big(U(t-D), Y(t)\big)^T = \begin{bmatrix} \begin{bmatrix} 0_{(n-1) \times 1} \\ 1 \end{bmatrix} U(t-D), & -I_n Y(t) \end{bmatrix}. \tag{4.6}
$$

Our control objectives are as follows:

- design a control scheme to compensate for the system uncertainty and actuator delay to ensure all the signals of the closed-loop system are globally bounded,
- make output $Y(t)$ asymptotically track a reference signal $Y_r(t)$.

To achieve the above control objectives, we have the following assumptions:

Assumption 4.3. *In the case of known θ, given a time-varying reference output trajectory $Y_r(t)$ that is known, bounded, and smooth, there exist known reference state signal $X^r(t,\theta)$ and reference input signal $U^r(t,\theta)$, which are bounded in t, continuously differentiable in the argument θ, and satisfy*

$$\dot{X}^r(t,\theta) = AX^r(t,\theta) + F\big(U^r(t,\theta), Y_r(t)\big)^T \theta$$
$$Y_r(t) = e_1^T X^r(t,\theta). \tag{4.7}$$

Remark 4.4. Assumption 4.3 is a mild variation of traditional Assumption 10.4 on page 418 of [89]. Suppose $Y_r(t)$ and its first n derivatives are known, bounded, and piecewise continuous. If we choose reference states as $X_1^r(t) = Y_r(t)$, $X_2^r(t,\theta) = \dot{Y}_r(t) + a_{n-1}Y_r(t)$, $\cdots$, $X_n^r(t,\theta) = Y_r^{(n-1)}(t) + a_{n-1}Y_r^{(n-2)}(t) + a_{n-2}Y_r^{(n-3)}(t) + \cdots + a_2\dot{Y}_r(t) + a_1Y_r(t)$, it is easy to find a known and bounded $U^r(t,\theta)$ to make the equality $Y_r^{(n)}(t) + a_{n-1}Y_r^{(n-1)}(t) + a_{n-2}Y_r^{(n-2)}(t) + \cdots + a_1\dot{Y}_r(t) + a_0Y_r(t) = b_0U^r(t,\theta)$ hold; thus (4.7) is satisfied.

Assumption 4.5. *There exist two known finite constants $\underline{D} > 0$ and $\bar{D} > 0$ such that $D \in [\underline{D}, \bar{D}]$. The sign of the high-frequency gain b_0, i.e., $\mathrm{sgn}(b_0)$, is known and there exist two known finite constants $\underline{b}_0$, $\bar{b}_0$ such that $0 < \underline{b}_0 \le |b_0| \le \bar{b}_0$. In addition, there exists a convex compact set $\mathscr{A} \subset \mathbb{R}^n$ such that $\exists \bar{a}, a^*, |a - a^*| \le \bar{a}$ for all $a \in \mathscr{A}$, where $a^* \in \mathbb{R}^n$ is a known constant vector, and $\bar{a} > 0$ is a known finite constant.*

Remark 4.6. Assumption 4.5 is consistent with Assumption 8.1 of [89] and Assumptions 3 and 4 of [132]. Though it requires some a priori knowledge regarding unknown delay and parameters that will be useful for projection-based update laws (4.96)–(4.103), it is still unrestrictive and allows a large uncertainty on unknown elements.

4.2 State Estimation with Kreisselmeier-Filters

In this section, Kreisselmeier-filters (K-filters) are introduced as follows to estimate the states of the system with exponential rates of convergence:

$$\dot{\eta}(t) = A_0\eta(t) + e_nY(t), \tag{4.8}$$

$$\dot{\lambda}(t) = A_0\lambda(t) + e_n U(t-D), \tag{4.9}$$

$$\xi(t) = -A_0^n \eta(t), \tag{4.10}$$

$$\Xi(t) = -[A_0^{n-1}\eta(t), \cdots, A_0\eta(t), \eta(t)], \tag{4.11}$$

$$\Omega(t)^T = [\lambda(t), \Xi(t)], \tag{4.12}$$

where the vector $k = [k_1, k_2, \cdots, k_n]^T$ is chosen so that the matrix $A_0 = A - ke_1^T$ is Hurwitz, that is,

$$A_0^T P + P A_0 = -I, \quad P = P^T > 0. \tag{4.13}$$

With the K-filters above, we virtually estimate the unmeasurable state $X(t)$ as $\hat{X}(t) = \xi(t) + \Omega(t)^T\theta$, and the state estimation error $\varepsilon(t) = X(t) - \hat{X}(t)$ decays exponentially because

$$\dot{\varepsilon}(t) = A_0\varepsilon(t). \tag{4.14}$$

Thus it is easy to show that

$$X(t) = \xi(t) + \Omega(t)^T\theta + \varepsilon(t), \tag{4.15}$$

$$= -A(A_0)\eta(t) + b_0\lambda(t) + \varepsilon(t), \tag{4.16}$$

where $A(A_0) = A_0^n + \sum_{i=0}^{n-1} a_i A_0^i$.

If θ is known, according to (4.7), we could readily produce the reference output $Y_r(t)$ with measurable reference K-filters $\eta^r(t)$ and $\lambda^r(t,\theta)$. Correspondingly, in the case of unknown θ, based on the certainty equivalence principle, we have

$$\dot{\eta}^r(t) = A_0\eta^r(t) + e_n Y_r(t), \tag{4.17}$$

$$\dot{\lambda}^r(t,\hat{\theta}) = A_0\lambda^r(t,\hat{\theta}) + e_n U^r(t,\hat{\theta}) + \frac{\partial\lambda^r(t,\hat{\theta})}{\partial\hat{\theta}}\dot{\hat{\theta}}, \tag{4.18}$$

$$\xi^r(t) = -A_0^n\eta^r(t), \tag{4.19}$$

$$\Xi^r(t) = -[A_0^{n-1}\eta^r(t), \cdots, A_0\eta^r(t), \eta^r(t)], \tag{4.20}$$

$$\Omega^r(t,\hat{\theta})^T = [\lambda^r(t,\hat{\theta}), \Xi^r(t)], \tag{4.21}$$

$$X^r(t,\hat{\theta}) = \xi^r(t) + \Omega^r(t,\hat{\theta})^T\hat{\theta}$$

$$= -\hat{A}(A_0)\eta^r(t) + \hat{b}_0\lambda^r(t,\hat{\theta}), \tag{4.22}$$

$$Y_r(t) = e_1^T X^r(t,\hat{\theta}), \tag{4.23}$$

where $\hat{\theta}$ is the estimate of unknown parameter θ with $\tilde{\theta} = \theta - \hat{\theta}$, and $\hat{A}(A_0) = A_0^n + \sum_{i=0}^{n-1} \hat{a}_i A_0^i$, $\tilde{A}(A_0) = \sum_{i=0}^{n-1} \tilde{a}_i A_0^i$, with $\hat{a}_i$ for $i = 0, 1, \cdots, n-1$ and $\hat{b}_0$ being estimates of a_i and b_0, $\tilde{a}_i = a_i - \hat{a}_i$, $\tilde{b}_0 = b_0 - \hat{b}_0$. Then we define a group of new error variables as follows:

$$z_1(t) = Y(t) - Y_r(t) \tag{4.24}$$

$$\tilde{U}(t-D) = U(t-D) - U^r(t,\hat{\theta}), \tag{4.25}$$

$$\tilde{\eta}(t) = \eta(t) - \eta^r(t), \tag{4.26}$$

$$\tilde{\lambda}(t) = \lambda(t) - \lambda^r(t,\hat{\theta}), \tag{4.27}$$

$$\tilde{\xi}(t) = \xi(t) - \xi^r(t), \tag{4.28}$$

$$\tilde{\Xi}(t) = \Xi(t) - \Xi^r(t), \tag{4.29}$$

$$\tilde{\Omega}(t)^T = \Omega(t)^T - \Omega^r(t,\hat{\theta})^T, \tag{4.30}$$

which are governed by the following dynamic equations:

$$\dot{\tilde{\eta}}(t) = A_0 \tilde{\eta}(t) + e_n z_1(t), \tag{4.31}$$

$$\dot{\tilde{\lambda}}(t) = A_0 \tilde{\lambda}(t) + e_n \tilde{U}(t-D) - \frac{\partial \lambda^r(t,\hat{\theta})}{\partial \hat{\theta}} \dot{\hat{\theta}}, \tag{4.32}$$

$$\dot{z}_1(t) = \tilde{\xi}_2(t) + \tilde{\omega}(t)^T \hat{\theta} + \varepsilon_2(t) + \omega(t)^T \tilde{\theta}, \tag{4.33}$$

$$= \hat{b}_0 \tilde{\lambda}_2(t) + \tilde{\xi}_2(t) + \bar{\omega}(t)^T \hat{\theta} + \varepsilon_2(t) + \omega(t)^T \tilde{\theta}, \tag{4.34}$$

where

$$\tilde{\omega}(t) = [\tilde{\lambda}_2(t), \tilde{\Xi}_2(t) - z_1(t)e_1^T]^T, \tag{4.35}$$

$$\bar{\omega}(t) = [0, \tilde{\Xi}_2(t) - z_1(t)e_1^T]^T, \tag{4.36}$$

$$\omega(t) = [\lambda_2(t), \Xi_2(t) - Y(t)e_1^T]^T. \tag{4.37}$$

If we further define a couple of new variables as

$$\chi(t) = \begin{bmatrix} Y(t) \\ \eta(t) \\ \lambda(t) \end{bmatrix} \in \mathbb{R}^{2n+1}, \quad \chi^r(t,\hat{\theta}) = \begin{bmatrix} Y_r(t) \\ \eta^r(t) \\ \lambda^r(t,\hat{\theta}) \end{bmatrix} \in \mathbb{R}^{2n+1}, \tag{4.38}$$

we can get a new error variable

$$\tilde{\chi}(t) = \chi(t) - \chi^r(t,\hat{\theta}) = \begin{bmatrix} z_1(t) \\ \tilde{\eta}(t) \\ \tilde{\lambda}(t) \end{bmatrix} \in \mathbb{R}^{2n+1}, \tag{4.39}$$

which is governed by

$$\dot{\chi}(t) = A_{\tilde{\chi}}(\hat{\theta})\tilde{\chi}(t) + e_{2n+1}\tilde{U}(t-D) + e_1\left(\varepsilon_2(t) + \omega(t)^T\tilde{\theta}\right)$$
$$- \frac{\partial \chi^r(t,\hat{\theta})}{\partial \hat{\theta}}\dot{\hat{\theta}}, \tag{4.40}$$

where

$$A_{\tilde{\chi}}(\hat{\theta}) = \begin{bmatrix} -\hat{a}_{n-1} & -e_2^T \hat{A}(A_0) & e_2^T \hat{b}_0 \\ e_n & A_0 & 0_{n\times n} \\ 0_{n\times 1} & 0_{n\times n} & A_0 \end{bmatrix}, \tag{4.41}$$

$$\frac{\partial \chi^r(t,\hat{\theta})}{\partial \hat{\theta}} = \begin{bmatrix} 0_{(1+n)\times p} \\ \frac{\partial \lambda^r(t,\hat{\theta})}{\partial \hat{\theta}} \end{bmatrix}. \tag{4.42}$$

4.3 Boundary Control with Adaptive Backstepping

We present a combination of prediction-based boundary control with adaptive backstepping to address unknown parameters and time delay in this section. Aiming at system

$$\dot{z}_1 = \hat{b}_0\tilde{\lambda}_2 + \tilde{\xi}_2 + \bar{\omega}^T\hat{\theta} + \varepsilon_2 + \omega^T\tilde{\theta}, \tag{4.43}$$

$$\dot{\tilde{\lambda}}_i = \tilde{\lambda}_{i+1} - k_i\tilde{\lambda}_1 - e_i^T \frac{\partial \lambda^r(t,\hat{\theta})}{\partial \hat{\theta}}\dot{\hat{\theta}},$$
$$i = 2, 3, \cdots, n-1, \tag{4.44}$$

$$\dot{\tilde{\lambda}}_n = \tilde{U}(t-D) - k_n\tilde{\lambda}_1 - e_n^T \frac{\partial \lambda^r(t,\hat{\theta})}{\partial \hat{\theta}}\dot{\hat{\theta}}, \tag{4.45}$$

the adaptive backstepping recursive control design is given as follows:
   Coordinate transformation:

$$z_1 = Y - Y_r, \tag{4.46}$$

$$z_i = \tilde{\lambda}_i - \alpha_{i-1}, \quad i = 2, 3, \cdots, n. \tag{4.47}$$

Stabilizing functions:

$$\alpha_1 = \frac{1}{\hat{b}_0}\left(-(c_1 + d_1)z_1 - \tilde{\xi}_2 - \bar{\omega}^T\hat{\theta}\right), \tag{4.48}$$

$$\alpha_2 = -\hat{b}_0 z_1 - \left(c_2 + d_2\left(\frac{\partial \alpha_1}{\partial z_1}\right)^2\right)z_2 + \beta_2, \tag{4.49}$$

$$\beta_2 = k_2\tilde{\lambda}_1 + \frac{\partial\alpha_1}{\partial z_1}(\tilde{\xi}_2 + \tilde{\omega}^T\hat{\theta}) + \frac{\partial\alpha_1}{\partial\tilde{\eta}}(A_0\tilde{\eta} + e_n z_1), \quad (4.50)$$

$$\alpha_i = -z_{i-1} - \left(c_i + d_i\left(\frac{\partial\alpha_{i-1}}{\partial z_1}\right)^2\right)z_i + \beta_i, \quad (4.51)$$

$$\beta_i = k_i\tilde{\lambda}_1 + \frac{\partial\alpha_{i-1}}{\partial z_1}(\tilde{\xi}_2 + \tilde{\omega}^T\hat{\theta}) + \frac{\partial\alpha_{i-1}}{\partial\tilde{\eta}}(A_0\tilde{\eta} + e_n z_1)$$

$$+ \sum_{j=1}^{i-1}\frac{\partial\alpha_{i-1}}{\partial\tilde{\lambda}}e_j e_j^T A_0\tilde{\lambda}, \quad i = 3, 4, \cdots, n, \quad (4.52)$$

where $c_i > 0$, $d_i > 0$ for $i = 1, 2, \cdots, n$ are design parameters.

Adaptive control law:

$$\tilde{U}(t - D) = \alpha_n \quad (4.53)$$

Through a recursive but straightforward calculation, we show the following equalities:

$$z_2 = K_{2,z_1}(\hat{\theta})z_1 + K_{2,\tilde{\eta}}(\hat{\theta})\tilde{\eta} + K_{2,\tilde{\lambda}}(\hat{\theta})\tilde{\lambda}, \quad (4.54)$$

$$z_3 = K_{3,z_1}(\hat{\theta})z_1 + K_{3,\tilde{\eta}}(\hat{\theta})\tilde{\eta} + K_{3,\tilde{\lambda}}(\hat{\theta})\tilde{\lambda}, \quad (4.55)$$

$$z_{i+1} = K_{i+1,z_1}(\hat{\theta})z_1 + K_{i+1,\tilde{\eta}}(\hat{\theta})\tilde{\eta} + K_{i+1,\tilde{\lambda}}(\hat{\theta})\tilde{\lambda}$$

$$i = 3, 4, \cdots, n-1, \quad (4.56)$$

$$\tilde{U}(t - D) = K_{z_1}(\hat{\theta})z_1(t) + K_{\tilde{\eta}}(\hat{\theta})\tilde{\eta}(t) + K_{\tilde{\lambda}}(\hat{\theta})\tilde{\lambda}(t), \quad (4.57)$$

$$= K_{\tilde{\chi}}(\hat{\theta})\tilde{\chi}(t), \quad (4.58)$$

where the explicit expressions of $K_{i,z_1}(\hat{\theta})$, $K_{i,\tilde{\eta}}(\hat{\theta})$, $K_{i,\tilde{\lambda}}(\hat{\theta})$ for $i = 2, 3, \cdots, n$ and $K_{z_1}(\hat{\theta})$, $K_{\tilde{\eta}}(\hat{\theta})$, $K_{\tilde{\lambda}}(\hat{\theta})$ are listed as follows:

$$K_{2,z_1}(\hat{\theta}) = \frac{1}{\hat{b}_0}((c_1 + d_1) - \hat{a}_{n-1}), \quad (4.59)$$

$$K_{2,\tilde{\eta}}(\hat{\theta}) = -\frac{1}{\hat{b}_0}e_2^T\hat{A}(A_0), \quad K_{2,\tilde{\lambda}}(\hat{\theta}) = e_2^T, \quad (4.60)$$

$$K_{3,z_1}(\hat{\theta}) = \hat{b}_0 + \left(c_2 + d_2\left(\frac{\partial\alpha_1}{\partial z_1}\right)^2\right)K_{2,z_1}(\hat{\theta}) + \frac{\partial\alpha_1}{\partial z_1}\hat{a}_{n-1} - \frac{\partial\alpha_1}{\partial\tilde{\eta}}e_n, \quad (4.61)$$

$$K_{3,\tilde{\eta}}(\hat{\theta}) = \left(c_2 + d_2\left(\frac{\partial\alpha_1}{\partial z_1}\right)^2\right)K_{2,\tilde{\eta}}(\hat{\theta}) + \frac{\partial\alpha_1}{\partial z_1}e_2^T\hat{A}(A_0) - \frac{\partial\alpha_1}{\partial\tilde{\eta}}A_0, \quad (4.62)$$

$$K_{3,\tilde{\lambda}}(\hat{\theta}) = \left(c_2 + d_2\left(\frac{\partial\alpha_1}{\partial z_1}\right)^2\right)K_{2,\tilde{\lambda}}(\hat{\theta}) + e_2^T A_0 - \frac{\partial\alpha_1}{\partial z_1}e_2^T\hat{b}_0, \quad (4.63)$$

$$K_{i+1,z_1}(\hat{\theta}) = K_{i-1,z_1}(\hat{\theta}) + \left(c_i + d_i \left(\frac{\partial \alpha_{i-1}}{\partial z_1}\right)^2\right) K_{i,z_1}(\hat{\theta})$$

$$+ \frac{\partial \alpha_{i-1}}{\partial z_1} \hat{a}_{n-1} - \frac{\partial \alpha_{i-1}}{\partial \tilde{\eta}} e_n, \tag{4.64}$$

$$K_{i+1,\tilde{\eta}}(\hat{\theta}) = K_{i-1,\tilde{\eta}}(\hat{\theta}) + \left(c_i + d_i \left(\frac{\partial \alpha_{i-1}}{\partial z_1}\right)^2\right) K_{i,\tilde{\eta}}(\hat{\theta})$$

$$+ \frac{\partial \alpha_{i-1}}{\partial z_1} e_2^T \hat{A}(A_0) - \frac{\partial \alpha_{i-1}}{\partial \tilde{\eta}} A_0, \tag{4.65}$$

$$K_{i+1,\tilde{\lambda}}(\hat{\theta}) = K_{i-1,\tilde{\lambda}}(\hat{\theta}) + \left(c_i + d_i \left(\frac{\partial \alpha_{i-1}}{\partial z_1}\right)^2\right) K_{i,\tilde{\lambda}}(\hat{\theta})$$

$$+ e_i^T A_0 - \frac{\partial \alpha_{i-1}}{\partial z_1} e_2^T \hat{b}_0 - \sum_{j=1}^{i-1} \frac{\partial \alpha_{i-1}}{\partial \tilde{\lambda}} e_j e_j^T A_0, \tag{4.66}$$

$$i = 3, 4, \cdots, n-1$$

$$K_{z_1}(\hat{\theta}) = -\left[K_{n-1,z_1}(\hat{\theta}) + \left(c_n + d_n \left(\frac{\partial \alpha_{n-1}}{\partial z_1}\right)^2\right) K_{n,z_1}(\hat{\theta})\right.$$

$$\left. + \frac{\partial \alpha_{n-1}}{\partial z_1} \hat{a}_{n-1} - \frac{\partial \alpha_{n-1}}{\partial \tilde{\eta}} e_n\right], \tag{4.67}$$

$$K_{\tilde{\eta}}(\hat{\theta}) = -\left[K_{n-1,\tilde{\eta}}(\hat{\theta}) + \left(c_n + d_n \left(\frac{\partial \alpha_{n-1}}{\partial z_1}\right)^2\right) K_{n,\tilde{\eta}}(\hat{\theta})\right.$$

$$\left. + \frac{\partial \alpha_{n-1}}{\partial z_1} e_2^T \hat{A}(A_0) - \frac{\partial \alpha_{n-1}}{\partial \tilde{\eta}} A_0\right], \tag{4.68}$$

$$K_{\tilde{\lambda}}(\hat{\theta}) = -\left[K_{n-1,\tilde{\lambda}}(\hat{\theta}) + \left(c_n + d_n \left(\frac{\partial \alpha_{n-1}}{\partial z_1}\right)^2\right) K_{n,\tilde{\lambda}}(\hat{\theta})\right.$$

$$\left. + e_n^T A_0 - \frac{\partial \alpha_{n-1}}{\partial z_1} e_2^T \hat{b}_0 - \sum_{j=1}^{n-1} \frac{\partial \alpha_{n-1}}{\partial \tilde{\lambda}} e_j e_j^T A_0\right] \tag{4.69}$$

and

$$K_{\tilde{\chi}}(\hat{\theta}) = [K_{z_1}(\hat{\theta}), K_{\tilde{\eta}}(\hat{\theta}), K_{\tilde{\lambda}}(\hat{\theta})] \in \mathbb{R}^{1 \times (2n+1)}. \tag{4.70}$$

If parameter θ and time delay D are known, replacing $\hat{\theta}$ with θ, we introduce a similar transport PDE proposed in section 2.2 of [80] and [27], which has the following form to represent the delay of input:

$$D u_t(x,t) = u_x(x,t), \quad x \in [0,1], \tag{4.71}$$

$$u(1,t) = U(t), \tag{4.72}$$

$$Du_t^r(x,t,\theta) = u_x^r(x,t,\theta), \quad x \in [0,1], \tag{4.73}$$

$$u^r(1,t,\theta) = U^r(t+D,\theta), \tag{4.74}$$

where $u(x,t) = U(t+D(x-1))$, $u^r(x,t,\theta) = U^r(t+Dx,\theta)$. Bearing (4.58) in mind, one can prove that the following prediction-based control law,

$$\begin{aligned} U(t) &= U^r(t+D,\theta) - K_{\tilde{\chi}}(\theta)\chi^r(t+D,\theta) \\ &\quad + K_{\tilde{\chi}}(\theta)e^{A_{\tilde{\chi}}(\theta)D}\chi(t) \\ &\quad + D\int_0^1 K_{\tilde{\chi}}(\theta)e^{A_{\tilde{\chi}}(\theta)D(1-y)}e_{2n+1}u(y,t)\,dy \end{aligned} \tag{4.75}$$

achieves our control objectives for system (4.3). To deal with the unknown plant parameters and the unknown actuator time delay, utilizing the certainty equivalence principle, we introduce a PDE error variable $\tilde{u}(x,t) = u(x,t) - u^r(x,t,\hat{\theta}) = U(t+D(x-1)) - U^r(t+Dx,\hat{\theta})$ satisfying

$$D\tilde{u}_t(x,t) = \tilde{u}_x(x,t) - D\frac{\partial u^r(x,t,\hat{\theta})}{\partial\hat{\theta}}\dot{\hat{\theta}}, \quad x \in [0,1], \tag{4.76}$$

$$\tilde{u}(1,t) = \tilde{U}(t) = U(t) - U^r(t+\hat{D},\hat{\theta}). \tag{4.77}$$

Here we employ a backstepping transformation similar to the one presented in [27] and section 2.2 of [80]:

$$\begin{aligned} w(x,t) &= \tilde{u}(x,t) - K_{\tilde{\chi}}(\hat{\theta})e^{A_{\tilde{\chi}}(\hat{\theta})\hat{D}x}\tilde{\chi}(t) \\ &\quad - \hat{D}\int_0^x K_{\tilde{\chi}}(\hat{\theta})e^{A_{\tilde{\chi}}(\hat{\theta})\hat{D}(x-y)}e_{2n+1}\tilde{u}(y,t)\,dy, \end{aligned} \tag{4.78}$$

$$\begin{aligned} \tilde{u}(x,t) &= w(x,t) + K_{\tilde{\chi}}(\hat{\theta})e^{(A_{\tilde{\chi}}+e_{2n+1}K_{\tilde{\chi}})(\hat{\theta})\hat{D}x}\tilde{\chi}(t) \\ &\quad + \hat{D}\int_0^x K_{\tilde{\chi}}(\hat{\theta})e^{(A_{\tilde{\chi}}+e_{2n+1}K_{\tilde{\chi}})(\hat{\theta})\hat{D}(x-y)}e_{2n+1} \\ &\quad \times w(y,t)\,dy, \end{aligned} \tag{4.79}$$

with which systems (4.14), (4.31), (4.43)–(4.45), and (4.76)–(4.77) are mapped into the target system as follows:

$$\begin{aligned} \dot{z} &= A_z(\hat{\theta})z + W_\varepsilon(\hat{\theta})(\varepsilon_2 + \omega^T\tilde{\theta}) + Q(z,t)^T\dot{\hat{\theta}} \\ &\quad + Q^r(t,\hat{\theta})^T\dot{\hat{\theta}} + e_n w(0,t), \end{aligned} \tag{4.80}$$

$$\dot{\tilde{\eta}} = A_0\tilde{\eta} + e_n z_1, \tag{4.81}$$

$$\dot{\varepsilon} = A_0\varepsilon, \tag{4.82}$$

$$Dw_t(x,t) = w_x(x,t) - D(\varepsilon_2 + \omega^T \tilde{\theta}) r_0(x,t) - \tilde{D} p_0(x,t)$$

$$- D\hat{\dot{D}} q_0(x,t) - D\hat{\dot{\theta}}^T q(x,t), \tag{4.83}$$

$$w(1,t) = 0, \tag{4.84}$$

where $\hat{D}$ is an estimate of D, $\tilde{D} = D - \hat{D}$, and $A_z(\hat{\theta})$, $W_\varepsilon(\hat{\theta})$, $Q(z,t)^T$, $Q^r(t,\hat{\theta})^T$, $r_0(x,t)$, $p_0(x,t)$, $q_0(x,t)$, and $q(x,t)$ are presented below:

$$A_z(\hat{\theta}) = \begin{bmatrix} -(c_1+d_1) & \hat{b}_0 & 0 \cdots & & 0 \\ -\hat{b}_0 & -\left(c_2+d_2\left(\frac{\partial\alpha_1}{\partial z_1}\right)^2\right) & 1 & \ddots & \vdots \\ 0 & -1 & \ddots & \ddots & 0 \\ \vdots & \ddots & \ddots & \ddots & 1 \\ 0 & \cdots & & 0 & -1 & -\left(c_n+d_n\left(\frac{\partial\alpha_{n-1}}{\partial z_1}\right)^2\right) \end{bmatrix}, \tag{4.85}$$

$$W_\varepsilon(\hat{\theta}) = \begin{bmatrix} 1 \\ -\frac{\partial\alpha_1}{\partial z_1} \\ \vdots \\ -\frac{\partial\alpha_{n-1}}{\partial z_1} \end{bmatrix}, \tag{4.86}$$

$$Q(z,t)^T = \begin{bmatrix} 0 \\ -\frac{\partial\alpha_1}{\partial\hat{\theta}} \\ \vdots \\ -\frac{\partial\alpha_{n-1}}{\partial\hat{\theta}} \end{bmatrix}, \quad Q^r(t,\hat{\theta})^T = \begin{bmatrix} 0 \\ -e_2^T \frac{\partial\lambda^r(t,\hat{\theta})}{\partial\hat{\theta}} \\ -\left(e_3^T - \sum_{j=1}^{2} \frac{\partial\alpha_2}{\partial\lambda} e_j e_j^T\right)\frac{\partial\lambda^r(t,\hat{\theta})}{\partial\hat{\theta}} \\ \vdots \\ -\left(e_n^T - \sum_{j=1}^{n-1} \frac{\partial\alpha_{n-1}}{\partial\lambda} e_j e_j^T\right)\frac{\partial\lambda^r(t,\hat{\theta})}{\partial\hat{\theta}} \end{bmatrix}, \tag{4.87}$$

$$r_0(x,t) = K_{\tilde{\chi}}(\hat{\theta}) e^{A_{\tilde{\chi}}(\hat{\theta})\hat{D}x} e_1, \tag{4.88}$$

$$p_0(x,t) = K_{\tilde{\chi}}(\hat{\theta}) e^{A_{\tilde{\chi}}(\hat{\theta})\hat{D}x} \left(A_{\tilde{\chi}}(\hat{\theta})\tilde{\chi}(t) + e_{2n+1}\tilde{u}(0,t) \right), \tag{4.89}$$

$$= K_{\tilde{\chi}}(\hat{\theta}) e^{A_{\tilde{\chi}}(\hat{\theta})\hat{D}x} \left[(A_{\tilde{\chi}} + e_{2n+1}K_{\tilde{\chi}})(\hat{\theta})\tilde{\chi}(t) + e_{2n+1}w(0,t) \right], \tag{4.90}$$

$$q_0(x,t) = K_{\tilde{\chi}} A_{\tilde{\chi}}(\hat{\theta}) x e^{A_{\tilde{\chi}}(\hat{\theta})\hat{D}x} \tilde{\chi}(t)$$

$$+ \int_0^x K_{\tilde{\chi}}(\hat{\theta})\left(I + A_{\tilde{\chi}}(\hat{\theta})\hat{D}(x-y)\right) e^{A_{\tilde{\chi}}(\hat{\theta})\hat{D}(x-y)} e_{2n+1}\tilde{u}(y,t)\, \mathrm{d}y, \tag{4.91}$$

$$= \left[K_{\tilde{\chi}} A_{\tilde{\chi}}(\hat{\theta}) x e^{A_{\tilde{\chi}}(\hat{\theta})\hat{D}x} + \int_0^x K_{\tilde{\chi}}(\hat{\theta}) \big(I + A_{\tilde{\chi}}(\hat{\theta})\hat{D}(x-y) \big) \right.$$

$$\left. \times e^{A_{\tilde{\chi}}(\hat{\theta})\hat{D}(x-y)} e_{2n+1} K_{\tilde{\chi}}(\hat{\theta}) e^{(A_{\tilde{\chi}}+e_{2n+1}K_{\tilde{\chi}})(\hat{\theta})\hat{D}y} \, \mathrm{d}y \right] \tilde{\chi}(t)$$

$$+ \int_0^x w(y,t) \left[K_{\tilde{\chi}}(\hat{\theta}) \big(I + A_{\tilde{\chi}}(\hat{\theta})\hat{D}(x-y) \big) e^{A_{\tilde{\chi}}(\hat{\theta})\hat{D}(x-y)} e_{2n+1} \right.$$

$$+ \hat{D} \int_y^x K_{\tilde{\chi}}(\hat{\theta}) \big(I + A_{\tilde{\chi}}(\hat{\theta})\hat{D}(x-s) \big) e^{A_{\tilde{\chi}}(\hat{\theta})\hat{D}(x-s)} e_{2n+1}$$

$$\left. \times K_{\tilde{\chi}}(\hat{\theta}) e^{(A_{\tilde{\chi}}+e_{2n+1}K_{\tilde{\chi}})(\hat{\theta})\hat{D}(s-y)} e_{2n+1} \, \mathrm{d}s \right] \mathrm{d}y, \tag{4.92}$$

$$q_i(x,t) = \left(\frac{\partial K_{\tilde{\chi}}(\hat{\theta})}{\partial \hat{\theta}_i} + K_{\tilde{\chi}}(\hat{\theta}) \frac{\partial A_{\tilde{\chi}}(\hat{\theta})}{\partial \hat{\theta}_i} \hat{D}x \right) e^{A_{\tilde{\chi}}(\hat{\theta})\hat{D}x} \tilde{\chi}(t)$$

$$+ \hat{D} \int_0^x \left(\frac{\partial K_{\tilde{\chi}}(\hat{\theta})}{\partial \hat{\theta}_i} + K_{\tilde{\chi}}(\hat{\theta}) \frac{\partial A_{\tilde{\chi}}(\hat{\theta})}{\partial \hat{\theta}_i} \hat{D}(x-y) \right) e^{A_{\tilde{\chi}}(\hat{\theta})\hat{D}(x-y)}$$

$$\times e_{2n+1} \tilde{u}(y,t) \, \mathrm{d}y - K_{\tilde{\chi}}(\hat{\theta}) e^{A_{\tilde{\chi}}(\hat{\theta})\hat{D}x} \frac{\partial \chi^r(t,\hat{\theta})}{\partial \hat{\theta}_i}$$

$$+ \frac{\partial u^r(x,t,\hat{\theta})}{\partial \hat{\theta}_i} - \hat{D} \int_0^x K_{\tilde{\chi}}(\hat{\theta}) e^{A_{\tilde{\chi}}(\hat{\theta})\hat{D}(x-y)} e_{2n+1} \frac{\partial u^r(y,t,\hat{\theta})}{\partial \hat{\theta}_i} \, \mathrm{d}y, \tag{4.93}$$

$$= \left[\left(\frac{\partial K_{\tilde{\chi}}(\hat{\theta})}{\partial \hat{\theta}_i} + K_{\tilde{\chi}}(\hat{\theta}) \frac{\partial A_{\tilde{\chi}}(\hat{\theta})}{\partial \hat{\theta}_i} \hat{D}x \right) e^{A_{\tilde{\chi}}(\hat{\theta})\hat{D}x} \right.$$

$$+ \hat{D} \int_0^x \left(\frac{\partial K_{\tilde{\chi}}(\hat{\theta})}{\partial \hat{\theta}_i} + K_{\tilde{\chi}}(\hat{\theta}) \frac{\partial A_{\tilde{\chi}}(\hat{\theta})}{\partial \hat{\theta}_i} \hat{D}(x-y) \right) e^{A_{\tilde{\chi}}(\hat{\theta})\hat{D}(x-y)} e_{2n+1}$$

$$\left. \times K_{\tilde{\chi}}(\hat{\theta}) e^{(A_{\tilde{\chi}}+e_{2n+1}K_{\tilde{\chi}})(\hat{\theta})\hat{D}y} \, \mathrm{d}y \right] \tilde{\chi}(t)$$

$$+ \hat{D} \int_0^x w(y,t) \left[\left(\frac{\partial K_{\tilde{\chi}}(\hat{\theta})}{\partial \hat{\theta}_i} + K_{\tilde{\chi}}(\hat{\theta}) \frac{\partial A_{\tilde{\chi}}(\hat{\theta})}{\partial \hat{\theta}_i} \hat{D}(x-y) \right) e^{A_{\tilde{\chi}}(\hat{\theta})\hat{D}(x-y)} e_{2n+1} \right.$$

$$+ \hat{D} \int_y^x \left(\frac{\partial K_{\tilde{\chi}}(\hat{\theta})}{\partial \hat{\theta}_i} + K_{\tilde{\chi}}(\hat{\theta}) \frac{\partial A_{\tilde{\chi}}(\hat{\theta})}{\partial \hat{\theta}_i} \hat{D}(x-s) \right) e^{A_{\tilde{\chi}}(\hat{\theta})\hat{D}(x-s)} e_{2n+1}$$

$$\left. \times K_{\tilde{\chi}}(\hat{\theta}) e^{(A_{\tilde{\chi}}+e_{2n+1}K_{\tilde{\chi}})(\hat{\theta})\hat{D}(s-y)} e_{2n+1} \, \mathrm{d}s \right] \mathrm{d}y$$

$$- K_{\tilde{\chi}}(\hat{\theta}) e^{A_{\tilde{\chi}}(\hat{\theta})\hat{D}x} \frac{\partial \chi^r(t,\hat{\theta})}{\partial \hat{\theta}_i} + \frac{\partial u^r(x,t,\hat{\theta})}{\partial \hat{\theta}_i}$$

$$- \hat{D} \int_0^x K_{\tilde{\chi}}(\hat{\theta}) e^{A_{\tilde{\chi}}(\hat{\theta})\hat{D}(x-y)} e_{2n+1} \frac{\partial u^r(y,t,\hat{\theta})}{\partial \hat{\theta}_i} \, \mathrm{d}y. \tag{4.94}$$

As a consequence, aiming at system (4.80)–(4.84), performing the certainty equivalence principle on (4.75), we design our control as follows to ensure that (4.84) holds:

$$U(t) = U^r(t + \hat{D}, \hat{\theta}) - K_{\tilde{\chi}}(\hat{\theta})\chi^r(t + \hat{D}, \hat{\theta})$$
$$+ K_{\tilde{\chi}}(\hat{\theta})e^{A_{\tilde{\chi}}(\hat{\theta})\hat{D}}\chi(t)$$
$$+ \hat{D}\int_0^1 K_{\tilde{\chi}}(\hat{\theta})e^{A_{\tilde{\chi}}(\hat{\theta})\hat{D}(1-y)}e_{2n+1}u(y,t)\,\mathrm{d}y. \qquad (4.95)$$

Remark 4.7. Please note that we do not utilize tuning functions or nonlinear damping terms like [150, 151] to deal with parameter estimation error in backstepping control design (4.46)–(4.53), so that we can obtain a linear controller (4.57)–(4.58) for employing prediction control (4.75) and (4.95) to address the input delay. To get a backstepping transformation of error variables in (4.78)–(4.79), we have made a preparation by introducing the reference signals in (4.17)–(4.23) to get the dynamic equation for $Y_r(t)$ and apply them in the controller design (4.46)–(4.53), which is different from standard adaptive backstepping feedback in [150, 151] and chapter 10 of [89], where $Y_r(t)$ and its first n derivatives are used.

4.4   Identification of Unknown Parameters and Delay

Two Lyapunov-based identifiers to estimate unknown plant parameters and actuator time delay are designed as follows:

The update law of D is designed as

$$\dot{\hat{D}} = \gamma_D \mathrm{Proj}_{[\underline{D},\bar{D}]}\{\tau_D\}, \quad \gamma_D > 0, \qquad (4.96)$$

$$\tau_D = \frac{1}{V_0}\left(-\int_0^1 (1+x)w(x,t)p_0(x,t)\,\mathrm{d}x\right), \qquad (4.97)$$

$$V_0 = 1 + \frac{1}{2}z^T z + \frac{1}{k_\eta}\tilde{\eta}^T P\tilde{\eta} + g\int_0^1 (1+x)w(x,t)^2\,\mathrm{d}x,$$

$$k_\eta > 0, \quad g > 0, \qquad (4.98)$$

where the projection operator is given by

$$\mathrm{Proj}_{[\underline{D},\bar{D}]}\{\tau_D\} = \tau_D \begin{cases} 0, & \hat{D} = \underline{D} \quad \text{and} \quad \tau_D < 0 \\ 0, & \hat{D} = \bar{D} \quad \text{and} \quad \tau_D > 0 \\ 1, & \text{else.} \end{cases} \qquad (4.99)$$

The update law of θ is designed as

$$\dot{\hat{\theta}} = \gamma_\theta \mathrm{Proj}_\Pi\{\tau_\theta\}, \quad \gamma_\theta > 0, \tag{4.100}$$

$$\tau_\theta = \frac{1}{V_0}\left(\frac{1}{2g}\omega W_\varepsilon(\hat{\theta})^T z \right.$$

$$\left. -\omega \int_0^1 (1+x)w(x,t)r_0(x,t)\,\mathrm{d}x\right), \tag{4.101}$$

where $\mathrm{Proj}_\Pi\{\cdot\}$ is a smooth projection operator employed to guarantee that $\hat{\theta} = [\hat{\theta}_1, \hat{\theta}_2, \cdots, \hat{\theta}_p]^T \in \Pi$, $\forall \hat{\theta}(0) \in \Pi$, with the set Π being defined as

$$\Pi = \left\{ \hat{\theta} \,\middle|\, \begin{array}{ll} |\hat{b}_0\mathrm{sgn}(b_0) - \sigma_0| < \sigma_1, & \hat{b}_0 = \hat{\theta}_1 \\ |\hat{a} - a^*| < \bar{a}, & \hat{a} = [\hat{\theta}_2, \hat{\theta}_3, ..., \hat{\theta}_p]^T \end{array} \right\} \tag{4.102}$$

and $\sigma_0 = (\underline{b}_0 + \bar{b}_0)/2$, $\sigma_1 = \sigma_0 - \underline{b}_0$. We now derive $\mathrm{Proj}_\Pi\{\cdot\}$ below. Firstly choose a smooth convex function $\mathscr{P}(\hat{\theta}) : \mathbb{R}^p \to \mathbb{R}$ as $\mathscr{P}(\hat{\theta}) = \left|\frac{\hat{b}_0\mathrm{sgn}(b_0)-\sigma_0}{\sigma_1}\right|^{\varsigma_2} + \left|\frac{\hat{a}-a^*}{\bar{a}}\right|^{\varsigma_2} - 1 + \varsigma_1$, where $0 < \varsigma_1 < 1$ and $\varsigma_2 \geq 2$ are two real numbers. Next we define two convex sets as $\Pi_0 = \{\hat{\theta} \in \mathbb{R}^p | \mathscr{P}(\hat{\theta}) \leq 0\}$ and $\Pi_\varsigma = \{\hat{\theta} \in \mathbb{R}^p | \mathscr{P}(\hat{\theta}) \leq \frac{\varsigma_1}{2}\}$. It is clear that Π_ς contains Π_0 and approaches Π as ς_1 decreases and ς_2 increases. Given the function $\mathscr{P}(\hat{\theta})$, we now obtain a smooth projection operator in (4.100) as

$$\mathrm{Proj}_\Pi\{\tau_\theta\} = \begin{cases} \tau_\theta, & \hat{\theta} \in \mathring{\Pi}_0 \quad \text{or} \quad \nabla_{\hat{\theta}}\mathscr{P}^T\tau_\theta \leq 0 \\ \left(I - \frac{\nabla_{\hat{\theta}}\mathscr{P}\nabla_{\hat{\theta}}\mathscr{P}^T}{\nabla_{\hat{\theta}}\mathscr{P}^T\nabla_{\hat{\theta}}\mathscr{P}}\right)\tau_\theta, & \hat{\theta} \in \Pi_\varsigma\backslash\mathring{\Pi}_0 \\ & \text{and} \quad \nabla_{\hat{\theta}}\mathscr{P}^T\tau_\theta > 0. \end{cases} \tag{4.103}$$

From (4.16), (4.22), and (4.54)–(4.56), through a recursive calculation similar to that on page 345 of [89], it is not difficult to express $\tilde{\lambda}$ as a smooth linear function of z, $\tilde{\eta}$, and ε such that

$$\tilde{\lambda} = H_z(\hat{\theta})z + H_{\tilde{\eta}}(\hat{\theta})\tilde{\eta} + H_\varepsilon(\hat{\theta})(\varepsilon_1 + \Omega_1^{rT}\tilde{\theta}), \tag{4.104}$$

where $H_z(\hat{\theta}) \in \mathbb{R}^{n \times n}$, $H_{\tilde{\eta}}(\hat{\theta}) \in \mathbb{R}^{n \times n}$, and $H_\varepsilon(\hat{\theta}) \in \mathbb{R}^n$ are bounded functions with respect to $\hat{\theta}$.

If we utilize projection algorithm, $\hat{\theta}$ is bounded for any $\hat{\theta}(0) \in \Pi_\varsigma$. Furthermore, we could show that ε is bounded from (4.14). As a consequence, employing Young's inequality and Cauchy-Schwartz inequality, through a lengthy but straightforward calculation, (4.88)–(4.94) imply that the following inequalities hold:

$$\int_0^1 (1+x)w(x,t)r_0(x,t)\,\mathrm{d}x \le M_{r_0}\|w(x,t)\|, \tag{4.105}$$

$$\int_0^1 (1+x)w(x,t)p_0(x,t)\,\mathrm{d}x$$
$$\le M_{p_0}\left(|z|^2 + |\tilde{\eta}|^2 + w(0,t)^2 + \|w(x,t)\|^2 + \|w(x,t)\|\right), \tag{4.106}$$

$$\int_0^1 (1+x)w(x,t)q_0(x,t)\,\mathrm{d}x$$
$$\le M_{q_0}\left(|z|^2 + |\tilde{\eta}|^2 + \|w(x,t)\|^2 + \|w(x,t)\|\right), \tag{4.107}$$

$$\int_0^1 (1+x)w(x,t)q_i(x,t)\,\mathrm{d}x$$
$$\le M_{q_i}\left(|z|^2 + |\tilde{\eta}|^2 + \|w(x,t)\|^2 + \|w(x,t)\|\right),$$
$$i = 1, 2, \cdots, p, \tag{4.108}$$

where $\|w(x,t)\| = \left(\int_0^1 w(x,t)^2\,\mathrm{d}x\right)^{\frac{1}{2}}$ and M_{r_0}, M_{p_0}, M_{q_0}, M_{q_i} are some positive constants. Moreover, the estimators (4.96)–(4.103) have the following properties:

$$|\dot{\hat{\theta}}_i| \le \frac{\gamma_\theta}{V_0} M_\theta\left(|z| + |z|^2 + |\tilde{\eta}|^2\right.$$
$$\left. +\|w(x,t)\|^2 + \|w(x,t)\|\right), \quad i = 1, 2, \cdots, p, \tag{4.109}$$
$$|\dot{\hat{D}}| \le \frac{\gamma_D}{V_0} M_D\left(|z|^2 + |\tilde{\eta}|^2 + w(0,t)^2\right.$$
$$\left. +\|w(x,t)\|^2 + \|w(x,t)\|\right), \tag{4.110}$$

where M_θ and M_D are some positive constants. These are derived from (4.105)–(4.106) such that

$$|\dot{\hat{\theta}}_i| = \gamma_\theta \left|e_i^T \mathrm{Proj}_\Pi\{\tau_\theta\}\right| \le \gamma_\theta \left|e_i^T \tau_\theta\right|$$
$$\le \frac{\gamma_\theta}{V_0}\left|\frac{1}{2g}e_i^T(\omega^r + \tilde{\omega})W_\varepsilon(\hat{\theta})^T z\right.$$
$$\left. -e_i^T(\omega^r + \tilde{\omega})\int_0^1 (1+x)w(x,t)r_0(x,t)\,\mathrm{d}x\right|$$
$$\le \frac{\gamma_\theta}{V_0}M_\theta\left(|z| + |z|^2 + |\tilde{\eta}|^2 + \|w(x,t)\|^2 + \|w(x,t)\|\right),$$
$$i = 1, 2, \cdots, p, \tag{4.111}$$
$$|\dot{\hat{D}}| = \gamma_D \left|\mathrm{Proj}_{[\underline{D},\bar{D}]}\{\tau_D\}\right| \le \gamma_D |\tau_D|$$
$$\le \frac{\gamma_D}{V_0}\left|\int_0^1 (1+x)w(x,t)p_0(x,t)\,\mathrm{d}x\right|$$
$$\le \frac{\gamma_D}{V_0}M_D\left(|z|^2 + |\tilde{\eta}|^2 + w(0,t)^2 + \|w(x,t)\|^2 + \|w(x,t)\|\right). \tag{4.112}$$

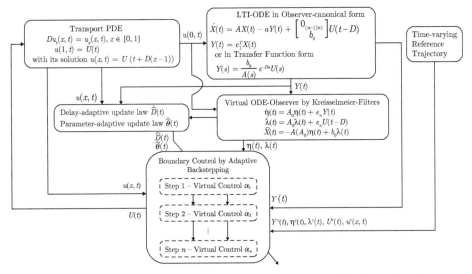

Figure 4.1. Adaptive control of single-input systems with full relative degree

4.5 Stability Analysis

The adaptive controller in this chapter is shown in figure 4.1 and we establish system stability summarized in the following theorem:

Theorem 4.8. *Consider the closed-loop system consisting of the plant (4.3), the ODE-state estimator (4.8)–(4.12), the adaptive controller (4.95), the time-delay identifier (4.96)–(4.99), and the parameter identifier (4.100)–(4.103). Under Assumptions 4.3–4.5, the following results hold:*

(i) all the signals of the closed-loop system are globally bounded,
(ii) the asymptotic tracking is achieved, i.e.,

$$\lim_{t\to\infty} z_1(t) = \lim_{t\to\infty} (Y(t) - Y_r(t)) = 0. \tag{4.113}$$

Proof. First of all, we build a nonnegative function to encompass major states of the closed-loop system as follows:

$$V(t) = D \ln \left(V_0(t) \right) + \frac{g}{\gamma_D} \tilde{D}(t)^2 + \frac{gD}{\gamma_\theta} \tilde{\theta}(t)^T \tilde{\theta}(t)$$

$$+ g_\varepsilon \sum_{i=1}^{n} \frac{D}{4d_i} \varepsilon(t)^T P \varepsilon(t), \tag{4.114}$$

where $V_0 = 1 + \frac{1}{2} z^T z + \frac{1}{k_\eta} \tilde{\eta}^T P \tilde{\eta} + g \int_0^1 (1+x) w(x,t)^2 \, dx$ has been given in (4.98) and $g_\varepsilon > 0$.

Taking the time derivative of (4.114) along the closed-loop target system (4.80)–(4.84), we obtain

$$
\dot{V}(t) = \frac{D}{V_0} \Big[z^T A_z(\hat{\theta}) z + z^T W_\epsilon(\hat{\theta})(\epsilon_2 + \omega^T \tilde{\theta}) + z^T Q(z,t)^T \dot{\hat{\theta}} + z^T Q^r(t,\hat{\theta})^T \dot{\hat{\theta}}
$$

$$
+ z^T e_n w(0,t) - \frac{1}{k_\eta}|\tilde{\eta}|^2 + \frac{2}{k_\eta}\tilde{\eta} P e_n z_1 + \frac{2g}{D}\int_0^1 (1+x)w(x,t)w_x(x,t)dx
$$

$$
- 2g(\epsilon_2 + \omega^T\tilde{\theta})\int_0^1 (1+x)w(x,t)r_0(x,t)dx - \frac{2g\tilde{D}}{D}\int_0^1 (1+x)w(x,t)p_0(x,t)dx
$$

$$
- 2g\dot{\tilde{D}}\int_0^1 (1+x)w(x,t)q_0(x,t)dx - 2g\dot{\hat{\theta}}^T\int_0^1 (1+x)w(x,t)q(x,t)dx \Big]
$$

$$
- \frac{2g}{\gamma_D}\tilde{D}\dot{\tilde{D}} - \frac{2gD}{\gamma_\theta}\tilde{\theta}^T\dot{\hat{\theta}} - g_\epsilon \sum_{i=1}^n \frac{D}{4d_i}|\epsilon|^2, \tag{4.115}
$$

where $z^T A_z(\hat{\theta}) z + z^T W_\epsilon(\hat{\theta})\epsilon_2$ is calculated by substituting (4.85)–(4.86) such that

$$
z^T A_z(\hat{\theta}) z + z^T W_\epsilon(\hat{\theta})\epsilon_2
$$

$$
= -\sum_{i=1}^n c_i z_i^2 - d_1 z_1^2 - \sum_{i=2}^n d_i \left(\frac{\partial \alpha_{i-1}}{\partial z_1}\right)^2 z_i^2 - z_1\epsilon_2 - \sum_{i=2}^n z_i \frac{\partial \alpha_{i-1}}{\partial z_1}\epsilon_2
$$

$$
= -\sum_{i=1}^n c_i z_i^2 - d_1 \left(z_1 + \frac{1}{2d_1}\epsilon_2\right)^2 - \sum_{i=2}^n d_i \left(z_i \frac{\partial \alpha_{i-1}}{\partial z_1} + \frac{1}{2d_i}\epsilon_2\right)^2 + \sum_{i=1}^n \frac{1}{4d_i}\epsilon_2^2
$$

$$
\leq -\sum_{i=1}^n c_i z_i^2 + \sum_{i=1}^n \frac{1}{4d_i}\epsilon_2^2 \tag{4.116}
$$

and $\frac{2g}{D}\int_0^1 (1+x)w(x,t)w_x(x,t)dx$ is calculated by using the integration by parts in x such that

$$
\frac{2g}{D}\int_0^1 (1+x)w(x,t)w_x(x,t)dx
$$

$$
= \frac{4g}{D}w(1,t)^2 - \frac{2g}{D}w(0,t)^2 - \frac{2g}{D}\|w(x,t)\|^2 - \frac{2g}{D}\int_0^1 (1+x)w_x(x,t)w(x,t)dx
$$

$$
= 0 - \frac{g}{D}w(0,t)^2 - \frac{g}{D}\|w(x,t)\|^2
$$

$$
\leq -\frac{g}{D}w(0,t)^2 - \frac{g}{D}\|w(x,t)\|^2. \tag{4.117}
$$

Substituting (4.116)–(4.117) into (4.115), and grouping like terms, we get

$$\dot{V}(t) = \frac{D}{V_0}\left[-\sum_{i=1}^{n}c_i z_i^2 + \sum_{i=1}^{n}\frac{1}{4d_i}\epsilon_2^2 + z^T Q(z,t)^T\dot{\hat{\theta}} + z^T Q^r(t,\hat{\theta})^T\dot{\hat{\theta}} + z_n w(0,t)\right.$$

$$-\frac{1}{k_\eta}|\tilde{\eta}|^2 + \frac{2}{k_\eta}\tilde{\eta}Pe_n z_1 - \frac{g}{D}w(0,t)^2 - \frac{g}{D}\|w(x,t)\|^2$$

$$-2g\epsilon_2\int_0^1 (1+x)w(x,t)r_0(x,t)dx$$

$$\left.-2g\dot{\hat{D}}\int_0^1 (1+x)w(x,t)q_0(x,t)dx - 2g\dot{\hat{\theta}}^T\int_0^1 (1+x)w(x,t)q(x,t)dx\right]$$

$$-\frac{2g}{\gamma_D}\tilde{D}\left(\dot{\hat{D}} + \gamma_D\frac{1}{V_0}\int_0^1 (1+x)w(x,t)p_0(x,t)dx\right)$$

$$-\frac{2gD}{\gamma_\theta}\tilde{\theta}^T\left(\dot{\hat{\theta}} - \gamma_\theta\frac{1}{V_0}\left(\frac{1}{2g}\omega W_\epsilon(\hat{\theta})^T z - \omega\int_0^1 (1+x)w(x,t)r_0(x,t)dx\right)\right)$$

$$-g_\epsilon\sum_{i=1}^{n}\frac{D}{4d_i}|\epsilon|^2. \tag{4.118}$$

Furthermore, considering update laws (4.97) and (4.101), we rewrite (4.118) as

$$\dot{V}(t) \le \frac{D}{V_0}\left[-\sum_{i=1}^{n}c_i z_i^2 + \sum_{i=1}^{n}\frac{1}{4d_i}\varepsilon_2^2\right.$$

$$+z^T Q(z,t)^T\dot{\hat{\theta}} + z^T Q^r(t,\hat{\theta})^T\dot{\hat{\theta}} + z_n w(0,t)$$

$$-\frac{1}{k_\eta}|\tilde{\eta}|^2 + \frac{2}{k_\eta}\tilde{\eta}^T Pe_n z_1$$

$$-\frac{g}{D}w(0,t)^2 - \frac{g}{D}\|w(x,t)\|^2$$

$$-2g\varepsilon_2\int_0^1 (1+x)w(x,t)r_0(x,t)\,dx$$

$$-2g\dot{\hat{D}}\int_0^1 (1+x)w(x,t)q_0(x,t)\,dx$$

$$\left.-2g\dot{\hat{\theta}}^T\int_0^1 (1+x)w(x,t)q(x,t)\,dx\right]$$

$$-\frac{2g}{\gamma_D}\tilde{D}(\dot{D}-\gamma_D\tau_D)-\frac{2gD}{\gamma_\theta}\tilde{\theta}^T(\dot{\hat{\theta}}-\gamma_\theta\tau_\theta)$$

$$-g_\varepsilon\sum_{i=1}^{n}\frac{D}{4d_i}|\varepsilon|^2. \tag{4.119}$$

From (4.47), (4.54)–(4.56), it is easy to show that $-\frac{\partial\alpha_{i-1}}{\partial\hat{\theta}_j}$ for $i=2,\cdots,n$ and $j=1,\cdots,p$ is a linear function of z, $\tilde{\eta}$, and ε. With (4.105)–(4.110), by using inequalities $0<\frac{x^2}{1+x^2}<1$ and $0<\frac{|x|}{1+x^2}<1$ to bound the cubic and biquadratic terms with only quadratic ones like those in [27], we have

$$z^TQ(z,t)^T\dot{\hat{\theta}}=-\sum_{i=2}^{n}\sum_{j=1}^{p}z_i\frac{\partial\alpha_{i-1}}{\partial\hat{\theta}_j}\dot{\hat{\theta}}_j$$

$$\leq\gamma_\theta M_Q\left(|z|^2+|\tilde{\eta}|^2+\|w(x,t)\|^2\right), \tag{4.120}$$

where

$$M_1=\sum_{i=2}^{n}\sum_{j=1}^{p}\left(\left\|\frac{\partial K_{i,z_1}(\hat{\theta})}{\partial\hat{\theta}_j}e_1^T+\frac{\partial K_{i,\tilde{\lambda}}(\hat{\theta})}{\partial\hat{\theta}_j}H_z(\hat{\theta})\right\|_\infty\right.$$

$$+\frac{1}{2}\left\|\frac{\partial K_{i,\tilde{\eta}}(\hat{\theta})}{\partial\hat{\theta}_j}+\frac{\partial K_{i,\tilde{\lambda}}(\hat{\theta})}{\partial\hat{\theta}_j}H_{\tilde{\eta}}(\hat{\theta})\right\|_\infty$$

$$\left.+\left\|\frac{\partial K_{i,\tilde{\lambda}}(\hat{\theta})}{\partial\hat{\theta}_j}H_\varepsilon(\hat{\theta})(\varepsilon_1+\Omega_1^{rT}\tilde{\theta})\right\|_\infty\right), \tag{4.121}$$

$$M_Q=M_\theta M_1\left(\frac{11}{2}+\frac{1}{\min\{1,g\}}+\frac{1}{\min\{\frac{1}{2},\frac{\lambda}{k_\eta}\}}\right), \tag{4.122}$$

$$\lambda=\min\{\lambda_{\min}(P),\ 1\}, \tag{4.123}$$

$$z^TQ^r(t,\hat{\theta})^T\dot{\hat{\theta}}=-\sum_{i=2}^{n}\sum_{k=1}^{p}z_i\left(e_i^T-\sum_{j=1}^{i-1}\frac{\partial\alpha_{i-1}}{\partial\tilde{\lambda}}e_je_j^T\right)$$

$$\times\frac{\partial\lambda^r(t,\hat{\theta})}{\partial\hat{\theta}_k}\dot{\hat{\theta}}_k$$

$$\leq\frac{\gamma_\theta}{V_0}M_\theta\sum_{i=2}^{n}\sum_{k=1}^{p}|z_i|\left|e_i^T-\sum_{j=1}^{i-1}\frac{\partial\alpha_{i-1}}{\partial\tilde{\lambda}}e_je_j^T\right|\left|\frac{\partial\lambda^r(t,\hat{\theta})}{\partial\hat{\theta}_k}\right|$$

$$\times\left(|z|+|z|^2+|\tilde{\eta}|^2+\|w(x,t)\|+\|w(x,t)\|^2\right)$$

$$\leq\gamma_\theta M_{Q^r}\left(|z|^2+|\tilde{\eta}|^2+\|w(x,t)\|^2\right), \tag{4.124}$$

where

$$M_2 = \sum_{i=2}^{n} \sum_{k=1}^{p} \left(\left\| e_i^T - \sum_{j=1}^{i-1} \frac{\partial \alpha_{i-1}}{\partial \tilde{\lambda}} e_j e_j^T \right\|_\infty \right.$$

$$\left. \times \left\| \frac{\partial \lambda^r(t, \hat{\theta})}{\partial \hat{\theta}_k} \right\|_\infty \right), \tag{4.125}$$

$$M_{Q^r} = \frac{7}{2} M_\theta M_2, \tag{4.126}$$

$$-2g\varepsilon_2 \int_0^1 (1+x) w(x,t) r_0(x,t) \, dx$$

$$\leq \frac{g}{2\bar{D}} \|w(x,t)\|^2 + g M_4 |\varepsilon|^2, \tag{4.127}$$

$$-2g\dot{D} \int_0^1 (1+x) w(x,t) q_0(x,t) \, dx$$

$$\leq \frac{2g\gamma_D}{V_0} M_D M_{q_0} (|z|^2 + |\tilde{\eta}|^2 + \|w(x,t)\|$$

$$+ \|w(x,t)\|^2 + w(0,t)^2)$$

$$\times (|z|^2 + |\tilde{\eta}|^2 + \|w(x,t)\|^2 + \|w(x,t)\|)$$

$$\leq g\gamma_D M_5 (|z|^2 + |\tilde{\eta}|^2 + \|w(x,t)\|^2 + w(0,t)^2), \tag{4.128}$$

$$-2g\dot{\hat{\theta}}^T \int_0^1 (1+x) w(x,t) q(x,t) \, dx$$

$$\leq \frac{2g\gamma_\theta}{V_0} M_\theta \sum_{i=1}^{p} (|z| + |z|^2 + |\tilde{\eta}|^2$$

$$+ \|w(x,t)\| + \|w(x,t)\|^2)$$

$$\times M_{q_i} (|z|^2 + |\tilde{\eta}|^2 + \|w(x,t)\| + \|w(x,t)\|^2)$$

$$\leq g\gamma_\theta M_6 (|z|^2 + |\tilde{\eta}|^2 + \|w(x,t)\|^2), \tag{4.129}$$

where M_4, M_5, M_6 are some positive constants that have explicit forms below.

$$M_4 = 2\bar{D} M_{r_0}^2, \tag{4.130}$$

$$M_5 = 2 M_D M_{q_0} \left(1 + \frac{2}{\min\{1, g\}} + \frac{1}{\min\{\frac{1}{2}, \frac{\Delta}{k_\eta}, g\}} \right), \tag{4.131}$$

$$M_6 = 2 M_\theta \sum_{i=1}^{p} M_{q_i} \left(\frac{5}{2} + \frac{2}{\min\{1, g\}} + \frac{1}{\min\{\frac{1}{2}, \frac{\Delta}{k_\eta}, g\}} \right). \tag{4.132}$$

Above all, (4.119) becomes

$$\dot{V}(t) \leq \frac{D}{V_0}\left[-\frac{c_0}{2}|z|^2 + \frac{1}{4d_0}|\varepsilon|^2 + \frac{1}{2c_0}w(0,t)^2 \right.$$

$$+\gamma_\theta M_z\left(|z|^2 + |\tilde{\eta}|^2 + \|w(x,t)\|^2\right)$$

$$-\frac{1}{2k_\eta}|\tilde{\eta}|^2 + \frac{2}{k_\eta}|Pe_n|^2 z_1^2$$

$$-\frac{g}{\bar{D}}w(0,t)^2 - \frac{g}{2\bar{D}}\|w(x,t)\|^2 + gM_4|\varepsilon|^2$$

$$+g\gamma_D M_5\left(|z|^2 + |\tilde{\eta}|^2 + \|w(x,t)\|^2 + w(0,t)^2\right)$$

$$\left. +g\gamma_\theta M_6\left(|z|^2 + |\tilde{\eta}|^2 + \|w(x,t)\|^2\right)\right]$$

$$-g_\varepsilon \frac{D}{4d_0}|\varepsilon|^2, \tag{4.133}$$

where $M_z = M_Q + M_{Q^r}$, $c_0 = \min\limits_{i=1,2,\cdots,n} c_i$, $d_0 = \left(\sum\limits_{i=1}^{n}\frac{1}{d_i}\right)^{-1}$.
If we design k_η, g, γ_θ, γ_D and g_ε appropriately such that

$$k_\eta \geq \frac{4}{c_0}|Pe_n|^2, \quad g \geq \frac{\bar{D}}{2c_0}, \quad g_\varepsilon \geq 1 + 4gM_4 d_0, \tag{4.134}$$

$$\gamma_\theta, \gamma_D \in [0, \gamma^*], \quad \gamma^* = \frac{m_1}{m_2}, \tag{4.135}$$

$$m_1 = \min\left\{ \frac{c_0}{2} - \frac{2}{k_\eta}|Pe_n|^2, \quad \frac{1}{2k_\eta}, \quad \frac{g}{\bar{D}} - \frac{1}{2c_0}, \quad \frac{g}{2\bar{D}} \right\}, \tag{4.136}$$

$$m_2 = 2\max\left\{ M_z + gM_6, \quad gM_5 \right\}, \tag{4.137}$$

it is obvious that we can make $\dot{V}$ nonpositive, and

$$\dot{V} \leq -\frac{M}{V_0}\left(|z|^2 + |\tilde{\eta}|^2 + \|w(x,t)\|^2 + w(0,t)^2\right), \tag{4.138}$$

where $M > 0$. Thus signals z, $\tilde{\eta}$, $\|w(x,t)\|$ are bounded. Then we can obtain boundedness of $\tilde{\lambda}$. Based on (4.78)–(4.79), it is easy to show that $\|\tilde{u}(x,t)\|$ is bounded. Bearing boundedness of ε, $\hat{\theta}$, Y_r, U^r, η^r, and λ^r and $u^r(x,t,\hat{\theta})$ in mind, we establish boundedness of η, λ, X, $u(x,t)$, and U from (4.26), (4.27), (4.16), and (4.95), respectively. Thus Theorem 4.8(i) has been proved.

By applying Barbalat's lemma to (4.138), it follows such that $z(t) \to 0$, as $t \to \infty$; based on (4.46), we have $\lim\limits_{t\to\infty}[Y(t) - Y_r(t)] = 0$. $\qquad\square$

4.6   Simulation

In this section, we illustrate the proposed scheme by applying it to the following potentially unstable plant which is similar to the example in [27] but has an unmeasured ODE system state:

$$\dot{X}_1(t) = X_2(t) - a_1 Y(t), \tag{4.139}$$

$$\dot{X}_2(t) = U(t - D), \tag{4.140}$$

$$Y(t) = X_1(t), \tag{4.141}$$

where $X_2(t)$ is the unmeasurable state, $Y(t)$ is the measured output, $D=1$ is the unknown constant time delay in control input, and $a_1 = -0.5$ is the unknown constant system parameter. The information including $\underline{D}=0.2$, $\bar{D}=1.5$, $a^* = [-0.5, 0]^T$, $\bar{a}=0.5$ is known. The reference signal $Y_r(t) = \sin t$ is known and it is easily produced by the following reference state and input:

$$X_1^r(t) = Y_r(t) = \sin t, \tag{4.142}$$

$$X_2^r(t, \hat{\theta}) = \dot{Y}_r(t) + \hat{a}_1 Y_r(t) = \cos t + \hat{a}_1 \sin t, \tag{4.143}$$

$$U^r(t, \hat{\theta}) = -\sin t + \hat{a}_1 \cos t \tag{4.144}$$

and the K-filters

$$\eta_1^r(t) = \frac{k_2 - 1}{k_1^2 + (k_2 - 1)^2} \sin t + \frac{-k_1}{k_1^2 + (k_2 - 1)^2} \cos t, \tag{4.145}$$

$$\eta_2^r(t) = \frac{k_1 k_2}{k_1^2 + (k_2 - 1)^2} \sin t + \frac{-k_1^2 + k_2 - 1}{k_1^2 + (k_2^2 - 1)^2} \cos t, \tag{4.146}$$

$$\lambda_1^r(t, \hat{\theta}) = \left(-\frac{k_2 - 1}{k_1^2 + (k_2 - 1)^2} + \hat{a}_1 \frac{k_1}{k_1^2 + (k_2 - 1)^2} \right) \sin t$$
$$+ \left(\frac{k_1}{k_1^2 + (k_2 - 1)^2} + \hat{a}_1 \frac{k_2 - 1}{k_1^2 + (k_2 - 1)^2} \right) \cos t, \tag{4.147}$$

$$\lambda_2^r(t, \hat{\theta}) = \left(-\frac{k_1 k_2}{k_1^2 + (k_2 - 1)^2} + \hat{a}_1 \frac{k_2 - k_2^2}{k_1^2 + (k_2 - 1)^2} + \hat{a}_1 \right) \sin t$$
$$+ \left(-\frac{k_2^2 - k_2}{k_1^2 + (k_2 - 1)^2} + \hat{a}_1 \frac{k_1 k_2}{k_1^2 + (k_2 - 1)^2} + 1 \right) \cos t. \tag{4.148}$$

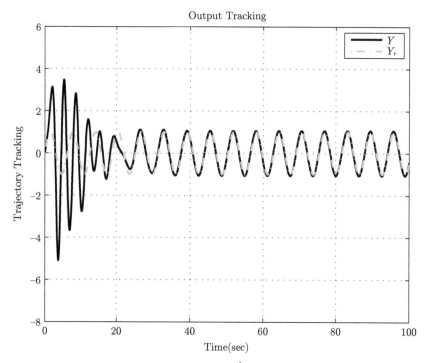

Figure 4.2. Output tracking for $\hat{D}(0) = 0$ and $\hat{a}_1(0) = -1$

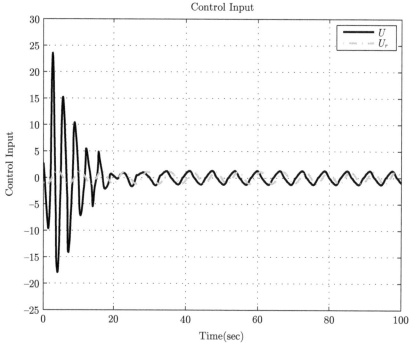

Figure 4.3. Control input for $\hat{D}(0) = 0$ and $\hat{a}_1(0) = -1$

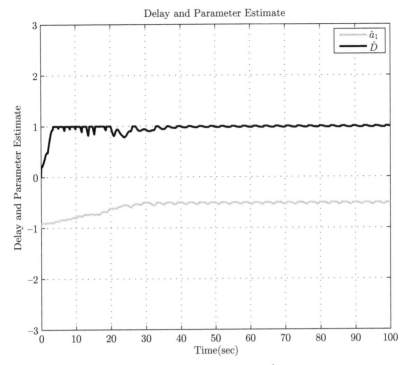

Figure 4.4. Delay and parameter estimates for $\hat{D}(0) = 0$ and $\hat{a}_1(0) = -1$

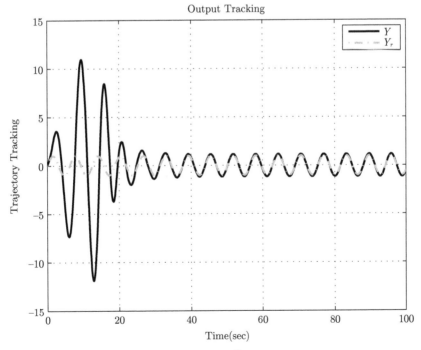

Figure 4.5. Output tracking for $\hat{D}(0) = 1.2$ and $\hat{a}_1(0) = 0$

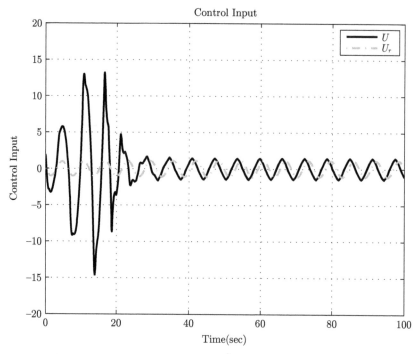

Figure 4.6. Control input for $\hat{D}(0) = 1.2$ and $\hat{a}_1(0) = 0$

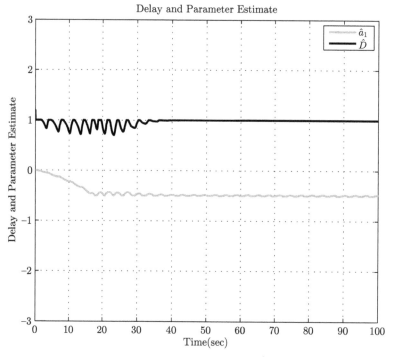

Figure 4.7. Delay and parameter estimates for $\hat{D}(0) = 1.2$ and $\hat{a}_1(0) = 0$

Furthermore, if we choose $k_1 = k_2 = 1$, the reference state (4.142)–(4.143) can be easily divided into the reference output filter and input filter as follows:

$$\eta_1^r(t) = -\cos t, \tag{4.149}$$

$$\eta_2^r(t) = \sin t - \cos t, \tag{4.150}$$

$$\lambda_1^r(t, \hat{\theta}) = \hat{a}_1 \sin t + \cos t, \tag{4.151}$$

$$\lambda_2^r(t, \hat{\theta}) = (\hat{a}_1 - 1) \sin t + (\hat{a}_1 + 1) \cos t. \tag{4.152}$$

It is obvious that Assumptions 4.3–4.5 are satisfied. Then we adopt the proposed boundary adaptive backstepping scheme to design controller and identifier. The initial values are set as $X_1(0) = 0$, $X_2(0) = 1$ and two different value combinations of $\hat{D}(0) = 0$ with $\hat{a}_1(0) = -1$ and $\hat{D}(0) = 1.2$ with $\hat{a}_1(0) = 0$, which means that there exists no predictor feedback or overcompensating predictor feedback initially, respectively. The design parameters are chosen as follows:

$$c_1 = c_2 = 1, \ d_1 = d_2 = 0.1, \ \varsigma_1 = 0.015, \ \varsigma_2 = 60, \ k_\eta = 300,$$

$$\gamma_\theta = 1 \times 10^{-3}, \ \gamma_D = 1 \times 10^{-2}, \ g = 2, \ P = \begin{bmatrix} 1 & -\frac{1}{2} \\ -\frac{1}{2} & \frac{3}{2} \end{bmatrix}$$

As a result, the boundedness and asymptotic tracking of system output and control input are shown in figures 4.2 and 4.3, though there exists no predictor feedback initially, that is, $\hat{D}(0) = 0$. Figures 4.5 and 4.6 show that system output and control input also have boundedness and asymptotic tracking despite an overcompensating predictor feedback at the beginning, that is, $\hat{D}(0) = 1.2$. Figures 4.4 and 4.7 exhibit estimations of θ and D when $\hat{D}(0) = 0$ and $\hat{D}(0) = 1.2$, respectively. Figures 4.4 and 4.7 exhibit an exact convergence of $\hat{\theta}$ to θ and an estimation of $\hat{D}$ to D when $\hat{D}(0) = 0$ and $\hat{D}(0) = 2$, respectively. $\hat{\theta}(t)$ converges to θ in both simulations because of a persistency of excitation, while this is not true for $\hat{D}(t)$.

Chapter Five

Single-Input Systems with Arbitrary Relative Degree

IN CHAPTER 4, we dealt with adaptive control for single-input LTI systems with uncertainty collection $(D, X(t), \theta)$ by assuming that the actuator state $u(x, t)$ is measurable. Furthermore, as stated in [88], the relative degree plays an important role in determining the difficulty of the output-feedback adaptive problems, namely, problems involving both unmeasurable ODE state $X(t)$ and uncertain parameter θ in system and input matrices $A(\theta)$ and $B(\theta)$. Thus, in chapter 4, the plants of interest were restricted to a class of systems with its relative degree being identical with its system dimension. In this chapter, we remove this restriction on relative degree and thus the results obtained are more general than in chapter 4. More importantly, we deal with the most challenging case of the coexistence of the four basic uncertainties of $(D, X(t), \theta, u(x, t))$.

5.1 Mathematical Model

Consider a general class of single-input single-output uncertain linear time-delay systems with discrete input delay

$$Y(s) = \frac{b_m s^m + \cdots + b_1 s + b_0}{s^n + a_{n-1} s^{n-1} + \cdots + a_1 s + a_0} e^{-Ds} U(s), \tag{5.1}$$

which can be represented as the following observer canonical form:

$$\dot{X}(t) = AX(t) - aY(t) + \begin{bmatrix} 0_{(\rho-1) \times 1} \\ b \end{bmatrix} U(t - D)$$

$$Y(t) = e_1^T X(t), \tag{5.2}$$

where

$$A = \begin{bmatrix} 0_{(n-1) \times 1} & I_{n-1} \\ 0 & 0_{1 \times (n-1)} \end{bmatrix}, \quad a = \begin{bmatrix} a_{n-1} \\ \vdots \\ a_0 \end{bmatrix}, \quad b = \begin{bmatrix} b_m \\ \vdots \\ b_0 \end{bmatrix}, \tag{5.3}$$

and e_i for $i = 1, 2, \cdots$ is the ith coordinate vector in corresponding space, ρ represents the relative degree satisfying $\rho = n - m$, $X(t) = [X_1(t), X_2(t), \cdots, X_n(t)]^T \in \mathbb{R}^n$ is the ODE state vector unavailable to measure, $Y(t) \in \mathbb{R}$ is the measurable output, and $U(t-D) \in \mathbb{R}$ is the actual input with an unknown constant time delay D, while $a_{n-1}, \cdots, a_0$ and $b_m, \cdots, b_0$ are unknown constant plant parameters and control coefficients, respectively.

The system (5.2) can be written compactly as

$$\dot{X}(t) = AX(t) + F\big(U(t-D), Y(t)\big)^T \theta$$
$$Y(t) = e_1^T X(t), \tag{5.4}$$

where $p = n + m + 1$ - dimensional parameter vector θ is defined by

$$\theta = \begin{bmatrix} b \\ a \end{bmatrix} \tag{5.5}$$

and

$$F\big(U(t-D), Y(t)\big)^T = \left[\begin{bmatrix} 0_{(\rho-1)\times(m+1)} \\ I_{m+1} \end{bmatrix} U(t-D), \quad -I_n Y(t) \right]. \tag{5.6}$$

Several traditional and mild assumptions concerning the system (5.1)–(5.2) are formulated. It is assumed that they hold throughout the chapter.

Assumption 5.1. *The plant (5.1) is minimum phase, i.e., the polynomial* $B(s) = b_m s^m + \cdots + b_1 s + b_0$ *is Hurwitz.*

Assumption 5.2. *There exist two known constants* $\underline{D} > 0$ *and* $\bar{D} > 0$ *such that* $D \in [\underline{D}, \bar{D}]$. *The high-frequency gain's sign* $\mathrm{sgn}(b_m)$ *is known and a constant* $\underline{b}_m$ *is known such that* $|b_m| \geq \underline{b}_m > 0$. *Furthermore,* θ *belongs to a convex compact set* $\Theta = \{\theta \in \mathbb{R}^p | \mathscr{P}(\theta) \leq 0\}$, *where* $\mathscr{P} : \mathbb{R}^p \to \mathbb{R}$ *is a smooth convex function.*

We start our design and analysis by introducing the distributed input

$$u(x,t) = U(t + D(x-1)), \quad x \in [0,1], \tag{5.7}$$

where the delayed actuator state is denoted by a transport PDE that is governed by the following dynamic PDE equation:

$$D u_t(x,t) = u_x(x,t), \tag{5.8}$$
$$u(1,t) = U(t). \tag{5.9}$$

In the rest of the chapter, the respective control schemes are developed for two different cases: measured or unmeasured actuator $u(x,t)$.

5.2 Trajectory Tracking by PDE Full-State Feedback

In this section, we consider the situation when the actuator state $u(x,t)$ is measurable. Our control objective is to design a control strategy to compensate for the system uncertainty as well as actuator delay to ensure all the signals of the closed-loop system are bounded, and to make output $Y(t)$ asymptotically track a time-varying reference signal $Y_r(t)$. To achieve this goal, we have one additional assumption.

Assumption 5.3. *In the case of known θ, given a time-varying reference output trajectory $Y_r(t)$ that is known, bounded, and smooth, there exist known reference state signal $X^r(t, \theta)$ and reference input signal $U^r(t, \theta)$, which are bounded in t, continuously differentiable in the argument θ, and satisfy*

$$\dot{X}^r(t, \theta) = AX^r(t, \theta) + F\big(U^r(t, \theta), Y_r(t)\big)^T \theta$$
$$Y_r(t) = e_1^T X^r(t, \theta). \tag{5.10}$$

5.2.1 State Observer with Kreisselmeier-Filters

Firstly, we employ the well-known K-filters as follows:

$$\dot{\eta}(t) = A_0 \eta(t) + e_n Y(t), \tag{5.11}$$
$$\dot{\lambda}(t) = A_0 \lambda(t) + e_n U(t - D), \tag{5.12}$$
$$\xi(t) = -A_0^n \eta(t), \tag{5.13}$$
$$\Xi(t) = -[A_0^{n-1} \eta(t), \cdots, A_0 \eta(t), \eta(t)], \tag{5.14}$$
$$v_j(t) = A_0^j \lambda(t), \quad j = 0, 1, \cdots, m, \tag{5.15}$$
$$\Omega(t)^T = [v_m(t), \cdots, v_1(t), v_0(t), \Xi(t)], \tag{5.16}$$

where the vector $k = [k_1, k_2, \cdots, k_n]^T$ is chosen so that the matrix $A_0 = A - k e_1^T$ is Hurwitz, that is, $A_0^T P + P A_0 = -I$, $P = P^T > 0$. The unmeasurable ODE system state $X(t)$ is virtually estimated as $\hat{X}(t) = \xi(t) + \Omega(t)^T \theta$ and the state estimation error $\varepsilon(t) = X(t) - \hat{X}(t)$ vanishes exponentially as

$$\dot{\varepsilon}(t) = A_0 \varepsilon(t). \tag{5.17}$$

Thus we get a static relationship such that

$$X(t) = \xi(t) + \Omega(t)^T \theta + \varepsilon(t), \tag{5.18}$$
$$= -A(A_0)\eta(t) + B(A_0)\lambda(t) + \varepsilon(t), \tag{5.19}$$

where $A(A_0) = A_0^n + \sum_{i=0}^{n-1} a_i A_0^i$, $B(A_0) = \sum_{i=0}^{m} b_i A_0^i$.

According to (5.10), when θ is known, we can use reference K-filters $\eta^r(t)$ and $\lambda^r(t, \theta)$ to produce the reference output $Y_r(t)$. When θ is unknown, by the certainty equivalence principle, we have

$$\dot{\eta}^r(t) = A_0 \eta^r(t) + e_n Y_r(t), \tag{5.20}$$
$$\dot{\lambda}^r(t, \hat{\theta}) = A_0 \lambda^r(t, \hat{\theta}) + e_n U^r(t, \hat{\theta}) + \frac{\partial \lambda^r(t, \hat{\theta})}{\partial \hat{\theta}} \dot{\hat{\theta}}, \tag{5.21}$$

$$\xi^r(t) = -A_0^n \eta^r(t), \tag{5.22}$$

$$\Xi^r(t) = -[A_0^{n-1}\eta^r(t), \cdots, A_0\eta^r(t), \eta^r(t)], \tag{5.23}$$

$$v_j^r(t, \hat\theta) = A_0^j \lambda^r(t, \hat\theta), \quad j = 0, 1, \cdots, m, \tag{5.24}$$

$$\Omega^r(t, \hat\theta)^T = [v_m^r(t, \hat\theta), \cdots, v_1^r(t, \hat\theta), v_0^r(t, \hat\theta), \Xi^r(t)], \tag{5.25}$$

$$X^r(t, \hat\theta) = \xi^r(t) + \Omega^r(t, \hat\theta)^T \hat\theta, \tag{5.26}$$

$$= -\hat{A}(A_0)\eta^r(t) + \hat{B}(A_0)\lambda^r(t, \hat\theta), \tag{5.27}$$

$$Y_r(t) = e_1^T X^r(t, \hat\theta), \tag{5.28}$$

where $\hat\theta$ is the estimate of unknown parameter θ with $\tilde\theta = \theta - \hat\theta$, and $\hat{A}(A_0) = A_0^n + \sum_{i=0}^{n-1} \hat{a}_i A_0^i$, $\hat{B}(A_0) = \sum_{i=0}^{m} \hat{b}_i A_0^i$.

Now a series of error variables are defined as follows:

$$z_1(t) = Y(t) - Y_r(t), \tag{5.29}$$

$$\tilde{U}(t - D) = U(t - D) - U^r(t, \hat\theta), \tag{5.30}$$

$$\tilde\eta(t) = \eta(t) - \eta^r(t), \tag{5.31}$$

$$\tilde\lambda(t) = \lambda(t) - \lambda^r(t, \hat\theta), \tag{5.32}$$

$$\tilde\xi(t) = \xi(t) - \xi^r(t), \tag{5.33}$$

$$\tilde\Xi(t) = \Xi(t) - \Xi^r(t), \tag{5.34}$$

$$\tilde{v}_j(t) = v_j(t) - v_j^r(t, \hat\theta), \quad j = 0, 1, \cdots, m, \tag{5.35}$$

$$\tilde\Omega(t)^T = \Omega(t)^T - \Omega^r(t, \hat\theta)^T, \tag{5.36}$$

which are governed by the following dynamic equations:

$$\dot{\tilde\eta}(t) = A_0 \tilde\eta(t) + e_n z_1(t), \tag{5.37}$$

$$\dot{\tilde\lambda}(t) = A_0 \tilde\lambda(t) + e_n \tilde{U}(t - D) - \frac{\partial \lambda^r(t, \hat\theta)}{\partial \hat\theta} \dot{\hat\theta}, \tag{5.38}$$

$$\dot{z}_1(t) = \tilde\xi_2(t) + \tilde\omega(t)^T \hat\theta + \varepsilon_2(t) + \omega(t)^T \tilde\theta, \tag{5.39}$$

$$= \hat{b}_m \tilde{v}_{m,2}(t) + \tilde\xi_2(t) + \bar{\tilde\omega}(t)^T \hat\theta$$

$$+ \varepsilon_2(t) + \omega(t)^T \tilde\theta, \tag{5.40}$$

where

$$\tilde\omega(t) = [\tilde{v}_{m,2}(t), \tilde{v}_{m-1,2}(t), \cdots, \tilde{v}_{0,2}(t), \tilde\Xi_2(t) - z_1(t)e_1^T]^T, \tag{5.41}$$

$$\bar{\tilde\omega}(t) = [0, \tilde{v}_{m-1,2}(t), \cdots, \tilde{v}_{0,2}(t), \tilde\Xi_2(t) - z_1(t)e_1^T]^T, \tag{5.42}$$

$$\omega(t) = [v_{m,2}(t), v_{m-1,2}(t), \cdots, v_{0,2}(t), \Xi_2(t) - Y(t)e_1^T]^T. \tag{5.43}$$

Then a couple of new variables are further defined as

$$\chi(t) = \begin{bmatrix} Y(t) \\ \eta(t) \\ \lambda(t) \end{bmatrix}, \quad \chi^r(t, \hat{\theta}) = \begin{bmatrix} Y_r(t) \\ \eta^r(t) \\ \lambda^r(t, \hat{\theta}) \end{bmatrix}. \tag{5.44}$$

Thus a new error variable is derived:

$$\tilde{\chi}(t) = \chi(t) - \chi^r(t, \hat{\theta}) = \begin{bmatrix} z_1(t) \\ \tilde{\eta}(t) \\ \tilde{\lambda}(t) \end{bmatrix} \in \mathbb{R}^{2n+1}, \tag{5.45}$$

which is driven by

$$\dot{\tilde{\chi}}(t) = A_{\tilde{\chi}}(\hat{\theta})\tilde{\chi}(t) + e_{2n+1}\tilde{U}(t - D) + e_1\left(\varepsilon_2(t) + \omega(t)^T \tilde{\theta}\right)$$
$$- \frac{\partial \chi^r(t, \hat{\theta})}{\partial \hat{\theta}}\dot{\hat{\theta}}, \tag{5.46}$$

where

$$A_{\tilde{\chi}}(\hat{\theta}) = \begin{bmatrix} -\hat{a}_{n-1} & -e_2^T \hat{A}(A_0) & e_2^T \hat{B}(A_0) \\ e_n & A_0 & 0_{n \times n} \\ 0_{n \times 1} & 0_{n \times n} & A_0 \end{bmatrix}, \tag{5.47}$$

$$\frac{\partial \chi^r(t, \hat{\theta})}{\partial \hat{\theta}} = \begin{bmatrix} 0_{(1+n) \times p} \\ \frac{\partial \lambda^r(t, \hat{\theta})}{\partial \hat{\theta}} \end{bmatrix}. \tag{5.48}$$

Next, similar to the transformation on pages 435–436 of [89], we bring in the dynamic equation for the m-dimensional inverse dynamics $\zeta(t) = TX(t)$ of (5.2) and its reference signal $\zeta^r(t)$ as

$$\dot{\zeta}(t) = A_b \zeta(t) + b_b Y(t), \tag{5.49}$$

$$\dot{\zeta}^r(t) = A_b \zeta^r(t) + b_b Y_r(t), \tag{5.50}$$

where

$$A_b = \begin{bmatrix} -\frac{b_{m-1}}{b_m} & & & \\ \vdots & & I_{m-1} & \\ -\frac{b_0}{b_m} & 0 & \cdots & 0 \end{bmatrix}, \tag{5.51}$$

$$b_b = T\left(A^\rho \begin{bmatrix} 0 \\ b \\ \frac{b}{b_m} \end{bmatrix} - a\right), \tag{5.52}$$

$$T = [A_b^\rho e_1, \cdots, A_b e_1, I_m]. \tag{5.53}$$

Thus the error state

$$\tilde{\zeta}(t) = \zeta(t) - \zeta^r(t) \tag{5.54}$$

is driven by

$$\dot{\tilde{\zeta}}(t) = A_b \tilde{\zeta}(t) + b_b z_1(t). \tag{5.55}$$

Under Assumption 5.1, we can see A_b is Hurwitz, that is, $P_b A_b + A_b^T P_b = -I$, $P_b = P_b^T > 0$.

5.2.2 Boundary Control with Adaptive Backstepping

Aiming at system

$$\dot{z}_1 = \hat{b}_m \tilde{v}_{m,2} + \tilde{\xi}_2 + \bar{\omega}^T \hat{\theta} + \varepsilon_2 + \omega^T \tilde{\theta}, \tag{5.56}$$

$$\dot{\tilde{v}}_{m,i} = \tilde{v}_{m,i+1} - k_i \tilde{v}_{m,1} - e_i^T A_0^m \frac{\partial \lambda^r(t, \hat{\theta})}{\partial \hat{\theta}} \dot{\hat{\theta}}, \tag{5.57}$$

$$i = 2, 3, \cdots, \rho - 1,$$

$$\dot{\tilde{v}}_{m,\rho} = \tilde{U}(t - D) + \tilde{v}_{m,\rho+1} - k_\rho \tilde{v}_{m,1} - e_\rho^T A_0^m \frac{\partial \lambda^r(t, \hat{\theta})}{\partial \hat{\theta}} \dot{\hat{\theta}}, \tag{5.58}$$

we present the adaptive backstepping recursive control design as follows:
    Coordinate transformation:

$$z_1 = Y - Y_r, \tag{5.59}$$

$$z_i = \tilde{v}_{m,i} - \alpha_{i-1}, \quad i = 2, 3, \cdots, \rho. \tag{5.60}$$

Stabilizing functions:

$$\alpha_1 = \frac{1}{\hat{b}_m} \left(-(c_1 + d_1) z_1 - \tilde{\xi}_2 - \bar{\omega}^T \hat{\theta} \right), \tag{5.61}$$

$$\alpha_2 = -\hat{b}_m z_1 - \left(c_2 + d_2 \left(\frac{\partial \alpha_1}{\partial z_1} \right)^2 \right) z_2 + \beta_2, \tag{5.62}$$

$$\alpha_i = -z_{i-1} - \left(c_i + d_i \left(\frac{\partial \alpha_{i-1}}{\partial z_1} \right)^2 \right) z_i + \beta_i, \tag{5.63}$$

$$i = 3, 4, \cdots, \rho,$$

$$\beta_i = k_i \tilde{v}_{m,1} + \frac{\partial \alpha_{i-1}}{\partial z_1} (\tilde{\xi}_2 + \tilde{\omega}^T \hat{\theta}) + \frac{\partial \alpha_{i-1}}{\partial \tilde{\eta}} (A_0 \tilde{\eta} + e_n z_1)$$

$$+ \sum_{j=1}^{m+i-1} \frac{\partial \alpha_{i-1}}{\partial \tilde{\lambda}_j} (\tilde{\lambda}_{j+1} - k_j \tilde{\lambda}_1), \quad i = 2, 3, \cdots, \rho, \tag{5.64}$$

where $c_i > 0$, $d_i > 0$ for $i = 1, 2, \cdots, \rho$ are design parameters.

Adaptive control law:

$$\tilde{U}(t-D) = -\tilde{v}_{m,\rho+1} + \alpha_\rho. \tag{5.65}$$

Here please note that α_i for $i = 1, 2, \cdots, \rho$ are linear in z_1, $\tilde{\eta}$, $\tilde{\lambda}$ but only nonlinear in $\hat{\theta}$. Thus through a recursive but straightforward calculation, we show the following equalities:

$$z_2 = K_{2,z_1}(\hat{\theta})z_1 + K_{2,\tilde{\eta}}(\hat{\theta})\tilde{\eta} + K_{2,\tilde{\lambda}}(\hat{\theta})\tilde{\lambda}, \tag{5.66}$$

$$z_3 = K_{3,z_1}(\hat{\theta})z_1 + K_{3,\tilde{\eta}}(\hat{\theta})\tilde{\eta} + K_{3,\tilde{\lambda}}(\hat{\theta})\tilde{\lambda}, \tag{5.67}$$

$$z_{i+1} = K_{i+1,z_1}(\hat{\theta})z_1 + K_{i+1,\tilde{\eta}}(\hat{\theta})\tilde{\eta} + K_{i+1,\tilde{\lambda}}(\hat{\theta})\tilde{\lambda}$$

$$i = 3, 4, \cdots, \rho - 1, \tag{5.68}$$

$$\tilde{U}(t-D) = K_{z_1}(\hat{\theta})z_1(t) + K_{\tilde{\eta}}(\hat{\theta})\tilde{\eta}(t) + K_{\tilde{\lambda}}(\hat{\theta})\tilde{\lambda}(t), \tag{5.69}$$

$$= K_{\tilde{\chi}}(\hat{\theta})\tilde{\chi}(t), \tag{5.70}$$

where the explicit expressions of $K_{i,z_1}(\hat{\theta})$, $K_{i,\tilde{\eta}}(\hat{\theta})$, $K_{i,\tilde{\lambda}}(\hat{\theta})$ for $i = 2, 3, \cdots, \rho$, and $K_{z_1}(\hat{\theta})$, $K_{\tilde{\eta}}(\hat{\theta})$, $K_{\tilde{\lambda}}(\hat{\theta})$ are as follows:

$$K_{2,z_1}(\hat{\theta}) = \frac{1}{\hat{b}_m}(c_1 + d_1 - \hat{a}_{n-1}), \tag{5.71}$$

$$K_{2,\tilde{\eta}}(\hat{\theta}) = -\frac{1}{\hat{b}_m}e_2^T \hat{A}(A_0), \tag{5.72}$$

$$K_{2,\tilde{\lambda}}(\hat{\theta}) = \frac{1}{\hat{b}_m}e_2^T \hat{B}(A_0), \tag{5.73}$$

$$K_{3,z_1}(\hat{\theta}) = \hat{b}_m + \left(c_2 + d_2\left(\frac{\partial \alpha_1}{\partial z_1}\right)^2\right)K_{2,z_1}(\hat{\theta})$$

$$+ \frac{\partial \alpha_1}{\partial z_1}\hat{a}_{n-1} - \frac{\partial \alpha_1}{\partial \tilde{\eta}}e_n, \tag{5.74}$$

$$K_{3,\tilde{\eta}}(\hat{\theta}) = \left(c_2 + d_2\left(\frac{\partial \alpha_1}{\partial z_1}\right)^2\right)K_{2,\tilde{\eta}}(\hat{\theta})$$

$$+ \frac{\partial \alpha_1}{\partial z_1}e_2^T \hat{A}(A_0) - \frac{\partial \alpha_1}{\partial \tilde{\eta}}A_0, \tag{5.75}$$

$$K_{3,\tilde{\lambda}}(\hat{\theta}) = \left(c_2 + d_2\left(\frac{\partial \alpha_1}{\partial z_1}\right)^2\right)K_{2,\tilde{\lambda}}(\hat{\theta}) + e_2^T A_0^{m+1}$$

$$- \frac{\partial \alpha_1}{\partial z_1}e_2^T \hat{B}(A_0) - \sum_{j=1}^{m+1}\frac{\partial \alpha_1}{\partial \tilde{\lambda}}e_j e_j^T A_0, \tag{5.76}$$

$$K_{i+1,z_1}(\hat{\theta}) = K_{i-1,z_1}(\hat{\theta}) + \left(c_i + d_i\left(\frac{\partial\alpha_{i-1}}{\partial z_1}\right)^2\right)K_{i,z_1}(\hat{\theta})$$

$$+ \frac{\partial\alpha_{i-1}}{\partial z_1}\hat{a}_{n-1} - \frac{\partial\alpha_{i-1}}{\partial\tilde{\eta}}e_n, \tag{5.77}$$

$$K_{i+1,\tilde{\eta}}(\hat{\theta}) = K_{i-1,\tilde{\eta}}(\hat{\theta}) + \left(c_i + d_i\left(\frac{\partial\alpha_{i-1}}{\partial z_1}\right)^2\right)K_{i,\tilde{\eta}}(\hat{\theta})$$

$$+ \frac{\partial\alpha_{i-1}}{\partial z_1}e_2^T\hat{A}(A_0) - \frac{\partial\alpha_{i-1}}{\partial\tilde{\eta}}A_0, \tag{5.78}$$

$$K_{i+1,\tilde{\lambda}}(\hat{\theta}) = K_{i-1,\tilde{\lambda}}(\hat{\theta}) + \left(c_i + d_i\left(\frac{\partial\alpha_{i-1}}{\partial z_1}\right)^2\right)K_{i,\tilde{\lambda}}(\hat{\theta})$$

$$+ e_i^T A_0^{m+1} - \frac{\partial\alpha_{i-1}}{\partial z_1}e_2^T\hat{B}(A_0) - \sum_{j=1}^{m+i-1}\frac{\partial\alpha_{i-1}}{\partial\tilde{\lambda}}e_j e_j^T A_0, \tag{5.79}$$

$$i = 3, 4, \cdots, \rho-1,$$

$$K_{z_1}(\hat{\theta}) = -\left[K_{\rho-1,z_1}(\hat{\theta}) + \left(c_\rho + d_\rho\left(\frac{\partial\alpha_{\rho-1}}{\partial z_1}\right)^2\right)K_{\rho,z_1}(\hat{\theta})\right.$$

$$\left. + \frac{\partial\alpha_{\rho-1}}{\partial z_1}\hat{a}_{n-1} - \frac{\partial\alpha_{\rho-1}}{\partial\tilde{\eta}}e_n\right], \tag{5.80}$$

$$K_{\tilde{\eta}}(\hat{\theta}) = -\left[K_{\rho-1,\tilde{\eta}}(\hat{\theta}) + \left(c_\rho + d_\rho\left(\frac{\partial\alpha_{\rho-1}}{\partial z_1}\right)^2\right)K_{\rho,\tilde{\eta}}(\hat{\theta})\right.$$

$$\left. + \frac{\partial\alpha_{\rho-1}}{\partial z_1}e_2^T\hat{A}(A_0) - \frac{\partial\alpha_{\rho-1}}{\partial\tilde{\eta}}A_0\right], \tag{5.81}$$

$$K_{\tilde{\lambda}}(\hat{\theta}) = -\left[K_{\rho-1,\tilde{\lambda}}(\hat{\theta}) + \left(c_\rho + d_\rho\left(\frac{\partial\alpha_{\rho-1}}{\partial z_1}\right)^2\right)K_{\rho,\tilde{\lambda}}(\hat{\theta})\right.$$

$$\left. + e_\rho^T A_0^{m+1} - \frac{\partial\alpha_{\rho-1}}{\partial z_1}e_2^T\hat{B}(A_0) - \sum_{j=1}^{m+\rho-1}\frac{\partial\alpha_{\rho-1}}{\partial\tilde{\lambda}}e_j e_j^T A_0\right], \tag{5.82}$$

and

$$K_{\tilde{\chi}}(\hat{\theta}) = [K_{z_1}(\hat{\theta}), K_{\tilde{\eta}}(\hat{\theta}), K_{\tilde{\lambda}}(\hat{\theta})] \in \mathbb{R}^{1\times(2n+1)}. \tag{5.83}$$

Provided a brief comparison between our backstepping control design (5.59)–(5.65) and that in [150, 151] and [89], we find that we do not utilize tuning functions or nonlinear damping terms to deal with parameter estimation error, so we can obtain a linear controller (5.70) for employing prediction control (5.84) and (5.104) to address the input delay. This is a further improvement for a traditional backstepping control methodology.

If parameter θ and time delay D are known, utilizing θ to replace $\hat{\theta}$, bearing (5.7) and (5.70) in mind, one can prove that the following prediction-based control law

$$U(t) = U^r(t + D, \theta) - K_{\tilde{\chi}}(\theta)\chi^r(t + D, \theta)$$
$$+ K_{\tilde{\chi}}(\theta)e^{A_{\tilde{\chi}}(\theta)D}\chi(t)$$
$$+ D\int_0^1 K_{\tilde{\chi}}(\theta)e^{A_{\tilde{\chi}}(\theta)D(1-y)}e_{2n+1}u(y,t)\,dy \qquad (5.84)$$

achieves our control objective for system (5.2).

To deal with the unknown plant parameters and the unknown actuator delay, utilizing the certainty equivalence principle, we bring in the reference transport PDE $u^r(x, t, \hat{\theta}) = U^r(t + Dx, \hat{\theta})$, $x \in [0, 1]$ and the corresponding PDE error variable $\tilde{u}(x, t) = u(x, t) - u^r(x, t, \hat{\theta})$ satisfying

$$D\tilde{u}_t(x, t) = \tilde{u}_x(x, t) - D\frac{\partial u^r(x, t, \hat{\theta})}{\partial \hat{\theta}}\dot{\hat{\theta}}, \quad x \in [0, 1], \qquad (5.85)$$

$$\tilde{u}(1, t) = \tilde{U}(t) = U(t) - U^r(t + \hat{D}, \hat{\theta}), \qquad (5.86)$$

where $\hat{D}$ is an estimate of D, $\tilde{D} = D - \hat{D}$. Here we further bring in a similar backstepping transformation presented in [27] and section 2.2 of [80],

$$w(x, t) = \tilde{u}(x, t) - K_{\tilde{\chi}}(\hat{\theta})e^{A_{\tilde{\chi}}(\hat{\theta})\hat{D}x}\tilde{\chi}(t)$$
$$- \hat{D}\int_0^x K_{\tilde{\chi}}(\hat{\theta})e^{A_{\tilde{\chi}}(\hat{\theta})\hat{D}(x-y)}e_{2n+1}\tilde{u}(y, t)\,dy, \qquad (5.87)$$

$$\tilde{u}(x, t) = w(x, t) + K_{\tilde{\chi}}(\hat{\theta})e^{(A_{\tilde{\chi}}+e_{2n+1}K_{\tilde{\chi}})(\hat{\theta})\hat{D}x}\tilde{\chi}(t)$$
$$+ \hat{D}\int_0^x K_{\tilde{\chi}}(\hat{\theta})e^{(A_{\tilde{\chi}}+e_{2n+1}K_{\tilde{\chi}})(\hat{\theta})\hat{D}(x-y)}e_{2n+1}$$
$$\times w(y, t)\,dy, \qquad (5.88)$$

with which systems (5.17), (5.37), (5.55), (5.56)–(5.58), and (5.85)–(5.86) are mapped into the closed-loop target error systems as follows:

$$\dot{z} = A_z(\hat{\theta})z + W_\varepsilon(\hat{\theta})(\varepsilon_2 + \omega^T\tilde{\theta}) + Q(z, t)^T\dot{\hat{\theta}}$$
$$+ Q^r(t, \hat{\theta})^T\dot{\hat{\theta}} + e_\rho w(0, t), \qquad (5.89)$$

$$\dot{\tilde{\eta}} = A_0\tilde{\eta} + e_n z_1, \qquad (5.90)$$

$$\dot{\tilde{\zeta}} = A_b\tilde{\zeta} + b_b z_1, \qquad (5.91)$$

$$\dot{\varepsilon} = A_0\varepsilon, \qquad (5.92)$$

$$Dw_t(x, t) = w_x(x, t) - D(\varepsilon_2 + \omega^T\tilde{\theta})r_0(x, t) - \tilde{D}p_0(x, t)$$
$$- D\dot{\hat{D}}q_0(x, t) - D\dot{\hat{\theta}}^T q(x, t), \qquad (5.93)$$

$$w(1, t) = 0, \qquad (5.94)$$

where $A_z(\hat{\theta})$, $W_\varepsilon(\hat{\theta})$, $Q(z,t)^T$, $Q^r(t,\hat{\theta})^T$, $r_0(x,t)$, $p_0(x,t)$, $q_0(x,t)$, and $q(x,t)$ are listed as follows:

$$
A_z(\hat{\theta}) =
\begin{bmatrix}
-(c_1+d_1) & \hat{b}_m & 0 & \cdots & & 0 \\
-\hat{b}_m & -\left(c_2+d_2\left(\frac{\partial \alpha_1}{\partial z_1}\right)^2\right) & 1 & \ddots & & \vdots \\
0 & -1 & \ddots & \ddots & & 0 \\
\vdots & \ddots & & \ddots & \ddots & 1 \\
0 & \cdots & & 0 & -1 & -\left(c_\rho+d_\rho\left(\frac{\partial \alpha_{\rho-1}}{\partial z_1}\right)^2\right)
\end{bmatrix},
$$

(5.95)

$$
W_\varepsilon(\hat{\theta}) =
\begin{bmatrix}
1 \\
-\frac{\partial \alpha_1}{\partial z_1} \\
\vdots \\
-\frac{\partial \alpha_{\rho-1}}{\partial z_1}
\end{bmatrix}, \quad
Q(z,t)^T =
\begin{bmatrix}
0 \\
-\frac{\partial \alpha_1}{\partial \hat{\theta}} \\
\vdots \\
-\frac{\partial \alpha_{\rho-1}}{\partial \hat{\theta}}
\end{bmatrix},
$$

$$
Q^r(t,\hat{\theta})^T =
\begin{bmatrix}
0 \\
-\left(e_2^T A_0^m - \sum_{j=1}^{m+1}\frac{\partial \alpha_1}{\partial \lambda_j}e_j^T\right)\frac{\partial \lambda^r(t,\hat{\theta})}{\partial \hat{\theta}} \\
\vdots \\
-\left(e_\rho^T A_0^m - \sum_{j=1}^{m+\rho-1}\frac{\partial \alpha_{\rho-1}}{\partial \lambda_j}e_j^T\right)\frac{\partial \lambda^r(t,\hat{\theta})}{\partial \hat{\theta}}
\end{bmatrix},
$$

(5.96)

$$
r_0(x,t) = K_{\tilde{\chi}}(\hat{\theta})e^{A_{\tilde{\chi}}(\hat{\theta})\hat{D}x}e_1,
$$

(5.97)

$$
p_0(x,t) = K_{\tilde{\chi}}(\hat{\theta})e^{A_{\tilde{\chi}}(\hat{\theta})\hat{D}x}\left(A_{\tilde{\chi}}(\hat{\theta})\tilde{\chi}(t) + e_{2n+1}\tilde{u}(0,t)\right),
$$

(5.98)

$$
= K_{\tilde{\chi}}(\hat{\theta})e^{A_{\tilde{\chi}}(\hat{\theta})\hat{D}x}\left[(A_{\tilde{\chi}} + e_{2n+1}K_{\tilde{\chi}})(\hat{\theta})\tilde{\chi}(t) + e_{2n+1}w(0,t)\right],
$$

(5.99)

$$
q_0(x,t) = K_{\tilde{\chi}}A_{\tilde{\chi}}(\hat{\theta})xe^{A_{\tilde{\chi}}(\hat{\theta})\hat{D}x}\tilde{\chi}(t)
$$
$$
+ \int_0^x K_{\tilde{\chi}}(\hat{\theta})\left(I + A_{\tilde{\chi}}(\hat{\theta})\hat{D}(x-y)\right)e^{A_{\tilde{\chi}}(\hat{\theta})\hat{D}(x-y)}e_{2n+1}\tilde{u}(y,t)\,dy, \quad (5.100)
$$
$$
= \left[K_{\tilde{\chi}}A_{\tilde{\chi}}(\hat{\theta})xe^{A_{\tilde{\chi}}(\hat{\theta})\hat{D}x} + \int_0^x K_{\tilde{\chi}}(\hat{\theta})\left(I + A_{\tilde{\chi}}(\hat{\theta})\hat{D}(x-y)\right)e^{A_{\tilde{\chi}}(\hat{\theta})\hat{D}(x-y)}e_{2n+1}\right.
$$
$$
\times K_{\tilde{\chi}}(\hat{\theta})e^{(A_{\tilde{\chi}}+e_{2n+1}K_{\tilde{\chi}})(\hat{\theta})\hat{D}y}\,dy\Big]\tilde{\chi}(t)
$$
$$
+ \int_0^x w(y,t)\left[K_{\tilde{\chi}}(\hat{\theta})\left(I + A_{\tilde{\chi}}(\hat{\theta})\hat{D}(x-y)\right)e^{A_{\tilde{\chi}}(\hat{\theta})\hat{D}(x-y)}e_{2n+1}\right.
$$
$$
+ \hat{D}\int_y^x K_{\tilde{\chi}}(\hat{\theta})\left(I + A_{\tilde{\chi}}(\hat{\theta})\hat{D}(x-s)\right)e^{A_{\tilde{\chi}}(\hat{\theta})\hat{D}(x-s)}e_{2n+1}
$$
$$
\times K_{\tilde{\chi}}(\hat{\theta})e^{(A_{\tilde{\chi}}+e_{2n+1}K_{\tilde{\chi}})(\hat{\theta})\hat{D}(s-y)}e_{2n+1}\,ds\Big]dy, \quad (5.101)
$$

$$q_i(x,t) = \left(\frac{\partial K_{\tilde{\chi}}(\hat{\theta})}{\partial \hat{\theta}_i} + K_{\tilde{\chi}}(\hat{\theta}) \frac{\partial A_{\tilde{\chi}}(\hat{\theta})}{\partial \hat{\theta}_i} \hat{D}x \right) e^{A_{\tilde{\chi}}(\hat{\theta})\hat{D}x} \tilde{\chi}(t)$$

$$+ \hat{D} \int_0^x \left(\frac{\partial K_{\tilde{\chi}}(\hat{\theta})}{\partial \hat{\theta}_i} + K_{\tilde{\chi}}(\hat{\theta}) \frac{\partial A_{\tilde{\chi}}(\hat{\theta})}{\partial \hat{\theta}_i} \hat{D}(x-y) \right)$$

$$\times e^{A_{\tilde{\chi}}(\hat{\theta})\hat{D}(x-y)} e_{2n+1} \tilde{u}(y,t)\, dy - K_{\tilde{\chi}}(\hat{\theta}) e^{A_{\tilde{\chi}}(\hat{\theta})\hat{D}x} \frac{\partial \chi^r(t,\hat{\theta})}{\partial \hat{\theta}_i} + \frac{\partial u^r(x,t,\hat{\theta})}{\partial \hat{\theta}_i}$$

$$- \hat{D} \int_0^x K_{\tilde{\chi}}(\hat{\theta}) e^{A_{\tilde{\chi}}(\hat{\theta})\hat{D}(x-y)} e_{2n+1} \frac{\partial u^r(y,t,\hat{\theta})}{\partial \hat{\theta}_i}\, dy, \tag{5.102}$$

$$= \Bigg[\left(\frac{\partial K_{\tilde{\chi}}(\hat{\theta})}{\partial \hat{\theta}_i} + K_{\tilde{\chi}}(\hat{\theta}) \frac{\partial A_{\tilde{\chi}}(\hat{\theta})}{\partial \hat{\theta}_i} \hat{D}x \right) e^{A_{\tilde{\chi}}(\hat{\theta})\hat{D}x}$$

$$+ \hat{D} \int_0^x \left(\frac{\partial K_{\tilde{\chi}}(\hat{\theta})}{\partial \hat{\theta}_i} + K_{\tilde{\chi}}(\hat{\theta}) \frac{\partial A_{\tilde{\chi}}(\hat{\theta})}{\partial \hat{\theta}_i} \hat{D}(x-y) \right) e^{A_{\tilde{\chi}}(\hat{\theta})\hat{D}(x-y)} e_{2n+1}$$

$$\times K_{\tilde{\chi}}(\hat{\theta}) e^{(A_{\tilde{\chi}} + e_{2n+1} K_{\tilde{\chi}})(\hat{\theta})\hat{D}y}\, dy \Bigg] \tilde{\chi}(t)$$

$$+ \hat{D} \int_0^x w(y,t) \Bigg[\left(\frac{\partial K_{\tilde{\chi}}(\hat{\theta})}{\partial \hat{\theta}_i} + K_{\tilde{\chi}}(\hat{\theta}) \frac{\partial A_{\tilde{\chi}}(\hat{\theta})}{\partial \hat{\theta}_i} \hat{D}(x-y) \right) e^{A_{\tilde{\chi}}(\hat{\theta})\hat{D}(x-y)} e_{2n+1}$$

$$+ \hat{D} \int_y^x \left(\frac{\partial K_{\tilde{\chi}}(\hat{\theta})}{\partial \hat{\theta}_i} + K_{\tilde{\chi}}(\hat{\theta}) \frac{\partial A_{\tilde{\chi}}(\hat{\theta})}{\partial \hat{\theta}_i} \hat{D}(x-s) \right) e^{A_{\tilde{\chi}}(\hat{\theta})\hat{D}(x-s)} e_{2n+1}$$

$$\times K_{\tilde{\chi}}(\hat{\theta}) e^{(A_{\tilde{\chi}} + e_{2n+1} K_{\tilde{\chi}})(\hat{\theta})\hat{D}(s-y)} e_{2n+1}\, ds \Bigg]\, dy$$

$$- K_{\tilde{\chi}}(\hat{\theta}) e^{A_{\tilde{\chi}}(\hat{\theta})\hat{D}x} \frac{\partial \chi^r(t,\hat{\theta})}{\partial \hat{\theta}_i} + \frac{\partial u^r(x,t,\hat{\theta})}{\partial \hat{\theta}_i}$$

$$- \hat{D} \int_0^x K_{\tilde{\chi}}(\hat{\theta}) e^{A_{\tilde{\chi}}(\hat{\theta})\hat{D}(x-y)} e_{2n+1} \frac{\partial u^r(y,t,\hat{\theta})}{\partial \hat{\theta}_i}\, dy. \tag{5.103}$$

As a consequence, we design our control below to ensure that (5.94) holds:

$$U(t) = U^r(t+\hat{D},\hat{\theta}) - K_{\tilde{\chi}}(\hat{\theta}) \chi^r(t+\hat{D},\hat{\theta})$$

$$+ K_{\tilde{\chi}}(\hat{\theta}) e^{A_{\tilde{\chi}}(\hat{\theta})\hat{D}} \chi(t)$$

$$+ \hat{D} \int_0^1 K_{\tilde{\chi}}(\hat{\theta}) e^{A_{\tilde{\chi}}(\hat{\theta})\hat{D}(1-y)} e_{2n+1} u(y,t)\, dy, \tag{5.104}$$

and two Lyapunov-based identifiers to identify unknown plant parameters and actuator time delay are presented as follows:

The update law of D is designed as

$$\dot{D} = \gamma_D \text{Proj}_{[\underline{D}, \bar{D}]}\{\tau_D\}, \quad \gamma_D > 0, \tag{5.105}$$

$$\tau_D = -\int_0^1 (1+x)w(x,t)p_0(x,t)\,\mathrm{d}x, \tag{5.106}$$

where $\text{Proj}_{[\underline{D}, \bar{D}]}\{\cdot\}$ is a standard projection operator defined on the interval $[\underline{D}, \bar{D}]$.

The update law of θ is designed as

$$\dot{\theta} = \gamma_\theta \text{Proj}_\Theta\{\tau_\theta\}, \quad \gamma_\theta > 0, \quad \hat{b}_m(0)\text{sgn}(b_m) \geq \underline{b}_m, \tag{5.107}$$

$$\tau_\theta = \frac{1}{2g} w W_\varepsilon(\hat{\theta})^T z - w \int_0^1 (1+x)w(x,t)r_0(x,t)\,\mathrm{d}x, \tag{5.108}$$

where $g > 0$ and $\text{Proj}_\Theta\{\cdot\}$ is a standard projection algorithm defined on the set Θ to guarantee that $|\hat{b}_m| \geq \underline{b}_m > 0$.

From (5.19), (5.32), and (5.59)–(5.60), based on a very similar derivation on pages 345 and 437 of [89], we could depict $\tilde{\lambda}$ as a linear function of z, $\tilde{\eta}$, $\tilde{\zeta}$, and ε such that

$$\tilde{\lambda} = H_z(\hat{\theta})z + H_\eta(\hat{\theta})\tilde{\eta} + H_\zeta(\hat{\theta})\tilde{\zeta} + H_\varepsilon(\hat{\theta})\varepsilon$$
$$+ H_{Y_r}(\hat{\theta})Y_r + H_{\eta^r}(\hat{\theta})\eta^r + H_{\zeta^r}(\hat{\theta})\zeta^r + H_{\lambda^r}(\hat{\theta})\lambda^r, \tag{5.109}$$

where $H_z(\hat{\theta}) \in \mathbb{R}^{n\times p}$, $H_\eta(\hat{\theta}) \in \mathbb{R}^{n\times n}$, $H_\zeta(\hat{\theta}) \in \mathbb{R}^{n\times m}$, $H_\varepsilon(\hat{\theta}) \in \mathbb{R}^{n\times n}$, $H_{Y_r}(\hat{\theta}) \in \mathbb{R}^n$, $H_{\eta^r}(\hat{\theta}) \in \mathbb{R}^{n\times n}$, $H_{\zeta^r}(\hat{\theta}) \in \mathbb{R}^{n\times m}$, $H_{\lambda^r}(\hat{\theta}) \in \mathbb{R}^{n\times n}$.

Remark 5.4. As analyzed on pages 345–346 and 437–438 of [89], one can notice that (5.109) is a one-to-one functional mapping if and only if $B(s)$ in Assumption 5.1 and $K(s) = s^n + k_1 s^{n-1} + k_2 s^{n-2} + \cdots + k_{n-1}s + k_n$ are coprime. This happens probably since $B(s)$ is unknown. However, since $K(s)$ is a designed polynomial that one can choose, we may satisfy the coprime condition by a continuously adjusting $K(s)$ through trial and error.

We establish the boundedness of ε, Y_r, η^r, ζ^r, λ^r, θ, and $\hat{\theta}$ from (5.17), Assumptions 5.1–5.2, and the projection algorithm. As a result, if $K(s)$ and $B(s)$ are assumed to satisfy the coprime condition, employing Young's inequality and Cauchy-Schwartz inequality, the formulas, (5.97)–(5.103) imply that the following inequalities hold:

$$\int_0^1 (1+x)w(x,t)r_0(x,t)\,\mathrm{d}x$$
$$\leq M_{r_0}\|w(x,t)\|, \tag{5.110}$$

$$\int_0^1 (1+x)w(x,t)p_0(x,t)\,\mathrm{d}x$$
$$\leq M_{p_0}\left(|z|^2 + |\tilde{\eta}|^2 + |\tilde{\zeta}|^2 + w(0,t)^2 + \|w(x,t)\|^2 + \|w(x,t)\|\right), \tag{5.111}$$

$$\int_0^1 (1+x)w(x,t)q_0(x,t)\,\mathrm{d}x$$
$$\leq M_{q_0}\left(|z|^2 + |\tilde{\eta}|^2 + |\tilde{\zeta}|^2 + \|w(x,t)\|^2 + \|w(x,t)\|\right), \tag{5.112}$$
$$\int_0^1 (1+x)w(x,t)q_i(x,t)\,\mathrm{d}x$$
$$\leq M_{q_i}\left(|z|^2 + |\tilde{\eta}|^2 + |\tilde{\zeta}|^2 + \|w(x,t)\|^2 + \|w(x,t)\|\right), \tag{5.113}$$
$$i=1,\cdots,p,$$

where M_{r_0}, M_{p_0}, M_{q_0}, M_{q_i} are some positive constants. Furthermore, the estimators (5.105)–(5.108) have the following properties:

$$|\dot{\hat{\theta}}_i| \leq \gamma_\theta M_\theta\left(|z| + |z|^2 + |\tilde{\eta}|^2 + |\tilde{\zeta}|^2\right.$$
$$\left. + \|w(x,t)\| + \|w(x,t)\|^2\right), \quad i=1,2,\cdots,p, \tag{5.114}$$
$$|\dot{\hat{D}}| \leq \gamma_D M_D\left(|z|^2 + |\tilde{\eta}|^2 + |\tilde{\zeta}|^2\right.$$
$$\left. + \|w(x,t)\| + \|w(x,t)\|^2 + w(0,t)^2\right), \tag{5.115}$$

where M_θ and M_D are some positive constants.

5.2.3 Stability Analysis

The adaptive controller in this section is shown in figure 5.1. Above all, the stability of the ODE-PDE cascade system of this section (in which the actuator state is measured) is summarized in the main theorem below.

Theorem 5.5. *Consider the closed-loop system consisting of the plant (5.2), the K-filters (5.11)–(5.16), the adaptive controller (5.104), the time-delay identifier (5.105)–(5.106), and the parameter identifier (5.107)–(5.108). Let Assumptions 5.1–5.3 hold and let $B(s)$ and $K(s)$ be coprime. There exists a constant $M>0$ such that if the initial error state satisfies the condition*

$$|z(0)|^2 + |\tilde{\eta}(0)|^2 + |\tilde{\zeta}(0)|^2$$
$$+ |\tilde{\theta}(0)|^2 + |\tilde{D}(0)|^2 + |\varepsilon(0)|^2 + \|w(x,0)\|^2 \leq M, \tag{5.116}$$

then all the signals of the closed-loop system are bounded and the asymptotic tracking is achieved, i.e.,

$$\lim_{t\to\infty} z_1(t) = \lim_{t\to\infty} (Y(t) - Y_r(t)) = 0 \tag{5.117}$$

Remark 5.6. Based on the equalities (5.19), (5.29)–(5.32), (5.54), (5.66)–(5.68), (5.87)–(5.88), $\tilde{\theta}=\theta-\hat{\theta}$, and $\tilde{D}=D-\hat{D}$, the initial condition of error state (5.116) can be transformed into the initial conditions of the states of the actual plants, filters, identifiers, and transport PDE, namely, $X(0)$, $\eta(0)$, $\lambda(0)$, $\hat{\theta}(0)$, $\hat{D}(0)$, and $u(x,0)$.

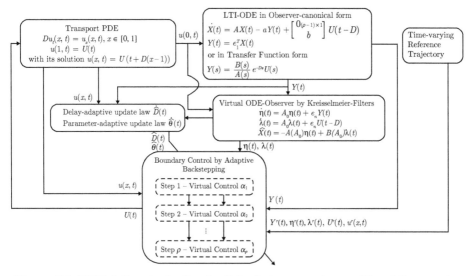

Figure 5.1. PDE full-state feedback of single-input systems with arbitrary relative degree

Proof. First of all, we build a nonnegative function to encompass major states of the closed-loop system as follows:

$$V = \frac{D}{2}z^T z + \frac{D}{k_\eta}\tilde{\eta}^T P\tilde{\eta} + \frac{D}{k_\zeta}\tilde{\zeta}^T P_b \tilde{\zeta}$$

$$+ gD\int_0^1 (1+x)w(x,t)^2 \, dx$$

$$+ \frac{g}{\gamma_D}\tilde{D}^2 + \frac{gD}{\gamma_\theta}\tilde{\theta}^T \tilde{\theta} + g_\varepsilon \sum_{i=1}^{\rho}\frac{D}{4d_i}\varepsilon^T P\varepsilon, \qquad (5.118)$$

where $k_\eta > 0$, $k_\zeta > 0$, $g > 0$, $g_\varepsilon > 0$.

Taking the time derivative of (5.118) along (5.89)–(5.94), we obtain

$$\dot{V} \leq D\left[-\sum_{i=1}^{\rho}c_i z_i^2 + \sum_{i=1}^{\rho}\frac{1}{4d_i}\varepsilon_2^2 + z^T W_\varepsilon(\hat{\theta})\omega^T \tilde{\theta} \right.$$

$$\left. + z^T Q(z,t)^T \dot{\hat{\theta}} + z^T Q^r(t,\hat{\theta})^T \dot{\hat{\theta}} + z_\rho w(0,t) \right]$$

$$+ D\left[-\frac{1}{k_\eta}|\tilde{\eta}|^2 + \frac{2}{k_\eta}\tilde{\eta}^T Pe_n z_1 \right]$$

$$+ D\left[-\frac{1}{k_\zeta}|\tilde{\zeta}|^2 + \frac{2}{k_\zeta}\tilde{\zeta}^T P_b b_b z_1 \right]$$

$$+\left[-gw(0,t)^2 - g\|w(x,t)\|^2\right.$$

$$-2gD\varepsilon_2 \int_0^1 (1+x)w(x,t)r_0(x,t)\,\mathrm{d}x$$

$$-2gD\tilde{\theta}^T\omega \int_0^1 (1+x)w(x,t)r_0(x,t)\,\mathrm{d}x$$

$$-2g\tilde{D} \int_0^1 (1+x)w(x,t)p_0(x,t)\,\mathrm{d}x$$

$$-2gD\dot{\tilde{D}} \int_0^1 (1+x)w(x,t)q_0(x,t)\,\mathrm{d}x$$

$$\left.-2gD\dot{\tilde{\theta}}^T \int_0^1 (1+x)w(x,t)q(x,t)\,\mathrm{d}x\right]$$

$$-\frac{2g}{\gamma_D}\tilde{D}\dot{\tilde{D}} - \frac{2gD}{\gamma_\theta}\tilde{\theta}^T\dot{\tilde{\theta}} - g_\varepsilon \sum_{i=1}^\rho \frac{D}{4d_i}|\varepsilon|^2, \qquad (5.119)$$

$$\leq D\left[-\frac{c_0}{2}|z|^2 + \frac{1}{4d_0}|\varepsilon|^2 + \frac{1}{2c_0}w(0,t)^2\right.$$

$$\left.+ z^T Q(z,t)^T\dot{\tilde{\theta}} + z^T Q^r(t,\hat{\theta})^T\dot{\tilde{\theta}}\right]$$

$$+D\left[-\frac{1}{2k_\eta}|\tilde{\eta}|^2 + \frac{2}{k_\eta}|Pe_n|^2 z_1^2\right]$$

$$+D\left[-\frac{1}{2k_\zeta}|\tilde{\zeta}|^2 + \frac{2}{k_\zeta}|P_b b_b|^2 z_1^2\right]$$

$$+\left[-gw(0,t)^2 - g\|w(x,t)\|^2\right.$$

$$-2gD\varepsilon_2 \int_0^1 (1+x)w(x,t)r_0(x,t)\,\mathrm{d}x$$

$$-2gD\dot{\tilde{D}} \int_0^1 (1+x)w(x,t)q_0(x,t)\,\mathrm{d}x$$

$$\left.-2gD\dot{\tilde{\theta}}^T \int_0^1 (1+x)w(x,t)q(x,t)\,\mathrm{d}x\right]$$

$$-\frac{2g}{\gamma_D}\tilde{D}(\dot{\tilde{D}} - \gamma_D\tau_D) - \frac{2gD}{\gamma_\theta}\tilde{\theta}^T(\dot{\tilde{\theta}} - \gamma_\theta\tau_\theta) - g_\varepsilon \frac{D}{4d_0}|\varepsilon|^2, \qquad (5.120)$$

where $c_0 = \min\limits_{i=1,2,\cdots,\rho} c_i, \quad d_0 = \left(\sum\limits_{i=1}^\rho \frac{1}{d_i}\right)^{-1}.$

From (5.60), (5.66)–(5.68), and (5.109), it is easy to show that $-\frac{\partial \alpha_{i-1}}{\partial \hat{\theta}_j}$ for $i = 2, \cdots, \rho$ and $j = 1, \cdots, p$ is a linear function of z, $\tilde{\eta}$, $\tilde{\zeta}$, and ε. If we denote

$$V_0 = |z|^2 + |\tilde{\eta}|^2 + |\tilde{\zeta}|^2 + \|w(x,t)\|^2, \qquad (5.121)$$

(5.110)–(5.115) imply that the following inequalities hold:

$$Dz^T Q(z,t)^T \dot{\hat{\theta}} = -D \sum_{i=2}^{\rho} \sum_{j=1}^{p} z_i \frac{\partial \alpha_{i-1}}{\partial \hat{\theta}_j} \dot{\hat{\theta}}_j$$
$$\leq \gamma_\theta M_Q \left(V_0 + V_0^2 \right), \qquad (5.122)$$

$$Dz^T Q^r(t, \hat{\theta})^T \dot{\hat{\theta}} = -D \sum_{i=2}^{\rho} \sum_{k=1}^{p} z_i \left(e_i^T A_0^m - \sum_{j=1}^{m+i-1} \frac{\partial \alpha_{i-1}}{\partial \tilde{\lambda}_j} e_j^T \right)$$
$$\times \frac{\partial \lambda^r(t, \hat{\theta})}{\partial \hat{\theta}_k} \dot{\hat{\theta}}_k$$
$$\leq \gamma_\theta M_{Q^r} \left(V_0 + V_0^2 \right) \qquad (5.123)$$

$$-2gD\varepsilon_2 \int_0^1 (1+x) w(x,t) r_0(x,t)\, dx$$
$$\leq \frac{g}{2} \|w(x,t)\|^2 + g M_1 |\varepsilon|^2 \qquad (5.124)$$

$$-2gD\dot{D} \int_0^1 (1+x) w(x,t) q_0(x,t)\, dx$$
$$\leq g \gamma_D M_2 \left(V_0 + V_0^2 \right) \qquad (5.125)$$

$$-2gD\dot{\hat{\theta}}^T \int_0^1 (1+x) w(x,t) q(x,t)\, dx$$
$$\leq g \gamma_\theta M_3 \left(V_0 + V_0^2 \right), \qquad (5.126)$$

where M_Q, M_{Q^r}, M_1, M_2, M_3 are some positive constants.

As a result, (5.120) becomes

$$\dot{V}(t) \leq -D \left(\frac{c_0}{2} - \frac{2}{k_\eta} |Pe_n|^2 - \frac{2}{k_\zeta} |P_b b_b|^2 \right) |z|^2$$
$$- \frac{D}{2k_\eta} |\tilde{\eta}|^2 - \frac{D}{2k_\zeta} |\tilde{\zeta}|^2$$
$$- \frac{g}{2} \|w(x,t)\|^2 - \left(g - \frac{D}{2c_0} \right) w(0,t)^2$$
$$- \left(g_\varepsilon \frac{D}{4d_0} - \frac{D}{4d_0} - g M_1 \right) |\varepsilon|^2$$

$$+ \gamma_\theta (M_Q + M_{Q^r} + gM_3)(V_0 + V_0^2)$$
$$+ \gamma_D gM_2(V_0 + V_0^2). \tag{5.127}$$

If we design k_η, k_ζ, g, and g_ε appropriately such that

$$k_\eta \geq \frac{8}{c_0}|Pe_n|^2, \quad k_\zeta \geq \frac{8}{c_0}|P_b b_b|^2,$$
$$g \geq \frac{\bar{D}}{2c_0}, \quad g_\varepsilon \geq \frac{4d_0 gM_1}{\underline{D}} + 1, \tag{5.128}$$

(5.127) becomes

$$\dot{V} \leq -M_4 V_0 + (\gamma_D + \gamma_\theta)M_5(V_0 + V_0^2), \tag{5.129}$$
$$\leq -M_4 V_0 + (\gamma_D + \gamma_\theta)M_5(V_0 + V_0^2)$$
$$+ \left(\frac{g}{\gamma_D}\tilde{D}^2 + \frac{gD}{\gamma_\theta}\tilde{\theta}^T\tilde{\theta} + g_\varepsilon \sum_{i=1}^{\rho} \frac{D}{4d_i}\varepsilon^T P\varepsilon \right) V_0, \tag{5.130}$$
$$\leq -M_4 V_0 + (\gamma_D + \gamma_\theta)M_5(V_0 + V_0^2)$$
$$+ (V - M_6 V_0)V_0, \tag{5.131}$$
$$\leq -\left(M_4 - (\gamma_D + \gamma_\theta)M_5 - V \right)V_0$$
$$- \left(M_6 - (\gamma_D + \gamma_\theta)M_5 \right)V_0^2, \tag{5.132}$$

where $M_4 = \min \left\{ \underline{D}\left(\frac{c_0}{2} - \frac{2}{k_\eta}|Pe_n|^2 - \frac{2}{k_\zeta}|P_b b_b|^2 \right), \frac{D}{2k_\eta}, \frac{D}{2k_\zeta}, \frac{g}{2} \right\}$, $M_5 > 0$ and $M_6 = \min \left\{ \frac{D}{2}, \frac{D\lambda_{\min}(P)}{k_\eta}, \frac{D\lambda_{\min}(P_b)}{k_\zeta}, g\underline{D} \right\}$. By designing γ_D and γ_θ and choosing initial state $V(0)$ appropriately such that

$$\gamma_D, \gamma_\theta \in (0, \gamma^*], \quad \gamma^* = \frac{1}{2}\min\left\{ \frac{M_4}{M_5}, \frac{M_6}{M_5} \right\}, \tag{5.133}$$
$$V(0) \leq M_4 - (\gamma_D + \gamma_\theta)M_5, \tag{5.134}$$

we have

$$\dot{V} \leq -M_7 V_0 - M_8 V_0^2, \tag{5.135}$$

where $M_7 = M_4 - (\gamma_D + \gamma_\theta)M_5 - V$, $M_8 = M_6 - (\gamma_D + \gamma_\theta)M_5$.

Thus signals $z, \tilde{\eta}, \tilde{\zeta}, \|w(x,t)\|$ are bounded. Bearing boundedness of ε, Y_r, U^r, η^r, and λ^r in mind, we show boundedness of X and $\|u(x,t)\|$ from (5.19), (5.27), (5.69), and (5.87)–(5.88), respectively. Thus all signals of the closed-loop system have been proved to be bounded. The inequality (5.135) shows the time derivative of the Lyapunov candidate is nonpositive, which implies that the Lyapunov function is nonincreasing such that $V(t) \leq V(0)$. From the definition of the

Lyapunov function (5.118), we know that all the states of the closed-loop system $z(t),\tilde{\eta}(t),\tilde{\zeta}(t),w(x,t),\tilde{D}(t),\tilde{\theta}(t),\epsilon(t)$ are bounded, and thus the update laws $\dot{\hat{D}}(t)$, $\dot{\hat{\theta}}(t)$ are bounded from (5.105)–(5.108). Using the dynamics (5.89), we find that $\frac{d|z(t)|^2}{dt}$ is bounded. From (5.135), we know that $|z(t)|$ is square integrable. Thus by Barbalat's lemma or its alternative in appendix of C.2 of this book (which can also be found in appendix A of [89] and Lemma 3.1 of [101]), we prove the asymptotic tracking property (5.117) such that $z(t)\to 0$, in other words, $\lim_{t\to\infty}[Y(t)-Y_r(t)]=0$.   □

5.3   Set-Point Regulation by PDE Output Feedback

In this section, the more complex situation of the unmeasured actuator state $u(x,t)$ is taken into account. To deal with this problem, we firstly bring in an infinite-dimensional input observer

$$\hat{u}(x,t)=U\big(t+\hat{D}(x-1)\big),\quad x\in[0,1],\qquad(5.136)$$

which satisfies the dynamic equation

$$\hat{D}\hat{u}_t(x,t)=\hat{u}_x(x,t)+\dot{\hat{D}}(x-1)\hat{u}_x(x,t),\qquad(5.137)$$
$$\hat{u}(1,t)=U(t).\qquad(5.138)$$

Our control objective in this part is to regulate output $Y(t)$ to a constant reference set-point Y_r. To achieve this, Assumption 5.3 is replaced by the following assumption:

Assumption 5.7. *In the case of known θ, provided a known and bounded constant set point Y_r, there exist known and bounded reference state $X^r(\theta)$ and reference input $U^r(\theta)$ continuously differentiable in the argument θ such that*

$$0=AX^r(\theta)+F\big(U^r(\theta),Y_r\big)^T\theta$$
$$Y_r=e_1^T X^r(\theta).\qquad(5.139)$$

By defining the reference signal $u^r(\hat{\theta})=U^r(\hat{\theta})$, we introduce several error variables

$$\tilde{u}(x,t)=u(x,t)-u^r(\hat{\theta}),\qquad(5.140)$$
$$e(x,t)=u(x,t)-\hat{u}(x,t),\qquad(5.141)$$
$$\hat{e}(x,t)=\hat{u}(x,t)-u^r(\hat{\theta}),\qquad(5.142)$$

where $\hat{e}(x,t)$ is governed by

$$\hat{D}\hat{e}_t(x,t) = \hat{e}_x(x,t) + \dot{\hat{D}}(x-1)\hat{e}_x(x,t)$$

$$- \hat{D}\frac{\partial u^r(\hat{\theta})}{\partial\hat{\theta}}\dot{\hat{\theta}}, \tag{5.143}$$

$$\hat{e}(1,t) = U(t) - U^r(\hat{\theta}). \tag{5.144}$$

5.3.1 State Observer with Kreisselmeier-Filters

Taking the place of unmeasured $U(t-D)$ by $\hat{u}(0,t)$, analogous to (5.11)–(5.16), K-filters are employed below to virtually estimate the unmeasured ODE state:

$$\dot{\eta}(t) = A_0\eta(t) + e_nY(t), \tag{5.145}$$

$$\dot{\lambda}(t) = A_0\lambda(t) + e_n\hat{u}(0,t), \tag{5.146}$$

$$\xi(t) = -A_0^n\eta(t), \tag{5.147}$$

$$\Xi(t) = -[A_0^{n-1}\eta(t), \cdots, A_0\eta(t), \eta(t)], \tag{5.148}$$

$$v_j(t) = A_0^j\lambda(t), \quad j = 0, 1, \cdots, m, \tag{5.149}$$

$$\Omega(t)^T = [v_m(t), \cdots, v_1(t), v_0(t), \Xi(t)]. \tag{5.150}$$

As $X(t) = \xi(t) + \Omega(t)^T\theta + \varepsilon(t) = -A(A_0)\eta(t) + B(A_0)\lambda(t) + \varepsilon(t)$ and $\hat{X}(t) = \xi(t) + \Omega(t)^T\theta$, obviously the state estimation error $\varepsilon(t) = X(t) - \hat{X}(t)$ is governed by

$$\dot{\varepsilon}(t) = A_0\varepsilon(t) + \bar{b}e(0,t), \tag{5.151}$$

where $\bar{b} = \begin{bmatrix} 0_{(\rho-1)\times 1} \\ b \end{bmatrix}$.

Based on Assumption 5.7, similar to (5.20)–(5.28), the reference output Y_r can be divided into the reference K-filters η^r and $\lambda^r(\hat{\theta})$.

$$0 = A_0\eta^r + e_nY_r, \tag{5.152}$$

$$0 = A_0\lambda^r(\hat{\theta}) + e_nU^r(\hat{\theta}), \tag{5.153}$$

$$\xi^r = -A_0^n\eta^r, \tag{5.154}$$

$$\Xi^r = -[A_0^{n-1}\eta^r, \cdots, A_0\eta^r, \eta^r], \tag{5.155}$$

$$v_j^r(\hat{\theta}) = A_0^j\lambda^r(\hat{\theta}), \quad j = 0, 1, \cdots, m, \tag{5.156}$$

$$\Omega^r(\hat{\theta})^T = [v_m^r(\hat{\theta}), \cdots, v_1^r(\hat{\theta}), v_0^r(\hat{\theta}), \Xi^r], \tag{5.157}$$

$$X^r(\hat{\theta}) = \xi^r + \Omega^r(\hat{\theta})^T\hat{\theta} \tag{5.158}$$

$$= -\hat{A}(A_0)\eta^r + \hat{B}(A_0)\lambda^r(\hat{\theta}), \tag{5.159}$$

$$Y_r = e_1^T X^r(\hat{\theta}). \tag{5.160}$$

Now we define a series of error variables below:

$$z_1(t) = Y(t) - Y_r, \tag{5.161}$$
$$\tilde{\eta}(t) = \eta(t) - \eta^r, \tag{5.162}$$
$$\tilde{\lambda}(t) = \lambda(t) - \lambda^r(\hat{\theta}), \tag{5.163}$$
$$\tilde{\xi}(t) = \xi(t) - \xi^r, \tag{5.164}$$
$$\tilde{\Xi}(t) = \Xi(t) - \Xi^r, \tag{5.165}$$
$$\tilde{v}_j(t) = v_j(t) - v_j^r(\hat{\theta}), \quad j = 0, 1, \cdots, m, \tag{5.166}$$
$$\tilde{\Omega}(t)^T = \Omega(t)^T - \Omega^r(\hat{\theta})^T, \tag{5.167}$$

which are governed by the following dynamic equations:

$$\dot{\tilde{\eta}}(t) = A_0 \tilde{\eta}(t) + e_n z_1(t), \tag{5.168}$$
$$\dot{\tilde{\lambda}}(t) = A_0 \tilde{\lambda}(t) + e_n \hat{e}(0, t) - \frac{\partial \lambda^r(\hat{\theta})}{\partial \hat{\theta}} \dot{\hat{\theta}}, \tag{5.169}$$
$$\dot{z}_1(t) = \tilde{\xi}_2(t) + \tilde{\omega}(t)^T \hat{\theta} + \varepsilon_2(t) + \omega(t)^T \tilde{\theta}, \tag{5.170}$$
$$= \hat{b}_m \tilde{v}_{m,2}(t) + \tilde{\xi}_2(t) + \bar{\tilde{\omega}}(t)^T \hat{\theta}$$
$$+ \varepsilon_2(t) + \omega(t)^T \tilde{\theta}, \tag{5.171}$$

where $\tilde{\omega}(t)$, $\bar{\tilde{\omega}}(t)$, and $\omega(t)$ have fully the same forms with (5.41)–(5.43).
Then we define a couple of new variables as

$$\chi(t) = \begin{bmatrix} Y(t) \\ \eta(t) \\ \lambda(t) \end{bmatrix}, \quad \chi^r(\hat{\theta}) = \begin{bmatrix} Y_r \\ \eta^r \\ \lambda^r(\hat{\theta}) \end{bmatrix}. \tag{5.172}$$

Thus a new error variable is derived:

$$\tilde{\chi}(t) = \chi(t) - \chi^r(\hat{\theta}) = \begin{bmatrix} z_1(t) \\ \tilde{\eta}(t) \\ \tilde{\lambda}(t) \end{bmatrix} \in \mathbb{R}^{2n+1}, \tag{5.173}$$

which is driven by

$$\dot{\tilde{\chi}}(t) = A_{\tilde{\chi}}(\hat{\theta}) \tilde{\chi}(t) + e_{2n+1} \hat{e}(0, t) + e_1 \left(\varepsilon_2(t) + \omega(t)^T \tilde{\theta} \right)$$
$$- \frac{\partial \chi^r(\hat{\theta})}{\partial \hat{\theta}} \dot{\hat{\theta}}, \tag{5.174}$$

where $A_{\tilde{\chi}}(\hat{\theta})$ has the same form as (5.47) while

$$\frac{\partial \chi^r(\hat{\theta})}{\partial \hat{\theta}} = \begin{bmatrix} 0_{(1+n) \times p} \\ \frac{\partial \lambda^r(\hat{\theta})}{\partial \hat{\theta}} \end{bmatrix}. \tag{5.175}$$

Next, similar to the transformation on pages 435–436 of [89], the dynamic equation for the m-dimensional inverse dynamics $\zeta(t) = TX(t)$ of (5.2) and its reference signal ζ^r are introduced as

$$\dot{\zeta}(t) = A_b \zeta(t) + b_b Y(t), \tag{5.176}$$

$$0 = A_b \zeta^r + b_b Y_r, \tag{5.177}$$

where A_b, b_b, and T have been defined by (5.51)–(5.53).

Thus the error state

$$\tilde{\zeta}(t) = \zeta(t) - \zeta^r \tag{5.178}$$

is driven by

$$\dot{\tilde{\zeta}}(t) = A_b \tilde{\zeta}(t) + b_b z_1(t). \tag{5.179}$$

5.3.2   Boundary Control with Adaptive Backstepping

Aiming at system

$$\dot{z}_1 = \hat{b}_m \tilde{v}_{m,2} + \tilde{\xi}_2 + \tilde{\bar{\omega}}^T \hat{\theta} + \varepsilon_2 + \omega^T \tilde{\theta}, \tag{5.180}$$

$$\dot{\tilde{v}}_{m,i} = \tilde{v}_{m,i+1} - k_i \tilde{v}_{m,1} - e_i^T A_0^m \frac{\partial \lambda^r(\hat{\theta})}{\partial \hat{\theta}} \dot{\hat{\theta}}, \tag{5.181}$$

$$i = 2, 3, \cdots, \rho - 1,$$

$$\dot{\tilde{v}}_{m,\rho} = \hat{e}(0,t) + \tilde{v}_{m,\rho+1} - k_\rho \tilde{v}_{m,1} - e_\rho^T A_0^m \frac{\partial \lambda^r(\hat{\theta})}{\partial \hat{\theta}} \dot{\hat{\theta}}, \tag{5.182}$$

an adaptive backstepping recursive control procedure very similar to (5.59)–(5.65) is depicted below,

Coordinate transformation:

$$z_1 = Y - Y_r, \tag{5.183}$$

$$z_i = \tilde{v}_{m,i} - \alpha_{i-1}, \quad i = 2, 3, \cdots, \rho. \tag{5.184}$$

Stabilizing functions:

$$\alpha_1 = \frac{1}{\hat{b}_m}\left(-(c_1 + d_1)z_1 - \tilde{\xi}_2 - \tilde{\bar{\omega}}^T \hat{\theta}\right), \tag{5.185}$$

$$\alpha_2 = -\hat{b}_m z_1 - \left(c_2 + d_2\left(\frac{\partial \alpha_1}{\partial z_1}\right)^2\right)z_2 + \beta_2, \tag{5.186}$$

$$\alpha_i = -z_{i-1} - \left(c_i + d_i\left(\frac{\partial \alpha_{i-1}}{\partial z_1}\right)^2\right)z_i + \beta_i, \tag{5.187}$$

$$i = 3, 4, \cdots, \rho$$

$$\beta_i = k_i \tilde{v}_{m,1} + \frac{\partial \alpha_{i-1}}{\partial z_1}(\tilde{\xi}_2 + \tilde{\omega}^T \hat{\theta}) + \frac{\partial \alpha_{i-1}}{\partial \tilde{\eta}}(A_0 \tilde{\eta} + e_n z_1)$$

$$+ \sum_{j=1}^{m+i-1} \frac{\partial \alpha_{i-1}}{\partial \tilde{\lambda}_j}(\tilde{\lambda}_{j+1} - k_j \tilde{\lambda}_1), \quad i = 2, 3, \cdots, \rho, \tag{5.188}$$

where $c_i > 0$, $d_i > 0$ for $i = 1, 2, \cdots, \rho$ are design parameters.

Adaptive control law:

$$\hat{e}(0, t) = U(t - \hat{D}) - U^r(\hat{\theta}) = -\tilde{v}_{m,\rho+1} + \alpha_\rho. \tag{5.189}$$

Based on a comparable calculation, $\hat{e}(0, t) = K_{\tilde{\chi}}(\hat{\theta}) \tilde{\chi}(t)$, with $K_{\tilde{\chi}}(\hat{\theta})$ being the same as (5.70)–(5.83).

In comparison with (5.87)–(5.88), by backstepping transformation

$$\hat{w}(x, t) = \hat{e}(x, t) - K_{\tilde{\chi}}(\hat{\theta}) e^{A_{\tilde{\chi}}(\hat{\theta}) \hat{D}x} \tilde{\chi}(t)$$
$$- \hat{D} \int_0^x K_{\tilde{\chi}}(\hat{\theta}) e^{A_{\tilde{\chi}}(\hat{\theta}) \hat{D}(x-y)} e_{2n+1} \hat{e}(y, t) \, dy, \tag{5.190}$$

$$\hat{e}(x, t) = \hat{w}(x, t) + K_{\tilde{\chi}}(\hat{\theta}) e^{(A_{\tilde{\chi}} + e_{2n+1} K_{\tilde{\chi}})(\hat{\theta}) \hat{D}x} \tilde{\chi}(t)$$
$$+ \hat{D} \int_0^x K_{\tilde{\chi}}(\hat{\theta}) e^{(A_{\tilde{\chi}} + e_{2n+1} K_{\tilde{\chi}})(\hat{\theta}) \hat{D}(x-y)} e_{2n+1}$$
$$\times \hat{w}(y, t) \, dy, \tag{5.191}$$

systems (5.8)–(5.9), (5.151), (5.168), (5.179), (5.180)–(5.182), and (5.143)–(5.144) are mapped into the closed-loop target error system below:

$$\dot{z} = A_z(\hat{\theta})z + W_\varepsilon(\hat{\theta})(\varepsilon_2 + \omega^T \tilde{\theta}) + Q(z, t)^T \dot{\hat{\theta}}$$
$$+ Q^r(\hat{\theta})^T \dot{\hat{\theta}} + e_\rho \hat{w}(0, t), \tag{5.192}$$

$$\dot{\tilde{\eta}} = A_0 \tilde{\eta} + e_n z_1, \tag{5.193}$$

$$\dot{\tilde{\zeta}} = A_b \tilde{\zeta} + b_b z_1, \tag{5.194}$$

$$\dot{\varepsilon} = A_0 \varepsilon + \bar{b} e(0, t), \tag{5.195}$$

$$De_t(x, t) = e_x(x, t) - \tilde{D} f(x, t) - D\hat{D}(x-1) f(x, t), \tag{5.196}$$

$$e(1, t) = 0, \tag{5.197}$$

$$\hat{D}\hat{w}_t(x, t) = \hat{w}_x(x, t) - \hat{D}\dot{\hat{D}} g(x, t) - \hat{D}\dot{\hat{\theta}}^T h(x, t)$$
$$- \hat{D}(\varepsilon_2 + \omega^T \tilde{\theta}) h_0(x, t), \tag{5.198}$$

$$\hat{w}(1, t) = 0, \tag{5.199}$$

$$\hat{D}\hat{w}_{xt}(x, t) = \hat{w}_{xx}(x, t) - \hat{D}\dot{\hat{D}} g_x(x, t) - \hat{D}\dot{\hat{\theta}}^T h_x(x, t)$$
$$- \hat{D}(\varepsilon_2 + \omega^T \tilde{\theta}) h_{0x}(x, t), \tag{5.200}$$

$$\hat{w}_x(1,t) = \hat{D}\dot{\hat{D}}g(1,t) + \hat{D}\dot{\hat{\theta}}^T h(1,t)$$
$$+ \hat{D}(\varepsilon_2 + \omega^T\tilde{\theta})h_0(1,t), \tag{5.201}$$

where $A_z(\hat{\theta})$, $W_\varepsilon(\hat{\theta})$, $Q(z,t)^T$, $Q^r(\hat{\theta})^T$ have almost the identical representations with those in (5.95)–(5.96), while $f(x,t)$, $h_0(x,t)$, $g(x,t)$, and $h_i(x,t)$ of $h(x,t)$ are presented below:

$$f(x,t) = \frac{\hat{w}_x(x,t)}{\hat{D}} + K_{\tilde{\chi}}(\hat{\theta})e_{2n+1}\hat{w}(x,t)$$
$$+ K_{\tilde{\chi}}(A_{\tilde{\chi}} + e_{2n+1}K_{\tilde{\chi}})(\hat{\theta})e^{(A_{\tilde{\chi}}+e_{2n+1}K_{\tilde{\chi}})(\hat{\theta})\hat{D}x}\tilde{\chi}(t)$$
$$+ \hat{D}\int_0^x K_{\tilde{\chi}}(A_{\tilde{\chi}} + e_{2n+1}K_{\tilde{\chi}})(\hat{\theta})e^{(A_{\tilde{\chi}}+e_{2n+1}K_{\tilde{\chi}})(\hat{\theta})\hat{D}(x-y)}e_{2n+1}\hat{w}(y,t)\,dy, \tag{5.202}$$

$$h_0(x,t) = K_{\tilde{\chi}}(\hat{\theta})e^{A_{\tilde{\chi}}(\hat{\theta})\hat{D}x}e_1, \tag{5.203}$$

$$g(x,t) = \frac{(1-x)}{\hat{D}}\hat{w}_x(x,t) + K_{\tilde{\chi}}(\hat{\theta})e^{A_{\tilde{\chi}}(\hat{\theta})\hat{D}x}e_{2n+1}\hat{w}(0,t)$$
$$+ K_{\tilde{\chi}}(\hat{\theta})e^{A_{\tilde{\chi}}(\hat{\theta})\hat{D}x}(A_{\tilde{\chi}} + e_{2n+1}K_{\tilde{\chi}})(\hat{\theta})\tilde{\chi}(t), \tag{5.204}$$

$$h_i(x,t) = \left[\left(\frac{\partial K_{\tilde{\chi}}(\hat{\theta})}{\partial\hat{\theta}_i} + K_{\tilde{\chi}}(\hat{\theta})\frac{\partial A_{\tilde{\chi}}(\hat{\theta})}{\partial\hat{\theta}_i}\hat{D}x\right)e^{A_{\tilde{\chi}}(\hat{\theta})\hat{D}x}\right.$$
$$+ \hat{D}\int_0^x\left(\frac{\partial K_{\tilde{\chi}}(\hat{\theta})}{\partial\hat{\theta}_i} + K_{\tilde{\chi}}(\hat{\theta})\frac{\partial A_{\tilde{\chi}}(\hat{\theta})}{\partial\hat{\theta}_i}\hat{D}(x-y)\right)e^{A_{\tilde{\chi}}(\hat{\theta})\hat{D}(x-y)}e_{2n+1}$$
$$\times K_{\tilde{\chi}}(\hat{\theta})e^{(A_{\tilde{\chi}}+e_{2n+1}K_{\tilde{\chi}})(\hat{\theta})\hat{D}y}\,dy\left.\right]\tilde{\chi}(t)$$
$$+ \hat{D}\int_0^x\hat{w}(y,t)\left[\left(\frac{\partial K_{\tilde{\chi}}(\hat{\theta})}{\partial\hat{\theta}_i} + K_{\tilde{\chi}}(\hat{\theta})\frac{\partial A_{\tilde{\chi}}(\hat{\theta})}{\partial\hat{\theta}_i}\hat{D}(x-y)\right)e^{A_{\tilde{\chi}}(\hat{\theta})\hat{D}(x-y)}e_{2n+1}\right.$$
$$+ \hat{D}\int_y^x\left(\frac{\partial K_{\tilde{\chi}}(\hat{\theta})}{\partial\hat{\theta}_i} + K_{\tilde{\chi}}(\hat{\theta})\frac{\partial A_{\tilde{\chi}}(\hat{\theta})}{\partial\hat{\theta}_i}\hat{D}(x-s)\right)e^{A_{\tilde{\chi}}(\hat{\theta})\hat{D}(x-s)}e_{2n+1}$$
$$\times K_{\tilde{\chi}}(\hat{\theta})e^{(A_{\tilde{\chi}}+e_{2n+1}K_{\tilde{\chi}})(\hat{\theta})\hat{D}(s-y)}e_{2n+1}\,ds\left.\right]dy$$
$$- K_{\tilde{\chi}}(\hat{\theta})e^{A_{\tilde{\chi}}(\hat{\theta})\hat{D}x}\frac{\partial\chi^r(\hat{\theta})}{\partial\hat{\theta}_i} + \frac{\partial u^r(\hat{\theta})}{\partial\hat{\theta}_i}$$
$$- \hat{D}\int_0^x K_{\tilde{\chi}}(\hat{\theta})e^{A_{\tilde{\chi}}(\hat{\theta})\hat{D}(x-y)}e_{2n+1}\frac{\partial u^r(\hat{\theta})}{\partial\hat{\theta}_i}\,dy. \tag{5.205}$$

As a consequence, by the certainty equivalence principle for (5.84), we design our control as follows to ensure that (5.199) holds:

$$U(t) = U^r(\hat\theta) - K_{\tilde\chi}(\hat\theta)\chi^r(\hat\theta)$$
$$+ K_{\tilde\chi}(\hat\theta)e^{A_{\tilde\chi}(\hat\theta)\hat{D}}\chi(t)$$
$$+ \hat{D}\int_0^1 K_{\tilde\chi}(\hat\theta)e^{A_{\tilde\chi}(\hat\theta)\hat{D}(1-y)}e_{2n+1}\hat{u}(y,t)\,\mathrm{d}y, \qquad (5.206)$$

where the update law of θ is chosen as

$$\dot\theta = \gamma_\theta \mathrm{Proj}_\Theta\{\tau_\theta\}, \quad \gamma_\theta > 0, \quad \hat{b}_m(0)\mathrm{sgn}(b_m) \geq \underline{b}_m, \qquad (5.207)$$
$$\tau_\theta = \frac{1}{2\kappa_2}\omega W_\varepsilon(\hat\theta)^T z$$
$$- \hat{D}\omega\int_0^1 (1+x)\big(\hat{w}(x,t)h_0(x,t) + \hat{w}_x(x,t)h_{0x}(x,t)\big)\,\mathrm{d}x, \qquad (5.208)$$

in which $\kappa_2 > 0$ and $\mathrm{Proj}_\Theta\{\cdot\}$ is a standard projection algorithm defined on the set Θ to guarantee that $|\hat{b}_m| \geq \underline{b}_m > 0$, and the update law of D is chosen such that

$$\dot{D} = \gamma_D \mathrm{Proj}_{[\underline{D},\bar{D}]}\{\tau_D\}, \quad \gamma_D > 0, \qquad (5.209)$$
$$|\tau_D| \leq M_D V_0^*, \qquad (5.210)$$
$$V_0^* = |z|^2 + |\tilde\eta|^2 + |\tilde\zeta|^2 + \|e(x,t)\|^2$$
$$+ \|\hat{w}(x,t)\|^2 + \|\hat{w}_x(x,t)\|^2 + |\varepsilon|^2, \qquad (5.211)$$

in which $M_D > 0$ and $\mathrm{Proj}_{[\underline{D},\bar{D}]}\{\cdot\}$ is a standard projection operator defined on the interval $[\underline{D},\bar{D}]$. One of the selections for the delay update law to satisfy (5.210)–(5.211) is a certainty-equivalence version of the update law (5.106) by replacing unmeasurable $u(x,t)$ and $w(x,t)$ with their estimates $\hat{u}(x,t)$ and $\hat{w}(x,t)$.

Remark 5.8. Keeping (5.139), (5.152)–(5.153), (5.177) in mind, based on a very similar derivation on pages 345–346 and 437–438 of [89], we could depict $\tilde\lambda$ as a one-to-one linear functional mapping of z, $\tilde\eta$, $\tilde\zeta$, and ε like (5.212), if and only if $B(s)$ in Assumption 5.1 and $K(s) = s^n + k_1 s^{n-1} + k_2 s^{n-2} + \cdots + k_{n-1}s + k_n$ are coprime. This appears possibly because $B(s)$ is not known. This coprime condition may be satisfied by a complete trial and error procedure.

$$\tilde\lambda = H_z(\hat\theta)z + H_\eta(\hat\theta)\tilde\eta + H_\zeta(\hat\theta)\tilde\zeta + H_\varepsilon(\hat\theta)\varepsilon$$
$$+ H_{Y_r}(\hat\theta)Y_r + H_{\eta^r}(\hat\theta)\eta^r + H_{\zeta^r}(\hat\theta)\zeta^r + H_{\lambda^r}(\hat\theta)\lambda^r, \qquad (5.212)$$

where $H_z(\hat\theta) \in \mathbb{R}^{n\times p}$, $H_\eta(\hat\theta) \in \mathbb{R}^{n\times n}$, $H_\zeta(\hat\theta) \in \mathbb{R}^{n\times m}$, $H_\varepsilon(\hat\theta) \in \mathbb{R}^{n\times n}$, $H_{Y_r}(\hat\theta) \in \mathbb{R}^n$, $H_{\eta^r}(\hat\theta) \in \mathbb{R}^{n\times n}$, $H_{\zeta^r}(\hat\theta) \in \mathbb{R}^{n\times m}$, $H_{\lambda^r}(\hat\theta) \in \mathbb{R}^{n\times n}$.

As a result, if $K(s)$ and $B(s)$ are assumed to satisfy the coprime condition, employing Young's inequality and Cauchy-Schwartz inequality, through

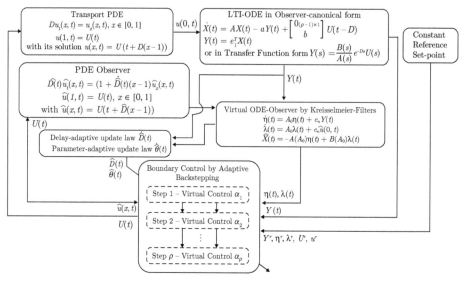

Figure 5.2. PDE output feedback of single-input systems with arbitrary relative degree

a lengthy but straightforward calculation, the parameter estimators (5.207)–(5.208) have the following property:

$$|\dot{\hat{\theta}}_i| \leq \gamma_\theta M_\theta \big(|z| + \|\hat{w}(x,t)\| + \|\hat{w}_x(x,t)\| + V_0^*\big),$$
$$i = 1, 2, \cdots, p, \tag{5.213}$$

where $M_\theta > 0$.

5.3.3 Stability Analysis

The adaptive controller in this section is shown in figure 5.2.

Next, the stability of the ODE-PDE cascade system in this section (in which the actuator state is unmeasured) is summarized in the main theorem below.

Theorem 5.9. *Consider the closed-loop system consisting of the plant (5.2), the PDE-observer (5.137)–(5.138), the ODE K-filters (5.145)–(5.150), the adaptive controller (5.206), the parameter identifier (5.207)–(5.208), and the time-delay identifier (5.209)–(5.211). Let Assumptions 5.1–5.2 and 5.7 hold, and let $B(s)$ and $K(s)$ be coprime. There exists a constant $M^* > 0$ such that if the initial error states satisfy the condition*

$$|z(0)|^2 + |\tilde{\eta}(0)|^2 + |\tilde{\zeta}(0)|^2$$
$$+ |\tilde{\theta}(0)|^2 + |\tilde{D}(0)|^2 + |\varepsilon(0)|^2 + \|e(x,0)\|^2$$
$$+ \|\hat{w}(x,0)\|^2 + \|\hat{w}_x(x,0)\|^2 \leq M^*, \tag{5.214}$$

*then all the signals of the closed-loop system are bounded and the set-point reg-
ulation is achieved, i.e.,*

$$\lim_{t\to\infty} z_1(t) = \lim_{t\to\infty} (Y(t) - Y_r) = 0. \tag{5.215}$$

Remark 5.10. Based on the equalities (5.19), (5.161)–(5.163), (5.178), (5.183)–
(5.184), (5.140)–(5.142), (5.190)–(5.191), $\tilde{\theta} = \theta - \hat{\theta}$, and $\tilde{D} = D - \hat{D}$, the initial
condition of error state (5.214) can be transformed into the initial conditions
of the actual ODE plants, filters, identifiers, and transport PDE, namely, $X(0)$,
$\eta(0)$, $\lambda(0)$, $\hat{\theta}(0)$, $\hat{D}(0)$, $u(x,0)$, and $\hat{u}(x,0)$.

Proof. First of all, a nonnegative function to contain major states of the closed-
loop system is built as follows:

$$V = \frac{1}{2} z^T z + \frac{1}{k_\eta} \tilde{\eta}^T P \tilde{\eta} + \frac{1}{k_\zeta} \tilde{\zeta}^T P_b \tilde{\zeta}$$

$$+ \kappa_1 D \int_0^1 (1+x) e(x,t)^2 \, \mathrm{d}x$$

$$+ \kappa_2 \hat{D} \int_0^1 (1+x) \big(\hat{w}(x,t)^2 + \hat{w}_x(x,t)^2 \big) \, \mathrm{d}x$$

$$+ \tilde{D}^2 + \frac{\kappa_2}{\gamma_\theta} \tilde{\theta}^T \tilde{\theta} + k_\varepsilon \sum_{i=1}^{\rho} \frac{1}{4d_i} \varepsilon^T P \varepsilon, \tag{5.216}$$

where $k_\eta > 0$, $k_\zeta > 0$, $\kappa_1 > 0$, $k_\varepsilon > 0$.

Taking the time derivative of (5.216) along (5.192)–(5.201), we obtain

$$\dot{V} \leq - \sum_{i=1}^{\rho} c_i z_i^2 + \sum_{i=1}^{\rho} \frac{1}{4d_i} \varepsilon_2^2 + z^T W_\varepsilon(\hat{\theta}) \omega^T \tilde{\theta}$$

$$+ z^T Q(z,t)^T \dot{\hat{\theta}} + z^T Q^r (\hat{\theta})^T \dot{\hat{\theta}}$$

$$+ z_\rho \hat{w}(0,t)$$

$$- \frac{1}{k_\eta} |\tilde{\eta}|^2 + \frac{2}{k_\eta} \tilde{\eta}^T P e_n z_1$$

$$- \frac{1}{k_\zeta} |\tilde{\zeta}|^2 + \frac{2}{k_\zeta} \tilde{\zeta}^T P_b b_b z_1$$

$$- \kappa_1 e(0,t)^2 - \kappa_1 \| e(x,t) \|^2$$

$$- 2\kappa_1 \tilde{D} \int_0^1 (1+x) e(x,t) f(x,t) \, \mathrm{d}x$$

$$- 2\kappa_1 D \dot{\hat{D}} \int_0^1 (x^2 - 1) e(x,t) f(x,t) \, \mathrm{d}x$$

$$+ \kappa_2 \dot{\hat{D}} \int_0^1 (1+x)\hat{w}(x,t)^2 \, \mathrm{d}x$$

$$- \kappa_2 \hat{w}(0,t)^2 - \kappa_2 \|\hat{w}(x,t)\|^2$$

$$- 2\kappa_2 \hat{D}\dot{\hat{D}} \int_0^1 (1+x)\hat{w}(x,t)g(x,t) \, \mathrm{d}x$$

$$- 2\kappa_2 \hat{D} \sum_{i=1}^p \dot{\hat{\theta}}_i \int_0^1 (1+x)\hat{w}(x,t)h_i(x,t) \, \mathrm{d}x$$

$$- 2\kappa_2 \hat{D}(\varepsilon_2 + \tilde{\theta}^T \omega) \int_0^1 (1+x)\hat{w}(x,t)h_0(x,t) \, \mathrm{d}x$$

$$+ \kappa_2 \dot{\hat{D}} \int_0^1 (1+x)\hat{w}_x(x,t)^2 \, \mathrm{d}x$$

$$+ 2\kappa_2 \hat{w}_x(1,t)^2 - \kappa_2 \hat{w}_x(0,t)^2 - \kappa_2 \|\hat{w}_x(x,t)\|^2$$

$$- 2\kappa_2 \hat{D}\dot{\hat{D}} \int_0^1 (1+x)\hat{w}_x(x,t)g_x(x,t) \, \mathrm{d}x$$

$$- 2\kappa_2 \hat{D} \sum_{i=1}^p \dot{\hat{\theta}}_i \int_0^1 (1+x)\hat{w}_x(x,t)h_{ix}(x,t) \, \mathrm{d}x$$

$$- 2\kappa_2 \hat{D}(\varepsilon_2 + \tilde{\theta}^T \omega) \int_0^1 (1+x)\hat{w}_x(x,t)h_{0x}(x,t) \, \mathrm{d}x$$

$$- \tilde{D}\dot{\hat{D}} - \frac{2\kappa_2}{\gamma_\theta}\tilde{\theta}^T \dot{\hat{\theta}} - k_\varepsilon \sum_{i=1}^\rho \frac{1}{4d_i}|\varepsilon|^2$$

$$+ k_\varepsilon \sum_{i=1}^\rho \frac{1}{2d_i}\varepsilon^T P\bar{b}e(0,t). \tag{5.217}$$

With the help of (5.199), Poincaré and Agmon's Inequalities imply that $\hat{w}(0,t)^2 \le 4\|\hat{w}_x(x,t)\|^2$. Bearing (5.139), (5.152)–(5.153), (5.177), (5.210)–(5.213) in mind, according to Young's and Cauchy-Schwartz inequalities, we have the following inequality:

$$z^T Q(z,t)^T \dot{\hat{\theta}} = -\sum_{i=2}^\rho \sum_{j=1}^p z_i \frac{\partial \alpha_{i-1}}{\partial \hat{\theta}_j}\dot{\hat{\theta}}_j$$

$$\le \gamma_\theta M_Q \left(V_0 + V_0^2\right) \tag{5.218}$$

$$z^T Q^r(\hat{\theta})^T \dot{\hat{\theta}} = -\sum_{i=2}^\rho \sum_{k=1}^p z_i \left(e_i^T A_0^m - \sum_{j=1}^{m+i-1} \frac{\partial \alpha_{i-1}}{\partial \tilde{\lambda}_j}e_j^T \right)$$

$$\times \frac{\partial \lambda^r(\hat{\theta})}{\partial \hat{\theta}_k}\dot{\hat{\theta}}_k$$

$$\le \gamma_\theta M_{Q^r}\left(V_0 + V_0^2\right) \tag{5.219}$$

$$-2\kappa_1\tilde{D}\int_0^1 (1+x)e(x,t)f(x,t)\,\mathrm{d}x$$

$$\leq |\tilde{D}|\kappa_1 M_1 V_0 \tag{5.220}$$

$$-2\kappa_1 D\dot{D}\int_0^1 (x^2-1)e(x,t)f(x,t)\,\mathrm{d}x$$

$$\leq \kappa_1\gamma_D M_2(V_0+V_0^2) \tag{5.221}$$

$$\kappa_2\dot{D}\int_0^1 (1+x)\hat{w}(x,t)^2\,\mathrm{d}x$$

$$\leq \kappa_2\gamma_D M_3(V_0+V_0^2) \tag{5.222}$$

$$-2\kappa_2\hat{D}\dot{D}\int_0^1 (1+x)\hat{w}(x,t)g(x,t)\,\mathrm{d}x$$

$$\leq \kappa_2\gamma_D M_4(V_0+V_0^2) \tag{5.223}$$

$$-2\kappa_2\hat{D}\sum_{i=1}^p \dot{\hat{\theta}}_i\int_0^1 (1+x)\hat{w}(x,t)h_i(x,t)\,\mathrm{d}x$$

$$\leq \kappa_2\gamma_\theta M_5(V_0+V_0^2) \tag{5.224}$$

$$-2\kappa_2\hat{D}\varepsilon_2\int_0^1 (1+x)\hat{w}(x,t)h_0(x,t)\,\mathrm{d}x$$

$$\leq \frac{\kappa_2}{4}\|\hat{w}(x,t)\|^2 + \kappa_2 M_6|\varepsilon|^2 \tag{5.225}$$

$$-2\kappa_2\hat{D}\int_0^1 (1+x)\hat{w}(x,t)K_{\tilde{\chi}}(\hat{\theta})e^{A_{\tilde{\chi}}(\hat{\theta})\hat{D}x}e_{2n+1}e(0,t)\,\mathrm{d}x$$

$$\leq \frac{\kappa_2}{4}\|\hat{w}(x,t)\|^2 + \kappa_2 M_7 e(0,t)^2 \tag{5.226}$$

$$\kappa_2\dot{D}\int_0^1 (1+x)\hat{w}_x(x,t)^2\,\mathrm{d}x$$

$$\leq \kappa_2\gamma_D M_8(V_0+V_0^2) \tag{5.227}$$

$$-2\kappa_2\hat{D}\dot{D}\int_0^1 (1+x)\hat{w}_x(x,t)g_x(x,t)\,\mathrm{d}x$$

$$\leq \kappa_2\gamma_D M_9(V_0+V_0^2) \tag{5.228}$$

$$-2\kappa_2\hat{D}\sum_{i=1}^p \dot{\hat{\theta}}_i\int_0^1 (1+x)\hat{w}_x(x,t)h_{ix}(x,t)\,\mathrm{d}x$$

$$\leq \kappa_2\gamma_\theta M_{10}(V_0+V_0^2) \tag{5.229}$$

$$-2\kappa_2\hat{D}\varepsilon_2\int_0^1 (1+x)\hat{w}_x(x,t)h_{0x}(x,t)\,\mathrm{d}x$$

$$\leq \frac{\kappa_2}{4}\|\hat{w}_x(x,t)\|^2 + \kappa_2 M_{11}|\varepsilon|^2 \tag{5.230}$$

$$- 2\kappa_2 \hat{D}^2 \int_0^1 (1+x)\hat{w}_x(x,t)K_{\tilde{\chi}}A_{\tilde{\chi}}(\hat{\theta})e^{A_{\tilde{\chi}}(\hat{\theta})\hat{D}x}e_{2n+1}e(0,t)\,\mathrm{d}x$$

$$\leq \frac{\kappa_2}{4}\|\hat{w}_x(x,t)\|^2 + \kappa_2 M_{12}e(0,t)^2 \tag{5.231}$$

$$2\kappa_2\hat{w}_x(1,t)^2$$

$$\leq \kappa_2(\gamma_D^2 + \gamma_\theta^2)M_{13}(V_0 + V_0^2)$$

$$+ \kappa_2 M_{14}|\varepsilon|^2 + \kappa_2|\tilde{\theta}|^2 M_{15}V_0 + \kappa_2 M_{16}e(0,t)^2. \tag{5.232}$$

Thus (5.217) becomes

$$\dot{V} \leq -\frac{c_0}{2}|z|^2 - \frac{k_\varepsilon - 2}{8d_0}|\varepsilon|^2 + \frac{1}{c_0}\hat{w}(0,t)^2 + \frac{k_\varepsilon}{2d_0}|P\bar{b}|^2 e(0,t)^2$$

$$+ \gamma_\theta M_Q(V_0^* + V_0^{*2}) + \gamma_\theta M_{Q^r}(V_0^* + V_0^{*2})$$

$$- \frac{1}{2k_\eta}|\tilde{\eta}|^2 + \frac{2}{k_\eta}|Pe_n|^2 z_1^2$$

$$- \frac{1}{2k_\zeta}|\tilde{\zeta}|^2 + \frac{2}{k_\zeta}|P_b b_b|^2 z_1^2$$

$$- \kappa_1 e(0,t)^2 - \kappa_1\|e(x,t)\|^2$$

$$+ |\tilde{D}|(\gamma_D M_D + \kappa_1 M_1)V_0^* + \kappa_1\gamma_D M_2(V_0^* + V_0^{*2})$$

$$- \kappa_2\hat{w}(0,t)^2 - \kappa_2\|\hat{w}(x,t)\|^2$$

$$+ \kappa_2\gamma_D M_3(V_0^* + V_0^{*2}) + \kappa_2\gamma_D M_4(V_0^* + V_0^{*2})$$

$$+ \kappa_2\gamma_\theta M_5(V_0^* + V_0^{*2})$$

$$+ \frac{\kappa_2}{4}\|\hat{w}(x,t)\|^2 + \kappa_2 M_6|\varepsilon|^2$$

$$- \kappa_2\hat{w}_x(0,t)^2 - \kappa_2\|\hat{w}_x(x,t)\|^2$$

$$+ \kappa_2\gamma_D M_7(V_0^* + V_0^{*2}) + \kappa_2\gamma_D M_8(V_0^* + V_0^{*2})$$

$$+ \kappa_2\gamma_\theta M_9(V_0^* + V_0^{*2})$$

$$+ \frac{\kappa_2}{4}\|\hat{w}_x(x,t)\|^2 + \kappa_2 M_{10}|\varepsilon|^2$$

$$+ \kappa_2(\gamma_D^2 + \gamma_\theta^2)M_{11}(V_0^{*2} + V_0^{*3}) + \kappa_2 M_{12}|\varepsilon|^2$$

$$+ \kappa_2|\tilde{\theta}|^2 M_{13}V_0^*$$

$$- \frac{2\kappa_2}{\gamma_\theta}\tilde{\theta}^T(\dot{\hat{\theta}} - \gamma_\theta\tau_\theta), \tag{5.233}$$

where $c_0 = \min\limits_{i=1,2,\cdots,\rho} c_i$, $d_0 = \left(\sum\limits_{i=1}^{\rho}\frac{1}{d_i}\right)^{-1}$, M_Q, M_{Q^r}, $M_1 \sim M_{13}$ are some positive constants.

A merger of similar items results in

$$\dot{V}(t) \leq -\left(\frac{c_0}{2} - \frac{2}{k_\eta}|Pe_n|^2 - \frac{2}{k_\zeta}|P_b b_b|^2\right)|z|^2$$

$$-\frac{1}{2k_\eta}|\tilde{\eta}|^2 - \frac{1}{2k_\zeta}|\tilde{\zeta}|^2 + \kappa_2|\tilde{\theta}|^2 M_{13}V_0^*$$

$$-\left(\frac{k_\varepsilon - 2}{8d_0} - \kappa_2(M_6 + M_{10} + M_{12})\right)|\varepsilon|^2$$

$$-\left(\kappa_1 - \frac{k_\varepsilon}{2d_0}|P\bar{b}|^2\right)e(0,t)^2$$

$$-\kappa_1\|e(x,t)\|^2 + |\tilde{D}|(\gamma_D M_D + \kappa_1 M_1)V_0^*$$

$$-\left(\kappa_2 - \frac{1}{c_0}\right)\hat{w}(0,t)^2 - \frac{\kappa_2}{2}\|\hat{w}(x,t)\|^2$$

$$-\kappa_2\hat{w}_x(0,t)^2 - \frac{\kappa_2}{2}\|\hat{w}_x(x,t)\|^2$$

$$+\phi_1(\gamma_D,\gamma_\theta)(V_0^* + V_0^{*2}) + \phi_2(\gamma_D,\gamma_\theta)(V_0^{*2} + V_0^{*3}), \qquad (5.234)$$

where ϕ_1 and ϕ_2 are linear and nonlinear functions with respect to γ_D and γ_θ, respectively.

Choose k_η, k_ζ, k_ε, and κ_1, κ_2 appropriately such that

$$k_\eta \geq \frac{8}{c_0}|Pe_n|^2, \quad k_\zeta \geq \frac{8}{c_0}|P_b b_b|^2, \quad \kappa_2 \geq \frac{1}{c_0}$$

$$k_\varepsilon \geq 8d_0\kappa_2(M_6 + M_{10} + M_{12}) + 2$$

$$\kappa_1 \geq \frac{k_\varepsilon}{2d_0}|P\bar{b}|^2. \qquad (5.235)$$

Thus (5.234) becomes

$$\dot{V} \leq -M_{14}V_0^* + |\tilde{D}|M_{15}V_0^* + |\tilde{\theta}|^2 M_{16}V_0^*$$

$$+\phi_1(\gamma_D,\gamma_\theta)(V_0^* + V_0^{*2}) + \phi_2(\gamma_D,\gamma_\theta)(V_0^{*2} + V_0^{*3})$$

$$\leq -M_{14}V_0^* + \left(\frac{\kappa_3}{2} + \frac{\tilde{D}^2}{2\kappa_3}\right)M_{15}V_0^* + |\tilde{\theta}|^2 M_{16}V_0^*$$

$$+\phi_1(\gamma_D,\gamma_\theta)(V_0^* + V_0^{*2}) + \phi_2(\gamma_D,\gamma_\theta)(V_0^{*2} + V_0^{*3})$$

$$\leq -M_{14}V_0^* + \frac{\kappa_3}{2}M_{15}V_0^* + (V - M_{17}V_0^*)V_0^*$$

$$+\phi_1(\gamma_D,\gamma_\theta)(V_0^* + V_0^{*2}) + \phi_2(\gamma_D,\gamma_\theta)(V_0^{*2} + V_0^{*3})$$

$$\leq -\left(M_{14} - \frac{\kappa_3}{2}M_{15} - \phi_1(\gamma_D,\gamma_\theta) - V\right)V_0^*$$

$$-\left(M_{17} - \phi_1(\gamma_D,\gamma_\theta) - \phi_2(\gamma_D,\gamma_\theta)\right.$$

$$\left. - \phi_2(\gamma_D,\gamma_\theta)V_0^*\right)V_0^{*2}, \tag{5.236}$$

where $M_{14} = \min\left\{\frac{c_0}{2} - \frac{2}{k_\eta}|Pe_n|^2 - \frac{2}{k_\zeta}|P_b b_b|^2, \frac{k_\varepsilon - 2}{8d_0} - \kappa_2(M_6 + M_{10} + M_{12}), \frac{1}{2k_\eta}, \right.$

$\left. \frac{1}{2k_\zeta}, \kappa_1, \frac{\kappa_2}{2}\right\}$, $M_{15} = \gamma_D M_D + \kappa_1 M_1$, $M_{16} = \kappa_2 M_{13}$, $\kappa_3 > 0$, $M_{17} > 0$.

By designing κ_3, γ_D, γ_θ and choosing initial state $V(0)$ suitably to satisfy

$$\frac{\kappa_3}{2}M_{15} + \phi_1(\gamma_D,\gamma_\theta) \leq M_{14}, \tag{5.237}$$

$$\phi_1(\gamma_D,\gamma_\theta) + \phi_2(\gamma_D,\gamma_\theta) \leq M_{17}, \tag{5.238}$$

$$V(0) \leq \min\left\{M_{14} - \frac{\kappa_3}{2}M_{15} - \phi_1(\gamma_D,\gamma_\theta),\right.$$

$$\left. M_{18}\frac{M_{17} - \phi_1(\gamma_D,\gamma_\theta) - \phi_2(\gamma_D,\gamma_\theta)}{\phi_2(\gamma_D,\gamma_\theta)}\right\}, \tag{5.239}$$

we have

$$\dot{V} \leq -M_{19}V_0^* - M_{20}V_0^{*2}, \tag{5.240}$$

where $M_{19} = M_{14} - \frac{\kappa_3}{2}M_{15} - \phi_1(\gamma_D,\gamma_\theta) - V$, $M_{20} = M_{17} - \phi_1(\gamma_D,\gamma_\theta) - \phi_2(\gamma_D,\gamma_\theta)$ $- \phi_2(\gamma_D,\gamma_\theta)V_0^*$, $M_{18} > 0$.

Thus signals z, $\tilde{\eta}$, $\tilde{\zeta}$, $e(x,t)$, $\hat{w}(x,t)$, $\hat{w}_x(x,t)$, and ε are bounded. Bearing the boundedness of Y_r, U^r, η^r, and λ^r in mind, we show the boundedness of X and $u(x,t)$ from (5.19), (5.159), and (5.190)–(5.191), respectively. Thus all signals of the closed-loop system have been proved to be bounded. Similar to the proof of Theorem 5.5, by applying Barbalat's lemma, we conclude that $z(t) \to 0$; in other words, $\lim_{t\to\infty}[Y(t) - Y_r] = 0$. □

Remark 5.11. Please note that the reference of (5.139) depends on neither time nor space. We stress that the result in this section cannot be extended directly to non-constant trajectory tracking. In that situation, on the one hand, (5.143)–(5.144) will lose its boundary symmetry, and correspondingly the closed-loop error system (5.196)–(5.201) will become too complicated to handle. On the other hand, as one can see in the previous calculation, (5.152)–(5.153), and (5.177) play their roles in the stability analysis.

5.4 Simulation

In this section, we illustrate the proposed scheme by applying it to the following three-dimensional plant with the relative degree being two:

$$\dot{X}_1(t) = X_2(t) - a_2 Y(t)$$
$$\dot{X}_2(t) = X_3(t) - a_1 Y(t) + b_1 U(t - D)$$
$$\dot{X}_3(t) = -a_0 Y(t) + b_0 U(t - D)$$
$$Y(t) = X_1(t), \tag{5.241}$$

where $X_2(t)$, $X_3(t)$ are the unmeasurable states, $Y(t)$ is the measured output, $D = 1$ is the unknown constant time delay in control input, and $a_2 = -3$ is the unknown constant system parameter. To show the core effectiveness of the developed control algorithm, without losing the generality, we simplify the simulation by assuming that $a_1 = a_0 = 0$, $b_1 = b_0 = 1$ are known and PDE state $u(x,t)$ is measurable. It is easy to check that this plant is potentially unstable due to the poles possessing nonnegative real parts. The prior information including $\underline{D} = 0.5$, $\bar{D} = 1.5$, $\underline{a}_2 = -4$, $\bar{a}_2 = -2$ is known to designers. $Y_r(t) = \sin t$ is the known reference signal to track and it is easily produced by the following reference state and input:

$$X_1^r(t) = \sin t, \tag{5.242}$$

$$X_2^r(t, \hat{\theta}) = \hat{a}_2 \sin t + \cos t, \tag{5.243}$$

$$X_3^r(t, \hat{\theta}) = \frac{\hat{a}_2 - 1}{2} \sin t + \frac{\hat{a}_2 + 1}{2} \cos t, \tag{5.244}$$

$$U^r(t, \hat{\theta}) = -\frac{\hat{a}_2 + 1}{2} \sin t + \frac{\hat{a}_2 - 1}{2} \cos t, \tag{5.245}$$

Furthermore, if we choose $k_1 = 6$, $k_2 = 12$, $k_3 = 8$, the reference state (5.242)–(5.245) can be readily divided into the reference output filter and input filter as follows:

$$\eta_1^r(t) = \frac{2}{125} \sin t - \frac{11}{125} \cos t, \tag{5.246}$$

$$\eta_2^r(t) = \frac{23}{125} \sin t - \frac{64}{125} \cos t, \tag{5.247}$$

$$\eta_3^r(t) = \frac{88}{125} \sin t - \frac{109}{125} \cos t, \tag{5.248}$$

$$\lambda_1^r(t, \hat{\theta}) = \left(-\frac{2}{125} \frac{\hat{a}_2 + 1}{2} + \frac{11}{125} \frac{\hat{a}_2 - 1}{2} \right) \sin t$$

$$+\left(\frac{11}{125}\frac{\hat{a}_2+1}{2}+\frac{2}{125}\frac{\hat{a}_2-1}{2}\right)\cos t, \tag{5.249}$$

$$\lambda_2^r(t,\hat{\theta})=\left(-\frac{23}{125}\frac{\hat{a}_2+1}{2}+\frac{64}{125}\frac{\hat{a}_2-1}{2}\right)\sin t$$

$$+\left(\frac{64}{125}\frac{\hat{a}_2+1}{2}+\frac{23}{125}\frac{\hat{a}_2-1}{2}\right)\cos t, \tag{5.250}$$

$$\lambda_3^r(t,\hat{\theta})=\left(-\frac{88}{125}\frac{\hat{a}_2+1}{2}+\frac{109}{125}\frac{\hat{a}_2-1}{2}\right)\sin t$$

$$+\left(\frac{109}{125}\frac{\hat{a}_2+1}{2}+\frac{88}{125}\frac{\hat{a}_2-1}{2}\right)\cos t. \tag{5.251}$$

It is obvious that Assumptions 5.1–5.3 are satisfied. Then we adopt the proposed adaptive control scheme to design the controller and the identifier. The initial values are set as $X(0)=[0,1,-1]^T$, $\eta(0)=[-11/125,-64/125,-109/125]^T$, $\lambda(0)=[-15/125,-110/125,-285/125]^T$, and $\hat{D}(0)=1.4$, $\hat{a}_2(0)=-3.4$. The design parameters are chosen as follows: $c_1=c_2=0.1$, $d_1=d_2=0.05$, $\gamma_\theta=5\times10^{-6}$, $\gamma_D=5\times10^{-7}$, $g=0.2$. The boundedness and the asymptotic tracking of system output and control input, the estimations of θ and D are shown in figures 5.3 and 5.4.

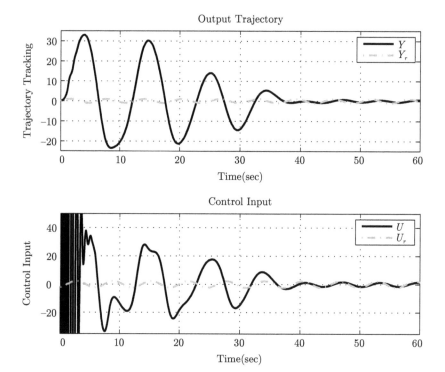

Figure 5.3. Output and input tracking

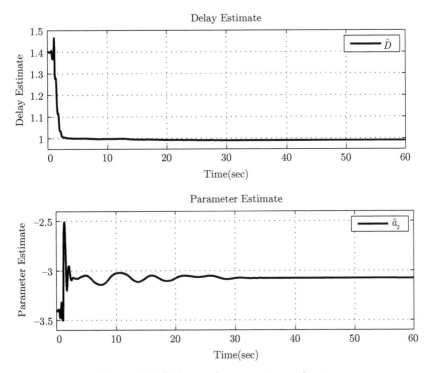

Figure 5.4. Delay and parameter estimate

Part II

Multi-Input Discrete Delays

Chapter Six

Exact Predictor Feedback for Multi-Input Systems

IN PART I, we investigated adaptive control problems of uncertain single-input LTI systems with discrete input delay, and the results obtained seem to easily (or even trivially) extend to systems with discrete multiple inputs when the delay is the same in all the input channels. In part II, we consider adaptive control problems of multi-input LTI systems with distinct discrete input delays; i.e., each delay length in several multi-input channels is not the same. The problem of distinct delays significantly complicates the prediction design, as it requires one to compute different future state values on time horizons, making this computation noncausal at first sight.

Before the description of adaptive control of uncertain systems, this chapter presents the exact predictor feedback for deterministic multi-input LTI systems with distinct discrete input delays, which offers a PDE-based framework for compensation for multi-input distinct delays and lays a foundation for adaptive control in later chapters. In other words, in this chapter, we assume that

- the finite-dimensional ODE plant state vector $X(t)$ is available for measurement,
- the infinite-dimensional PDE actuator state $u_i(x, t)$ for $i = 1, \cdots, m$, which represents the distinct delayed input channel $U_i(\eta)$ over the time window $\eta \in [t - D_i, t]$, is available for measurement,
- the ODE system matrix A and input matrices b_i for $i = 1, \cdots, m$ are fully known (every element of the matrices is certain and known),
- the length value of the arbitrarily large input delay D_i in distinct input channel for $i = 1, \cdots, m$ is certain and known.

In the PDE-based framework to deal with the discrete multi-input delays, the problem is formulated as a cascade between several first-order transport PDEs (accounting for the distinct delayed inputs) and an ODE (accounting for the original plant) and solved using backstepping to design a prediction-based controller. The key idea of this prediction technique is to replace future inputs involved in the state predictions computation by their closed-loop expressions.

6.1 Basic Idea of Predictor Feedback of Multi-Input LTI Systems with Distinct Input Delays

Consider a general class of multi-input LTI systems with distinct discrete input delays as follows:

$$\dot{X}(t) = AX(t) + \sum_{i=1}^{m} b_i U_i(t - D_i), \tag{6.1}$$

where $X(t) \in \mathbb{R}^n$ is the state vector of an ODE, and $U_i(t) \in \mathbb{R}$ is the ith scalar control input with a constant delay $D_i \in \mathbb{R}^+$ for $i = 1, 2, \cdots, m$. Without loss of generality, the length sequence satisfies $0 < D_1 \le D_2 \le \cdots \le D_m$. The matrices $A \in \mathbb{R}^{n \times n}$ and $b_i \in \mathbb{R}^n$ are the system matrix and the input matrix, respectively.

Our objective is to develop a control scheme to stabilize the potentially unstable LTI systems (6.1). To achieve the stabilization target, the following assumption about (6.1) is supposed to hold throughout the chapter.

Assumption 6.1. *The pair* (A, B) *is stabilizable with matrix* $B = (b_1, b_2, \cdots, b_m) \in \mathbb{R}^{n \times m}$. *In other words, there exists a matrix* $K = (k_1^T, k_2^T, \cdots, k_m^T)^T \in \mathbb{R}^{m \times n}$ *in which* $k_i \in \mathbb{R}^{1 \times n}$ *for* $i = 1, 2, \cdots, m$ *such that* $A + BK = A + \sum_{i=1}^{m} b_i k_i$ *is Hurwitz, namely,*

$$(A + BK)^T P + P(A + BK) = -Q, \tag{6.2}$$

with $P = P^T > 0$ *and* $Q = Q^T > 0$.

To address the delayed input, similar to the single-input case in chapter 2, a transport PDE is introduced as

$$u_i(\sigma, t) = U_i(t + \sigma - D_i), \sigma \in [0, D_i] \tag{6.3}$$

by which the LTI system (6.1) is transformed into the ODE-PDE cascade such that

$$\dot{X}(t) = AX(t) + \sum_{i=1}^{m} b_i u_i(0, t), \tag{6.4}$$

$$\partial_t u_i(\sigma, t) = \partial_\sigma u_i(\sigma, t), \quad \sigma \in [0, D_i], \tag{6.5}$$

$$u_i(D_i, t) = U_i(t), \quad i \in \{1, 2, \cdots, m\}. \tag{6.6}$$

Recent publications [127, 128] provide an exact predictor feedback law to stabilize deterministic LTI systems (6.1) and (6.4)–(6.6), in which multi-input delays D_i and matrices A and b_i are totally known, and the ODE state $X(t)$ and the PDE state $u_i(\sigma, t)$ are fully measurable.

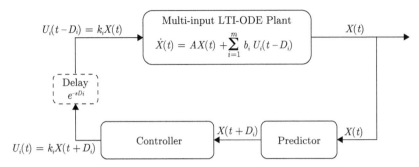

Figure 6.1. Exact predictor feedback for multi-input systems in ODE notation

For $i = 1, 2, \cdots, m$, as shown in figure 6.1, the idea of exact predictor feedback in [127, 128] is to realize

$$U_i(t - D_i) = u_i(0, t) = k_i X(t),$$

$$\text{equivalently,} \quad U_i(t) = u_i(D_i, t) = k_i X(t + D_i), \tag{6.7}$$

where k_i is defined in Assumption 6.1.

In parallel with the single-input case in chapter 2, the purpose is to find a state transformation and a state feedback control law that convert the system (6.4)–(6.6) into a stable target system satisfying

$$\dot{X}(t) = (A + BK)X(t) + \sum_{i=1}^{m} b_i w_i(0, t), \tag{6.8}$$

$$\partial_t w_i(\sigma, t) = \partial_\sigma w_i(\sigma, t), \quad \sigma \in [0, D_i], \tag{6.9}$$

$$w_i(D_i, t) = 0, \quad i \in \{1, 2, \cdots, m\}. \tag{6.10}$$

For the target system, the solution to (6.9)–(6.10) satisfies $w_i(\sigma, t) = 0$ for any $\sigma \in [0, D_i]$ after $t = D_i$. Hence, the state $X(t)$ satisfies

$$\dot{X}(t) = (A + BK)X(t), \quad t \geq D_m. \tag{6.11}$$

Thus, the target plant obeys the nominal closed-loop equation after $t = D_m$.

In order to achieve the above conversion $(X(t), u_i(\sigma, t)) \to (X(t), w_i(\sigma, t))$, the invertible backstepping-like transformations for $i = 1, 2, \cdots, m$ are expressed as follows:

$$w_i(\sigma, t) = u_i(\sigma, t) - k_i \Phi(\sigma, 0) X(t)$$

$$- \sum_{j=1}^{m} \int_0^{\phi_j(\sigma)} k_i \Phi(\sigma, \delta) b_j u_j(\delta, t) d\delta, \tag{6.12}$$

$$u_i(\sigma, t) = w_i(\sigma, t) + k_i e^{A_m \sigma} X(t)$$

$$+ \sum_{j=1}^{m} \int_0^{\phi_j(\sigma)} k_i e^{A_m(\sigma-\delta)} b_j w_j(\delta, t) d\delta, \tag{6.13}$$

where for $j \in \{1, 2, \cdots, m\}$,

$$\phi_j(\sigma) = \begin{cases} \sigma, & \sigma \in [0, D_j] \\ D_j, & \sigma \in [D_j, D_m], \end{cases} \tag{6.14}$$

and for $D_{i-1} \le \delta \le \sigma \le D_i$, $i \in \{1, 2, \cdots, m\}$,

$$\Phi(\sigma, \delta) = e^{A_{i-1}(\sigma-\delta)}, \tag{6.15}$$

for $\sigma \in [D_{i-1}, D_i]$, $\delta \in [D_{j-1}, D_j]$, $i, j \in \{1, 2, \cdots, m\}$, $j < i$,

$$\Phi(\sigma, \delta) = e^{A_{i-1}(\sigma-D_{i-1})} e^{A_{i-2}(D_{i-1}-D_{i-2})}$$

$$\cdots e^{A_j(D_{j+1}-D_j)} e^{A_{j-1}(D_j-\delta)}, \tag{6.16}$$

and specifically for (6.16), if $j = i - 1$, then

$$\Phi(\sigma, \delta) = e^{A_{i-1}(\sigma-D_{i-1})} e^{A_{i-2}(D_{i-1}-\delta)} \tag{6.17}$$

with

$$D_0 = 0, \quad \begin{cases} A_0 = A, \\ A_i = A_{i-1} + b_i k_i, i \in \{1, 2, \cdots, m\} \\ A_m = A + BK = A + \sum_{i=1}^{m} b_i k_i. \end{cases} \tag{6.18}$$

Readers may wonder how to derive the above transformations (6.12)–(6.13). Similar to the single-input case in chapter 2 of this book and [80], the invertible transformations are guided by a general backstepping design procedure, and the detailed calculation is found in [127].

Then, by substituting $\sigma = D_i$ into (6.12) to make (6.10) hold, the predictor-based PDE boundary controllers for $i = 1, 2, \cdots, m$ have expressions such that

$$U_i(t) = u_i(D_i, t) = k_i \Phi(D_i, 0) X(t)$$

$$+ \sum_{j=1}^{m} \int_0^{\phi_j(D_i)} k_i \Phi(D_i, \delta) b_j u_j(\delta, t) d\delta. \tag{6.19}$$

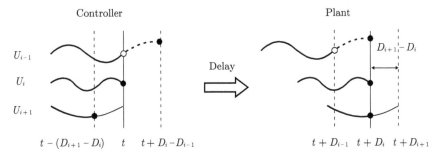

Figure 6.2. Difference in arrival time of the control inputs

The backstepping transformation (6.12)–(6.13) and the boundary control (6.19) may be somewhat abstract to readers who are not familiar with the PDE-based notation. Next, we give a concrete example to illustrate the above control scheme and discuss its relationship with the traditional predictor feedback form. As illustrated in figure 6.2, the problem of distinct delays significantly complicates the prediction design, as it requires one to compute different future state values on time horizons, making this computation noncausal at first sight. The key idea of the backstepping prediction technique is to replace future inputs involved in the state predictions computation by their closed-loop expressions.

Example 6.2. Let us consider an LTI system with two-input distinct delays as follows:

$$\dot{X}(t) = AX(t) + b_1 U_1(t - D_1) + b_2 U_2(t - D_2), \quad t \geq 0, \tag{6.20}$$

where $0 < D_1 \leq D_2$ is assumed without loss of generality.

In the standard delay notation, regarding $X(t)$ as the current state and employing the predictor state $X(t + D_1)$, we expect the first input channel of the exact predictor feedback design to be

$$U_1(t) = k_1 X(t + D_1), \tag{6.21}$$

$$= k_1 \left[e^{A(t+D_1-t)} X(t) + \int_t^{t+D_1} e^{A(t+D_1-s)} b_1 U_1(s - D_1) ds \right.$$

$$\left. + \int_t^{t+D_1} e^{A(t+D_1-s)} b_2 U_2(s - D_2) ds \right], \tag{6.22}$$

$$= k_1 \left[e^{AD_1} X(t) + \int_{t-D_1}^t e^{A(t-\theta)} b_1 U_1(\theta) d\theta \right.$$

$$\left. + \int_{t-D_2}^{t+D_1-D_2} e^{A(t+D_1-D_2-\theta)} b_2 U_2(\theta) d\theta \right]. \tag{6.23}$$

Thus after $t \geq D_1$, the first delay D_1 is compensated for and (6.20) becomes the closed-loop system such that

$$\dot{X}(t) = (A + b_1 k_1) X(t) + b_2 U_2(t - D_2)$$
$$= A_1 X(t) + b_2 U_2(t - D_2), \quad t \geq D_1, \tag{6.24}$$

where $A_1 = A + b_1 k_1$.

Concentrating on the system (6.24), regarding $X(t + D_1)$ as the current state and employing the predictor state $X(t + D_2)$, in the standard delay notation, we select the second input channel of the exact predictor feedback design as

$$U_2(t) = k_2 X(t + D_2), \tag{6.25}$$

$$= k_2 \left[e^{A_1(t + D_2 - t - D_1)} X(t + D_1) + \int_{t+D_1}^{t+D_2} e^{A_1(t+D_2-s)} b_2 U_2(s - D_2) ds \right]$$

$$= k_2 \left[e^{A_1(D_2 - D_1)} X(t + D_1) + \int_{t+D_1-D_2}^{t} e^{A_1(t-\theta)} b_2 U_2(\theta) d\theta \right], \tag{6.26}$$

$$= k_2 \left[e^{A_1(D_2 - D_1)} e^{A D_1} X(t) + \int_{t-D_1}^{t} e^{A_1(D_2 - D_1)} e^{A(t-\theta)} b_1 U_1(\theta) d\theta \right.$$

$$+ \int_{t-D_2}^{t+D_1-D_2} e^{A_1(D_2 - D_1)} e^{A(t+D_1-D_2-\theta)} b_2 U_2(\theta) d\theta$$

$$+ \left. \int_{t+D_1-D_2}^{t} e^{A_1(t-\theta)} b_2 U_2(\theta) d\theta \right]. \tag{6.27}$$

Then after $t \geq D_2$, the delay D_2 is compensated for and the system (6.24) becomes

$$\dot{X}(t) = (A_1 + b_2 k_2) X(t)$$
$$= A_2 X(t), \quad t \geq D_2, \tag{6.28}$$

where $A_2 = A_1 + b_2 k_2 = A + b_1 k_1 + b_2 k_2$. Keep the exact predictor feedback design (6.23) and (6.27) in the standard delay notation in mind.

Now we consider the PDE-based approach. The transport PDEs representing the different input delays are

$$u_1(\sigma, t) = U_1(t + \sigma - D_1), \quad \sigma \in [0, D_1], \tag{6.29}$$
$$u_2(\sigma, t) = U_2(t + \sigma - D_2), \quad \sigma \in [0, D_2]. \tag{6.30}$$

Then the invertible backstepping transformations (6.12)–(6.13) have the concrete forms as follows:

- $i = 1, \sigma \in [0, D_1]$

$$w_1(\sigma, t) = u_1(\sigma, t) - k_1 e^{A\sigma} X(t) - \int_0^\sigma k_1 e^{A(\sigma - \delta)} b_1 u_1(\delta, t) d\delta$$

$$- \int_0^\sigma k_1 e^{A(\sigma - \delta)} b_2 u_2(\delta, t) d\delta, \tag{6.31}$$

$$u_1(\sigma, t) = w_1(\sigma, t) + k_1 e^{A_2\sigma} X(t) + \int_0^\sigma k_1 e^{A_2(\sigma - \delta)} b_1 w_1(\delta, t) d\delta$$

$$+ \int_0^\sigma k_1 e^{A_2(\sigma - \delta)} b_2 w_2(\delta, t) d\delta. \tag{6.32}$$

- $i = 2, \sigma \in [0, D_1]$

$$w_2(\sigma, t) = u_2(\sigma, t) - k_2 e^{A\sigma} X(t) - \int_0^\sigma k_2 e^{A(\sigma - \delta)} b_1 u_1(\delta, t) d\delta$$

$$- \int_0^\sigma k_2 e^{A(\sigma - \delta)} b_2 u_2(\delta, t) d\delta, \tag{6.33}$$

$$u_2(\sigma, t) = w_2(\sigma, t) + k_2 e^{A_2\sigma} X(t) + \int_0^\sigma k_2 e^{A_2(\sigma - \delta)} b_1 w_1(\delta, t) d\delta$$

$$+ \int_0^\sigma k_2 e^{A_2(\sigma - \delta)} b_2 w_2(\delta, t) d\delta. \tag{6.34}$$

- $i = 2, \sigma \in [D_1, D_2]$

$$w_2(\sigma, t) = u_2(\sigma, t) - k_2 e^{A_1(\sigma - D_1)} e^{AD_1} X(t)$$

$$- \int_0^{D_1} k_2 e^{A_1(\sigma - D_1)} e^{A(D_1 - \delta)} b_1 u_1(\delta, t) d\delta$$

$$- \int_0^{D_1} k_2 e^{A_1(\sigma - D_1)} e^{A(D_1 - \delta)} b_2 u_2(\delta, t) d\delta$$

$$- \int_{D_1}^\sigma k_2 e^{A_1(\sigma - \delta)} b_2 u_2(\delta, t) d\delta, \tag{6.35}$$

$$u_2(\sigma, t) = w_2(\sigma, t) + k_2 e^{A_2\sigma} X(t) + \int_0^{D_1} k_2 e^{A_2(\sigma - \delta)} b_1 w_1(\delta, t) d\delta$$

$$+ \int_0^\sigma k_2 e^{A_2(\sigma - \delta)} b_2 w_2(\delta, t) d\delta, \tag{6.36}$$

where $A_2 = A_1 + b_2 k_2$, $A_1 = A + b_1 k_1$.

By substituting $\sigma = D_1$ into (6.31) and $\sigma = D_2$ into (6.35) to meet boundary conditions $w_1(D_1, t) = 0$ and $w_2(D_2, t) = 0$, the boundary control laws have the concrete forms

$$U_1(t) = u_1(D_1, t)$$

$$= k_1 e^{AD_1} X(t) + \int_0^{D_1} k_1 e^{A(D_1-\delta)} b_1 u_1(\delta, t) d\delta$$

$$+ \int_0^{D_1} k_1 e^{A(D_1-\delta)} b_2 u_2(\delta, t) d\delta, \tag{6.37}$$

$$U_2(t) = u_2(D_2, t)$$

$$= k_2 e^{A_1(D_2-D_1)} e^{AD_1} X(t) + \int_0^{D_1} k_2 e^{A_1(D_2-D_1)} e^{A(D_1-\delta)} b_1 u_1(\delta, t) d\delta$$

$$+ \int_0^{D_1} k_2 e^{A_1(D_2-D_1)} e^{A(D_1-\delta)} b_2 u_2(\delta, t) d\delta + \int_{D_1}^{D_2} k_2 e^{A_1(D_2-\delta)} b_2 u_2(\delta, t) d\delta.$$
$$\tag{6.38}$$

If we compare the exact predictor feedback laws (6.23) and (6.27) in standard ODE delay notation with the boundary control laws (6.37)–(6.38) in transport PDE notation, it is evident they are equivalent by the variable changes below:

$$u_1(\sigma, t) = U_1(t + \sigma - D_1) = U_1(\theta), \tag{6.39}$$

$$u_1(\delta, t) = U_1(t + \delta - D_1) = U_1(\theta), \tag{6.40}$$

$$u_2(\sigma, t) = U_2(t + \sigma - D_2) = U_2(\theta), \tag{6.41}$$

$$u_2(\delta, t) = U_2(t + \delta - D_2) = U_2(\theta). \tag{6.42}$$

As a result, the PDE-based boundary control recovers the classic predictor feedback law in standard delay notation.

For the above predictor feedback in PDE notation, which is shown in figure 6.3, the following theorem holds.

Theorem 6.3. *The closed-loop system consisting of the ODE-PDE cascade (6.4)–(6.6) representing the multi-input LTI system with distinct input delays (6.1) and the control law (6.19) is exponentially stable at the origin in the sense of the norm*

$$|X(t)| + \left(\int_0^{D_1} u_1(\sigma, t)^2 d\sigma \right)^{\frac{1}{2}} \cdots + \left(\int_0^{D_m} u_m(\sigma, t)^2 d\sigma \right)^{\frac{1}{2}}. \tag{6.43}$$

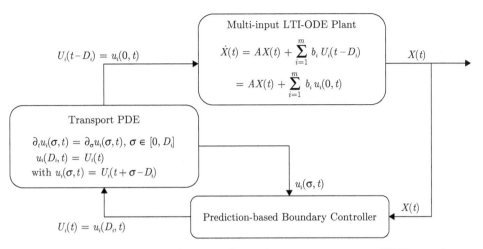

$U_i(t-D_i) = u_i(0,t)$

Multi-input LTI-ODE Plant

$$\dot{X}(t) = AX(t) + \sum_{i=1}^{m} b_i\, U_i(t-D_i)$$

$$= AX(t) + \sum_{i=1}^{m} b_i\, u_i(0,t)$$

$X(t)$

Transport PDE

$\partial_t u_i(\sigma,t) = \partial_\sigma u_i(\sigma,t),\ \sigma \in [0, D_i]$
$u_i(D_i, t) = U_i(t)$
with $u_i(\sigma,t) = U_i(t+\sigma-D_i)$

$u_i(\sigma, t)$

$X(t)$

Prediction-based Boundary Controller

$U_i(t) = u_i(D_i, t)$

Figure 6.3. Exact predictor feedback for multi-input systems in PDE notation

Proof. Consider a Lyapunov functional:

$$V(t) = X^T(t)PX(t) + \frac{a}{2}\sum_{i=1}^{m}\int_0^{D_i}(1+\sigma)w_i(\sigma,t)^2 d\sigma, \qquad (6.44)$$

where P is the solution of the equation (6.2) and $a > 0$ is an analysis coefficient to be determined later. Note that $V(t)$ in (6.44) is a multi-input counterpart of the Lyapunov functional of the single-input case in chapter 2. In addition, there exist positive constants $\underline{c}$ and $\bar{c}$ such that

$$\underline{c}\left\|\big(X(t), w(\cdot, t)\big)\right\|^2 \le V(t) \le \bar{c}\left\|\big(X(t), w(\cdot, t)\big)\right\|^2. \qquad (6.45)$$

Taking the time derivative along the target system (6.8)–(6.10), we have

$$\dot{V}(t) = -X^T(t)QX(t) + 2\sum_{i=1}^{m}X^T(t)Pb_i w_i(0,t)$$

$$- \sum_{i=1}^{m}\frac{a}{2}\left(w_i(0,t)^2 + \int_0^{D_i}w_i(\sigma,t)^2 d\sigma\right), \qquad (6.46)$$

$$\le -X^T(t)QX(t) + \frac{2}{a}X^T(t)PBB^T PX(t) - \sum_{i=1}^{m}\frac{a}{2}\int_0^{D_i}w_i(\sigma,t)^2 d\sigma. \qquad (6.47)$$

Setting $a = \frac{4\lambda_{\max}(PBB^T P)}{\lambda_{\min}(Q)}$, where $\lambda_{\max}(\cdot)$ and $\lambda_{\min}(\cdot)$ stand for the maximum and minimum eigenvalues of the argument, we get

$$\dot{V}(t) \leq -\frac{\lambda_{\min}(Q)}{\lambda_{\max}(P)} X(t)^T P X(t) - \frac{a}{2(1+D_m)} \sum_{i=1}^{m} \int_0^{D_m} (1+\sigma) w_i(\sigma, t)^2 d\sigma$$

$$\leq -2\omega V(t), \tag{6.48}$$

where $\omega = \min \left\{ \frac{\lambda_{\min}(Q)}{4\lambda_{\max}(P)}, \frac{1}{2(1+D_m)} \right\}$. Thus we get

$$V(t) \leq e^{-2\omega t} V(0) \leq e^{-\omega t} V(0). \tag{6.49}$$

Combining (6.45) with (6.49) yields

$$\|(X(t), w(\cdot, t))\| \leq M_\omega e^{-\omega t} \|(X(0), w(\cdot, 0))\|, \quad t \geq 0, \tag{6.50}$$

where $M_\omega = \left(\frac{\bar{c}}{\underline{c}}\right)^{\frac{1}{2}}$. Utilizing invertible backstepping transformations (6.12)–(6.13), we get the exponential stability (6.43) in terms of the original states $(X(t), u(\cdot, t))$ from (6.50). Hence, we have shown that under the backstepping transformation (6.12)–(6.13) and the PDE control scheme (6.19), the closed-loop target ODE-PDE cascade (6.8)–(6.10) is stable. Thus, stabilization of the original system (6.1) and (6.4)–(6.6) is achieved.   □

6.2 Unity-Interval Rescaling for Multi-Input LTI Systems with Input Delays

To develop a certainty-equivalence-based adaptive control for uncertain systems, we first need to rescale the ODE-PDE cascade (6.4)–(6.6) and (6.8)–(6.10) on the spatial interval $\sigma \in [0, D_i]$ into the rescaled systems on the unity interval $[0, 1]$. The unity-interval rescaling is helpful for the Lyapunov-based design. We still assume that $X(t)$ and $u_i(\cdot, t)$ are measurable and D_i, A, and b_i are known.

By the change of variable $\sigma = D_i x$, the transport PDE (6.3) representing the input with delay is turned into the following form:

$$u_i(x, t) = U_i(t + D_i(x - 1)), x \in [0, 1]. \tag{6.51}$$

As a consequence, the initial system (6.4)–(6.6) is converted to the following rescaled unity-interval system:

$$\dot{X}(t) = AX(t) + \sum_{i=1}^{m} b_i u_i(0, t), \tag{6.52}$$

$$D_i \partial_t u_i(x, t) = \partial_x u_i(x, t), \quad x \in [0, 1], \tag{6.53}$$

$$u_i(1, t) = U_i(t), \quad i \in \{1, 2, \cdots, m\}, \tag{6.54}$$

while the closed-loop system (6.8)–(6.10) is transformed into the following rescaled unity-interval system:

$$\dot{X}(t) = (A + BK)X(t) + \sum_{i=1}^{m} b_i w_i(0, t), \tag{6.55}$$

$$D_i \partial_t w_i(x, t) = \partial_x w_i(x, t), \quad x \in [0, 1], \tag{6.56}$$

$$w_i(1, t) = 0, \quad i \in \{1, 2, \cdots, m\}. \tag{6.57}$$

To achieve the conversion of (6.52)–(6.54) into (6.55)–(6.57), according to the method of substitution $\sigma = D_i x$ and $\delta = D_j y$, the backstepping transformations $\big(X(t), u_i(\sigma, t)\big) \to \big(X(t), w_i(\sigma, t)\big)$ in (6.12)–(6.13) are replaced by the rescaled unity-interval backstpping transformations $\big(X(t), u_i(x, t)\big) \to \big(X(t), w_i(x, t)\big)$ as follows:

$$w_i(x, t) = u_i(x, t) - k_i \Phi(D_i x, D_l^v, 0) X(t)$$
$$- \sum_{j=1}^{m} D_j \int_0^{\phi_j(D_i x, D_j)} k_i \Phi(D_i x, D_l^v, D_j y) b_j u_j(y, t) dy, \tag{6.58}$$

$$u_i(x, t) = w_i(x, t) + k_i e^{A_m D_i x} X(t),$$
$$+ \sum_{j=1}^{m} D_j \int_0^{\phi_j(D_i x, D_j)} k_i e^{A_m(D_i x - D_j y)} b_j w_j(y, t) dy, \tag{6.59}$$

where for $j \in \{1, 2, \cdots, m\}$,

$$\phi_j(D_i x, D_j) = \begin{cases} \dfrac{D_i}{D_j} x, & x \in \left[0, \dfrac{D_j}{D_i}\right] \\ 1, & x \in \left[\dfrac{D_j}{D_i}, \dfrac{D_m}{D_i}\right] \end{cases} \tag{6.60}$$

and $D_l^v = (D_{v-1}, D_{v-2}, \cdots, D_{l+1}, D_l)$ for $\frac{D_{v-1}}{D_j} \leq y \leq \frac{D_i}{D_j} x \leq \frac{D_v}{D_j}$, $i, j \in \{1, 2, \cdots, m\}$, $v \in \{1, 2, \cdots, i\}$,

$$\Phi(D_i x, D_l^v, D_j y) = e^{A_{v-1}(D_i x - D_j y)}, \tag{6.61}$$

for $\frac{D_{v-1}}{D_i} \leq x \leq \frac{D_v}{D_i}$, $\frac{D_{l-1}}{D_j} \leq y \leq \frac{D_l}{D_j}$, $i, j \in \{1, 2, \cdots, m\}$, $v, l \in \{1, 2, \cdots, i\}$, $l < v$,

$$\Phi(D_i x, D_l^v, D_j y) = e^{A_{v-1}(D_i x - D_{v-1})} e^{A_{v-2}(D_{v-1} - D_{v-2})}$$
$$\cdots e^{A_l(D_{l+1} - D_l)} e^{A_{l-1}(D_l - D_j y)}, \tag{6.62}$$

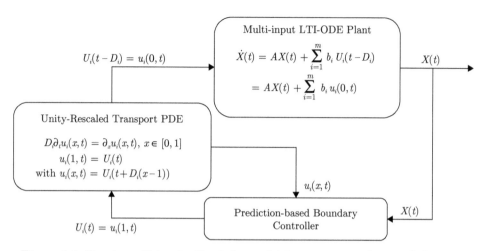

Figure 6.4. Exact predictor feedback for multi-input systems in rescaled unity-interval PDE notation

and specifically for (6.62), if $l = v - 1$,

$$\Phi(D_i x, D_l^v, D_j y) = e^{A_{v-1}(D_i x - D_{v-1})} e^{A_{v-2}(D_{v-1} - D_j y)}, \qquad (6.63)$$

where D_0 and A_i for $i = 0, 1, \cdots, m$ are defined the same as those in (6.18) such that

$$D_0 = 0, \quad \begin{cases} A_0 = A, \\ A_i = A_{i-1} + b_i k_i, i \in \{1, 2, \cdots, m\} \\ A_m = A + BK = A + \sum_{i=1}^{m} b_i k_i. \end{cases} \qquad (6.64)$$

Subsequently, If we substitute $x = 1$ into (6.58) to meet (6.57), the predictor-based PDE boundary controller for $i = 1, 2, \cdots, m$ in (6.19) is replaced by the rescaled unity-interval expression such that

$$U_i(t) = u_i(1, t) = k_i \Phi(D_i, D_l^v, 0) X(t)$$

$$+ \sum_{j=1}^{m} D_j \int_0^{\phi_j(D_i, D_j)} k_i \Phi(D_i, D_l^v, D_j y) b_j u_j(y, t) dy. \qquad (6.65)$$

For the above predictor feedback in rescaled unity-interval PDE notation, which is shown in figure 6.4, the following result holds.

Theorem 6.4. *The closed-loop system consisting of the ODE-PDE cascade (6.52)–(6.54) representing the multi-input LTI systems with distinct input delays (6.1) and the control law (6.65) is exponentially stable at the origin in the sense of the norm*

$$|X(t)| + \left(\int_0^1 u_1(x,t)^2 dx \right)^{\frac{1}{2}} \cdots + \left(\int_0^1 u_m(x,t)^2 dx \right)^{\frac{1}{2}}. \qquad (6.66)$$

Proof. The proof is almost the same as for the nonrescaled unity-interval case and it is omitted here.   □

If we compare (6.3)–(6.19) with (6.51)–(6.65), for the system (6.1), it is obvious that the unity-interval rescaling from the spatial interval $[0, D_i]$ to the spatial interval $[0, 1]$ is accomplished by the change of variables $\sigma = D_i x$, $\sigma \in [0, D_i]$, $x \in [0, 1]$, and $\delta = D_j y$, $\delta \in [0, D_j]$, $y \in [0, 1]$. More importantly, it is evident that formulas (6.3)–(6.19) and (6.51)–(6.65) are equivalent, and both of them achieve stabilization of (6.1).

Chapter Seven

Full-State Feedback of Uncertain Multi-Input Systems

IN CHAPTER 6, under the assumption of a perfect prior knowledge of the system, a PDE-based framework of exact predictor feedback for multi-input LTI systems with distinct discrete input delays was proposed. On the basis of that, in this chapter, we investigate adaptive control for uncertain multi-input LTI systems with distinct discrete actuator delays. In parallel with the single-input case in chapter 3, four types of basic uncertainties come with multi-input LTI time-delay systems:

- the finite-dimensional ODE plant state vector $X(t)$ is unavailable for measurement,
- the infinite-dimensional PDE actuator state $u_i(x,t)$ for $i = 1, \cdots, m$, which represents the distinct delayed input channel $U_i(\eta)$ over the time window $\eta \in [t - D_i, t]$, is unavailable for measurement,
- the ODE system matrix $A(\theta)$ and input matrices $b_i(\theta)$ for $i = 1, \cdots, m$ contain the unknown parameter vector θ representing both plant parameters and control coefficients,
- the length value of the arbitrarily large actuator delay D_i in distinct input channel for $i = 1, \cdots, m$ is unknown.

Different combinations of the four uncertainties above result in different design difficulties. For example, when the full-state measurement of the transport PDE state is available, the global stabilization is acquired, whereas when the actuator state is not measurable and the delay value is unknown at the same time, the problem is not solvable globally, since the problem is not linearly parameterized.

To clearly describe the organization of part II of this book, in parallel with table 3.1 for the single-input case in section 3.1, different collections of uncertainties for the multi-input case are summarized in table 7.1. The interested readers and practitioners can simply make their own selections to address a vast class of relevant problems.

When some of the four variables above are unknown or unmeasured, the basic idea of certainty-equivalence-based adaptive control is to use an estimator (a parameter estimator or a state estimator) to replace the unknown variables in the PDE-based framework in chapter 6, and carefully select their adaptive update laws based on Lyapunov-based analysis, as shown in figure 7.1.

Table 7.1. Uncertainty Collections of Multi-Input LTI Systems with Input Delays

Organization of Part II	Multi-Input Delays D_i	Parameter θ	ODE State $X(t)$	PDE States $u_i(x,t)$
Section 6	known	known	known	known
Section 7.2	*unknown*	known	known	known
Section 7.3	*unknown*	known	*unknown*	known
Section 7.4	*unknown*	*unknown*	known	known
Chapter 8	*unknown*	known	known	*unknown*
Section 9.2	*unknown*	known	*unknown*	*unknown*
Section 9.3	*unknown*	*unknown*	known	*unknown*

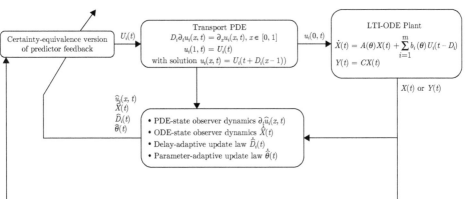

Figure 7.1. Adaptive control framework of uncertain multi-input LTI systems with distinct input delays

7.1 Problem Statement

Consider a general class of LTI systems with distinct discrete multi-input time delay as follows:

$$\dot{X}(t) = A(\theta)X(t) + \sum_{i=1}^{m} b_i(\theta)U_i(t - D_i), \qquad (7.1)$$

$$Y(t) = CX(t), \qquad (7.2)$$

where $X(t) \in \mathbb{R}^n$ is the state vector of ODE, $Y(t) \in \mathbb{R}^q$ is the output vector, and $U_i(t) \in \mathbb{R}$ is the ith scalar control input with a constant delay $D_i \in \mathbb{R}^+$ for $i = 1, 2, \cdots, m$. Without loss of generality, the length sequence satisfies

$0 < D_1 \leq D_2 \leq \cdots \leq D_m$. The matrices $A(\theta) \in \mathbb{R}^{n \times n}$, $b_i(\theta) \in \mathbb{R}^n$, and $C \in \mathbb{R}^{q \times n}$ are the system matrix, the input matrix, and the output matrix, respectively. The constant parameter vector $\theta \in \mathbb{R}^p$ includes both plant parameters and control coefficients, and can be linearly parameterized such that

$$A(\theta) = A_0 + \sum_{h=1}^{p} \theta_h A_h, \quad b_i(\theta) = b_{i0} + \sum_{h=1}^{p} \theta_h b_{ih}. \tag{7.3}$$

Please note that the above linearly parameterized condition (7.3) is not restrictive. We give a concrete example to explain this.

Example 7.1. Consider the following three-dimensional LTI systems with two-input delays:

$$\dot{X}(t) = A(\theta)X(t) + b_1(\theta)U_1(t - D_1) + b_2(\theta)U_2(t - D_2), \tag{7.4}$$

which has a structure such that

$$\begin{bmatrix} \dot{X}_1(t) \\ \dot{X}_2(t) \\ \dot{X}_3(t) \end{bmatrix} = \begin{bmatrix} \alpha_{11} & \alpha_{12} & \alpha_{13} \\ \alpha_{21} & \alpha_{22} & \alpha_{23} \\ \alpha_{31} & \alpha_{32} & \alpha_{33} \end{bmatrix} \begin{bmatrix} X_1(t) \\ X_2(t) \\ X_3(t) \end{bmatrix} + \begin{bmatrix} \beta_{11} \\ \beta_{21} \\ \beta_{31} \end{bmatrix} U_1(t - D_1) + \begin{bmatrix} \beta_{12} \\ \beta_{22} \\ \beta_{32} \end{bmatrix} U_2(t - D_2),$$
$$\tag{7.5}$$

where $\alpha_{ij} \in \mathbb{R}$ and $\beta_{ij} \in \mathbb{R}$ are elements of the system matrix $A(\theta)$ and the input matrices $b_1(\theta)$ and $b_2(\theta)$, respectively. The p-dimensional parameter vector is randomly selected as

$$\theta = \begin{bmatrix} \theta_1 \\ \theta_2 \\ \theta_3 \\ \theta_4 \end{bmatrix} = \begin{bmatrix} \alpha_{12} \\ \alpha_{23} \\ \beta_{21} \\ \beta_{32} \end{bmatrix}. \tag{7.6}$$

Then $A(\theta)$, $b_1(\theta)$, and $b_2(\theta)$ are linearly parameterized with respect to four above parameters as follows:

$$\begin{bmatrix} \alpha_{11} & \alpha_{12} & \alpha_{13} \\ \alpha_{21} & \alpha_{22} & \alpha_{23} \\ \alpha_{31} & \alpha_{32} & \alpha_{33} \end{bmatrix} = \begin{bmatrix} \alpha_{11} & 0 & \alpha_{13} \\ \alpha_{21} & \alpha_{22} & 0 \\ \alpha_{31} & \alpha_{32} & \alpha_{33} \end{bmatrix} + \alpha_{12} \begin{bmatrix} 0 & 1 & 0 \\ 0 & 0 & 0 \\ 0 & 0 & 0 \end{bmatrix} + \alpha_{23} \begin{bmatrix} 0 & 0 & 0 \\ 0 & 0 & 1 \\ 0 & 0 & 0 \end{bmatrix}$$
$$+ \beta_{21} \begin{bmatrix} 0 & 0 & 0 \\ 0 & 0 & 0 \\ 0 & 0 & 0 \end{bmatrix} + \beta_{32} \begin{bmatrix} 0 & 0 & 0 \\ 0 & 0 & 0 \\ 0 & 0 & 0 \end{bmatrix}, \tag{7.7}$$

$$\begin{bmatrix} \beta_{11} \\ \beta_{21} \\ \beta_{31} \end{bmatrix} = \begin{bmatrix} \beta_{11} \\ 0 \\ \beta_{31} \end{bmatrix} + \alpha_{12} \begin{bmatrix} 0 \\ 0 \\ 0 \end{bmatrix} + \alpha_{23} \begin{bmatrix} 0 \\ 0 \\ 0 \end{bmatrix} + \beta_{21} \begin{bmatrix} 0 \\ 1 \\ 0 \end{bmatrix} + \beta_{32} \begin{bmatrix} 0 \\ 0 \\ 0 \end{bmatrix}, \tag{7.8}$$

$$\begin{bmatrix} \beta_{12} \\ \beta_{22} \\ \beta_{32} \end{bmatrix} = \begin{bmatrix} \beta_{12} \\ \beta_{22} \\ 0 \end{bmatrix} + \alpha_{12} \begin{bmatrix} 0 \\ 0 \\ 0 \end{bmatrix} + \alpha_{23} \begin{bmatrix} 0 \\ 0 \\ 0 \end{bmatrix} + \beta_{21} \begin{bmatrix} 0 \\ 0 \\ 0 \end{bmatrix} + \beta_{32} \begin{bmatrix} 0 \\ 0 \\ 1 \end{bmatrix}. \tag{7.9}$$

As a result, the condition (7.3) is satisfied. This three-dimensional two-input example can be straightforwardly extended to the general n-dimensional LTI systems with m-input delays.

Our objective is to develop a control scheme to stabilize the potentially unstable LTI system (7.1). To achieve the stabilization target, the following assumption about (7.1) is supposed to hold throughout the whole chapter:

Assumption 7.2. *The pair* $\big(A(\theta), B(\theta)\big)$ *is stabilizable with matrix* $B(\theta) = \big(b_1(\theta), b_2(\theta), \cdots, b_m(\theta)\big) \in \mathbb{R}^{n \times m}$. *In other words, there exists a matrix* $K(\theta) = \big(k_1^T(\theta), k_2^T(\theta), \cdots, k_m^T(\theta)\big)^T \in \mathbb{R}^{m \times n}$ *in which* $k_i(\theta) \in \mathbb{R}^{1 \times n}$ *for* $i = 1, 2,$ $\cdots, m$ *such that* $A(\theta) + B(\theta)K(\theta) = A(\theta) + \sum_{i=1}^{m} b_i(\theta)k_i(\theta)$ *is Hurwitz, namely,*

$$(A + BK)^T(\theta)P(\theta) + P(\theta)(A + BK)(\theta) = -Q(\theta) \tag{7.10}$$

with $P(\theta) = P(\theta)^T > 0$ *and* $Q(\theta) = Q(\theta)^T > 0$.

Assumption 7.2 is not restrictive, as illustrated by the following example in the general strict-feedback form.

Example 7.3. Consider a potentially unstable linear time-delay system with two inputs as follows:

$$\dot{X}_1(t) = \theta X_1(t) + X_2(t), \tag{7.11}$$

$$\dot{X}_2(t) = U_1(t - D_1) + U_2(t - D_2), \tag{7.12}$$

$$Y(t) = X_1(t). \tag{7.13}$$

If we contrast (7.11)–(7.13) with (7.1)–(7.2), it is obvious that (7.3) has the concrete form below:

$$A(\theta) = A_0 + \theta A_1 = \begin{bmatrix} 0 & 1 \\ 0 & 0 \end{bmatrix} + \theta \begin{bmatrix} 1 & 0 \\ 0 & 0 \end{bmatrix} = \begin{bmatrix} \theta & 1 \\ 0 & 0 \end{bmatrix}, \tag{7.14}$$

$$b_1(\theta) = b_{10} = \begin{bmatrix} 0 \\ 1 \end{bmatrix}, \quad b_2(\theta) = b_{20} = \begin{bmatrix} 0 \\ 1 \end{bmatrix}. \tag{7.15}$$

If gain matrices $k_1(\theta)$ and $k_2(\theta)$ are designed as

$$k_1(\theta) = [-1 - (\theta + 1)^2, 0], \quad k_2(\theta) = [0, -\theta - 2], \tag{7.16}$$

then it is not hard to calculate that there exist positive definite and symmetric matrices $P(\theta)$ and $Q(\theta)$ having the forms

$$P(\theta) = \frac{1}{2}Q(\theta) = \begin{bmatrix} 1+(1+\theta)^2 & 1+\theta \\ 1+\theta & 1 \end{bmatrix} \tag{7.17}$$

to let (7.10) hold.

To address the delayed input, a transport PDE is introduced as

$$u_i(x,t) = U_i(t + D_i(x-1)), x \in [0,1]. \tag{7.18}$$

As a consequence, the initial system (7.1) is converted to the following rescaled unity-interval ODE-PDE cascade:

$$\dot{X}(t) = A(\theta)X(t) + \sum_{i=1}^{m} b_i(\theta)u_i(0,t), \tag{7.19}$$

$$Y(t) = CX(t), \tag{7.20}$$

$$D_i\partial_t u_i(x,t) = \partial_x u_i(x,t), \quad x \in [0,1], \tag{7.21}$$

$$u_i(1,t) = U_i(t), \quad i \in \{1,2,\cdots,m\}. \tag{7.22}$$

Based on the state-space model (7.19)–(7.22), four types of uncertainties that need to be addressed in adaptive control methodology are outlined below:

- unknown and distinct multi-input delays D_i,
- unknown plant parameter vector θ,
- unmeasurable finite-dimensional ODE plant state $X(t)$,
- unmeasurable infinite-dimensional PDE actuator state $u_i(x,t)$.

As stated in sections 6.1 and 6.2 of this book, where the four variables above are assumed to be known and available, for $i = 1, 2, \cdots, m$, the idea of exact predictor feedback is to realize

$$U_i(t-D_i) = k_i(\theta)X(t), \text{equivalently}, U_i(t) = k_i(\theta)X(t+D_i), \tag{7.23}$$

where $k_i(\theta)$ in Assumption 7.2 renders $(A+BK)(\theta)$ Hurwitz. Furthermore, chapter 6 provides a PDE-based exact predictor feedback framework to stabilize the uncertainty-free LTI system (7.1) by assuming multi-input delay D_i and parameter vector θ are completely known, as well as the ODE plant state $X(t)$ and PDE actuator state $u_i(x,t)$ being fully measured.

Adaptive control problems for single-input delay $m=1$ or identical multi-input delay $D_1 = D_2 = \cdots = D_m$ have been presented in [27–32], [154–159], and part I of this book. However, adaptive control for distinct multi-input delays are nontrivial and challenging problems. When some of the above four variables are unknown or unmeasured, the basic idea of certainty-equivalence-based adaptive control is to use an estimator (a parameter estimator or a state estimator) to replace the unknown variables in the PDE-based framework in chapter 6, and carefully select their adaptive update laws based on Lyapunov-based analysis, as shown in figure 7.1.

As shown in table 7.1, in later sections of chapter 7, we deal with adaptive global stabilization for delay uncertainty (D_i) and a couple of uncertainty combinations $(D_i, X(t))$ and (D_i, θ), respectively.

7.2   Global Stabilization under Unknown Delays

Starting with this section, we deal with adaptive global stabilization for a variety of the aforementioned uncertainties like D_i, θ, and $X(t)$. Furthermore, throughout chapter 7, we assume that the actuator state $u_i(x,t)$ of (7.18) driven by PDE dynamics (7.21)–(7.22) is measurable. Due to the available measurement of the infinite-dimensional PDE actuator state $u_i(x,t)$, similar to the single-input case of part I, the global stabilization is achievable by making use of the normalization-based update law and the log function–based Lyapunov candidate. Firstly, as shown in the second row of table 7.1, we focus on the delay uncertainty in this section.

As $A(\theta)$ and $b_i(\theta)$ of (7.1) and (7.19) are assumed to be known in this section, for the sake of brevity, we denote

$$A = A(\theta), \quad b_i = b_i(\theta). \tag{7.24}$$

Thus (7.19)–(7.22) is rewritten as

$$\dot{X}(t) = AX(t) + \sum_{i=1}^{m} b_i u_i(0,t), \tag{7.25}$$

$$D_i \partial_t u_i(x,t) = \partial_x u_i(x,t), \quad x \in [0,1], \tag{7.26}$$

$$u_i(1,t) = U_i(t), \quad i \in \{1, 2, \cdots, m\}. \tag{7.27}$$

Remark 7.4. Some readers may wonder whether the assumption that the actuator state $u_i(x,t)$ is measurable but the delay D_i is unknown is reasonable. They may argue that if the infinite-dimensional variable $u_i(x,t)$ can be measured, then the delay can be derived from (7.26) as $D_i = \frac{\partial_x u_i(x,t)}{\partial_t u_i(x,t)}$; thus the assumption that the distributed input $u_i(x,t)$ is measurable while the delay D_i is unknown may be a problem that does not make sense. This seems true mathematically. However, in practice, measurements are noisy spatially and temporally. As the previous calculus involves time and space derivatives of the measurement of the PDE state, this methodology should not give an accurate delay estimate. Thus adaptive control by PDE full-state feedback under unknown delay can still be of importance. Furthermore, some plants in practice do fit in this kind of structure, which possesses a certain setup to measure the distributed input despite the delay being unknown. They correspond to a big family of process control plants in chemical and biological engineering where delay arises from a physical transport process, rather than computational or communication-based delay, and in which the distributed input is measured but not the speed of propagation. In fact, this case of unknown delay but measurable actuator state for single-input or identical multi-input systems has been considered in [149], chapter 7 and 9 of [80], and also [27, 29]. Please see the fifth paragraph of its introduction section in [29] and the realistic examples in section III of [29] and section 5 of [149].

To achieve stabilization under unknown delays, another assumption is made.

Assumption 7.5. *Distinct and unknown multi-input delays satisfy*

$$0 < \underline{D}_1 \leq D_1 \leq \overline{D}_{1,2} \leq D_2 \leq \overline{D}_{2,3} \leq D_3 \leq \overline{D}_{3,4} \leq \cdots$$
$$\leq \overline{D}_{i-1,i} \leq D_i \leq \overline{D}_{i,i+1} \leq \cdots \leq \overline{D}_{m-1,m} \leq D_m \leq \overline{D}_m, \qquad (7.28)$$

where $\overline{D}_{i-1,i} > 0$, $\overline{D}_{i,i+1} > 0$ *for* $i = 1, 2, \cdots, m$, *with* $\overline{D}_{0,1} = \underline{D}_1$, $\overline{D}_{m,m+1} = \overline{D}_m$ *are known constants, representing lower and upper bounds of* D_i, *respectively.*

Remark 7.6. In comparison with the requirement of the exact value of the delay in every control channel, though Assumption 7.5 requires some a priori knowledge regarding the boundedness of unknown delays that will be used in the delay update law, and the length sequence of unknown delays that is important to develop a predictor feedback law, it is unrestrictive and still allows a large uncertainty among the interval $[\underline{D}_{i-1,i}, \overline{D}_{i,i+1}]$. On the other hand, some readers may suspect that the assumption of a priori knowledge about the length order of delays is strong, especially when the delay value in each input channel is unknown. This concern is valid and a switched technique to overcome this shortcoming is given in the simulation in the next chapter.

The estimate $\hat{D}_i(t)$ with estimation error $\tilde{D}_i(t) = D_i - \hat{D}_i(t)$ is brought in for unknown D_i. Referring to (6.58) and (6.59) in section 6.2 of the book, by using the certainty-equivalence principle (i.e., replacing unknown D_i with its estimate $\hat{D}_i(t)$), we present the invertible backstepping transformation $\big(X(t), u_i(x,t)\big) \leftrightarrow \big(X(t), w_i(x,t)\big)$ in the delay-adaptive scheme as follows:

$$w_i(x,t) = u_i(x,t) - k_i\Phi(\hat{D}_i x, \hat{D}_l^v, 0)X(t)$$
$$- \sum_{j=1}^{m} \hat{D}_j \int_0^{\phi_j(\hat{D}_i x, \hat{D}_j)} k_i\Phi(\hat{D}_i x, \hat{D}_l^v, \hat{D}_j y)b_j u_j(y,t)dy, \qquad (7.29)$$

$$u_i(x,t) = w_i(x,t) + k_i e^{A_m \hat{D}_i x}X(t)$$
$$+ \sum_{j=1}^{m} \hat{D}_j \int_0^{\phi_j(\hat{D}_i x, \hat{D}_j)} k_i e^{A_m(\hat{D}_i x - \hat{D}_j y)}b_j w_j(y,t)dy, \qquad (7.30)$$

where for $j \in \{1, 2, \cdots, m\}$,

$$\phi_j(\hat{D}_i x, \hat{D}_j) = \begin{cases} \frac{\hat{D}_i}{\hat{D}_j} x, & x \in \left[0, \frac{\hat{D}_j}{\hat{D}_i}\right] \\ 1, & x \in \left[\frac{\hat{D}_j}{\hat{D}_i}, \frac{\hat{D}_m}{\hat{D}_i}\right] \end{cases} \tag{7.31}$$

and $\hat{D}_l^v = (\hat{D}_{v-1}, \hat{D}_{v-2}, \cdots, \hat{D}_{l+1}, \hat{D}_l)$, for $\frac{\hat{D}_{v-1}}{\hat{D}_j} \leq y \leq \frac{\hat{D}_i}{\hat{D}_j} x \leq \frac{\hat{D}_v}{\hat{D}_j}$, $i, j \in \{1, 2, \cdots, m\}$, $v \in \{1, 2, \cdots, i\}$,

$$\Phi(\hat{D}_i x, \hat{D}_l^v, \hat{D}_j y) = e^{A_{v-1}(\hat{D}_i x - \hat{D}_j y)}, \tag{7.32}$$

for $\frac{\hat{D}_{v-1}}{\hat{D}_i} \leq x \leq \frac{\hat{D}_v}{\hat{D}_i}$, $\frac{\hat{D}_{l-1}}{\hat{D}_j} \leq y \leq \frac{\hat{D}_l}{\hat{D}_j}$, $i, j \in \{1, 2, \cdots, m\}$, $v, l \in \{1, 2, \cdots, i\}$, $l < v$,

$$\Phi(\hat{D}_i x, \hat{D}_l^v, \hat{D}_j y) = e^{A_{v-1}(\hat{D}_i x - \hat{D}_{v-1})} e^{A_{v-2}(\hat{D}_{v-1} - \hat{D}_{v-2})}$$
$$\cdots e^{A_l(\hat{D}_{l+1} - \hat{D}_l)} e^{A_{l-1}(\hat{D}_l - \hat{D}_j y)}, \tag{7.33}$$

and specifically for (7.33), if $l = v - 1$,

$$\Phi(\hat{D}_i x, \hat{D}_l^v, \hat{D}_j y) = e^{A_{v-1}(\hat{D}_i x - \hat{D}_{v-1})} e^{A_{v-2}(\hat{D}_{v-1} - \hat{D}_j y)} \tag{7.34}$$

with

$$\hat{D}_0 = 0, \quad \begin{cases} A_0 = A, \\ A_i = A_{i-1} + b_i k_i, i \in \{1, 2, \cdots, m\} \\ A_m = A + BK = A + \sum_{i=1}^{m} b_i k_i. \end{cases} \tag{7.35}$$

Subsequently, for $i = 1, 2, \cdots, m$, if we substitute $x = 1$ into (7.29) to meet boundary condition $w_i(1, t) = 0$ of the target system, the prediction-based PDE boundary control in the delay-adaptive feedback scheme has the expression such that

$$U_i(t) = u_i(1, t) = k_i \Phi(\hat{D}_i, \hat{D}_l^v, 0) X(t)$$
$$+ \sum_{j=1}^{m} \hat{D}_j \int_0^{\phi_j(\hat{D}_i, \hat{D}_j)} k_i \Phi(\hat{D}_i, \hat{D}_l^v, \hat{D}_j y) b_j u_j(y, t) dy. \tag{7.36}$$

After a lengthy and intricate calculation, the ODE-PDE cascade (7.25)–(7.27) is converted into the following closed-loop system:

$$\dot{X}(t) = (A + BK) X(t) + \sum_{i=1}^{m} b_i w_i(0, t), \tag{7.37}$$

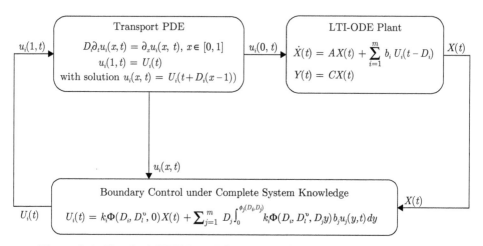

Figure 7.2. Nominal PDE-based framework for uncertainty-free system

$$D_i \partial_t w_i(x,t) = \partial_x w_i(x,t) - \sum_{h=1}^{m} \frac{D_i}{D_h} \tilde{D}_h p_{ih}(x,t) - D_i q_i(x,t,\dot{\hat{D}}),$$

$$x \in [0,1], \tag{7.38}$$

$$w_i(1,t) = 0, \quad i \in \{1,2,\cdots,m\}, \tag{7.39}$$

where $p_{ih}(x,t)$ and $q_i(x,t,\dot{\hat{D}})$ are shown as (7.136) and (7.145) in section 7.6, respectively.

Following (7.38), as shown later in the proof of Theorem 7.7, we design the delay-adaptive update law to cancel the terms containing estimation error $\tilde{D}_i$ in the Lyapunov-based stability analysis. It has the form as follows: for $i = \{1,2,\cdots,m\}$,

$$\dot{\hat{D}}_i(t) = \gamma_D \text{Proj}_{[\underline{D}_{i-1,i}, \overline{D}_{i,i+1}]} \tau_{D_i}(t), \quad \hat{D}_i(0) \in [\underline{D}_{i-1,i}, \overline{D}_{i,i+1}], \tag{7.40}$$

$$\tau_{D_i}(t) = -\frac{\sum_{h=1}^{m} \int_0^1 (1+x) w_h(x,t) p_{hi}(x,t) dx}{1 + X(t)^T P X(t) + g \sum_{i=1}^{m} \int_0^1 (1+x) w_i(x,t)^2 dx}, \tag{7.41}$$

where $\gamma_D > 0$ and $g > 0$ are design coefficients, P is a symmetric and positive matrix defined in Assumption 7.2, and $\text{Proj}_{[\underline{D}_{i-1,i}, \overline{D}_{i,i+1}]}(\cdot)$ is a standard projection operator defined on the interval $[\underline{D}_{i-1,i}, \overline{D}_{i,i+1}]$.

Referring to the nominal PDE-based framework for uncertainty-free systems in chapter 6, which is shown in figure 7.2, we show the adaptive controller in this section in figure 7.3 and the results are summarized as a main theorem below and followed by its proof subsequently.

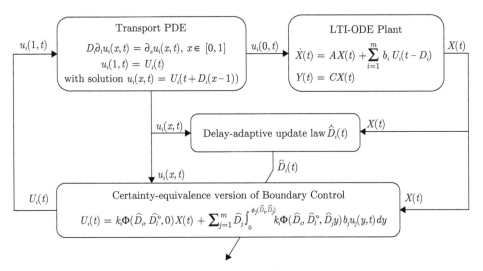

Figure 7.3. Adaptive control under uncertain delays

Theorem 7.7. *Under Assumptions 7.2 and 7.5 and assuming that the actuator states $u_i(x,t)$ in (7.26)–(7.27) are measurable, consider the closed-loop system consisting of the ODE-PDE cascade (7.25)–(7.27), the backstepping transformation (7.29)–(7.30), the PDE boundary control (7.36), and the delay-adaptive update law (7.40)–(7.41). All the signals of the closed-loop system are globally bounded and regulation is achieved, i.e., $\lim_{t\to\infty} X(t) = 0$.*

Proof. We build a Lyapunov candidate encompassing all states of the closed-loop system (7.37)–(7.39):

$$V(t) = \left(\prod_{h=1}^{m} D_h \right) \log V_0(t) + \frac{g}{\gamma_D} \sum_{i=1}^{m} \left(\left(\prod_{\substack{h=1 \\ h\neq i}}^{m} D_h \right) \tilde{D}_i(t)^2 \right), \qquad (7.42)$$

where $V_0(t)$ is the denominator of (7.41) such that

$$V_0(t) = 1 + X(t)^T P X(t) + g \sum_{i=1}^{m} \int_0^1 (1+x) w_i(x,t)^2 dx. \qquad (7.43)$$

Taking the time derivative of $V(t)$ along the dynamics of the target closed-loop system (7.37)–(7.39), we have

$$\dot{V}(t) = \frac{1}{V_0(t)} \left[-\left(\prod_{h=1}^{m} D_h \right) X(t)^T Q X(t) \right.$$
$$\left. + 2 \left(\prod_{h=1}^{m} D_h \right) \sum_{i=1}^{m} X(t)^T P b_i w_i(0,t) \right.$$

$$
+ 2g \left(\prod_{h=1}^{m} D_h \right) \sum_{i=1}^{m} \int_0^1 (1+x) w_i(x,t) \partial_t w_i(x,t) dx \Bigg]
$$

$$
- \frac{2g}{\gamma_D} \sum_{i=1}^{m} \left(\left(\prod_{\substack{h=1 \\ h \neq i}}^{m} D_h \right) \tilde{D}_i \dot{\hat{D}}_i \right), \tag{7.44}
$$

$$
= \frac{1}{V_0(t)} \Bigg[- \left(\prod_{h=1}^{m} D_h \right) X(t)^T Q X(t)
$$

$$
+ 2 \left(\prod_{h=1}^{m} D_h \right) \sum_{i=1}^{m} X(t)^T P b_i w_i(0,t)
$$

$$
+ 2g \sum_{i=1}^{m} \left(\left(\prod_{\substack{h=1 \\ h \neq i}}^{m} D_h \right) \int_0^1 (1+x) w_i(x,t) \partial_x w_i(x,t) dx \right)
$$

$$
- 2g \sum_{i=1}^{m} \left(\left(\prod_{\substack{h=1 \\ h \neq i}}^{m} D_h \right) \int_0^1 (1+x) w_i(x,t) \right.
$$

$$
\left. \times \left(\sum_{h=1}^{m} \frac{D_i}{D_h} \tilde{D}_h p_{ih}(x,t) \right) dx \right)
$$

$$
- 2g \sum_{i=1}^{m} \left(\left(\prod_{h=1}^{m} D_h \right) \int_0^1 (1+x) w_i(x,t) q_i(x,t,\hat{D}) dx \right) \Bigg]
$$

$$
- \frac{2g}{\gamma_D} \sum_{i=1}^{m} \left(\left(\prod_{\substack{h=1 \\ h \neq i}}^{m} D_h \right) \tilde{D}_i \dot{\hat{D}}_i \right). \tag{7.45}
$$

According to (7.29)–(7.30), applying Young's and Cauchy-Schwartz inequalities, for $i \in \{1, 2, \cdots, m\}$, we have the following inequality:

$$
\| u_i(x,t) \|^2 \le N_0 \left(|X(t)|^2 + \sum_{i=1}^{m} \| w_i(x,t) \|^2 \right), \tag{7.46}
$$

where $N_0 > 0$ is a constant independent of initial conditions. In parallel with [127, 128], if we bear the boundary condition (7.39) in mind, according to Young's and Cauchy-Schwartz inequalities, the following inequality and equality hold:

$$
- X(t)^T Q X(t) + 2 X(t)^T P \sum_{i=1}^{m} b_i w_i(0,t)
$$

$$
\le - \frac{\lambda_{\min}(Q)}{2} |X(t)|^2
$$

$$
- \frac{\lambda_{\min}(Q)}{2m} \sum_{i=1}^{m} \left(X(t) - \frac{2m P b_i}{\lambda_{\min}(Q)} w_i(0,t) \right)^2
$$

$$+ \sum_{i=1}^{m} \frac{2m|Pb_i|^2}{\lambda_{\min}(Q)} w_i(0,t)^2$$

$$\leq -\frac{\lambda_{\min}(Q)}{2}|X(t)|^2 + \sum_{i=1}^{m} \frac{2m|Pb_i|^2}{\lambda_{\min}(Q)} w_i(0,t)^2, \tag{7.47}$$

$$2g \sum_{i=1}^{m} \int_{0}^{1} (1+x)w_i(x,t)\partial_x w_i(x,t)dx$$

$$= g \sum_{i=1}^{m} \left(w_i(1,t)^2 - w_i(0,t)^2 - \|w_i(x,t)\|^2 \right)$$

$$= -g \sum_{i=1}^{m} \left(w_i(0,t)^2 + \|w_i(x,t)\|^2 \right). \tag{7.48}$$

Recalling (7.40)–(7.41), (7.145), and (7.46), if we utilize Young's and Cauchy-Schwartz inequalities, as well as inequalities $0 \leq \frac{|x|}{1+x^2} \leq 1$ and $0 \leq \frac{x^2}{1+x^2} \leq 1$, the following inequality holds:

$$-2g \sum_{i=1}^{m} \int_{0}^{1} (1+x)w_i(x,t)q_i(x,t,\dot{\hat{D}})dx$$

$$\leq g\gamma_D N_1 \left(|X(t)|^2 + \sum_{i=1}^{m} \left(\|w_i(x,t)\|^2 + \|w_i(0,t)\|^2 \right) \right), \tag{7.49}$$

where $N_1 > 0$ is a constant independent of initial conditions.

If we substitute (7.47), (7.48), and (7.49) into (7.45) and combine similar terms, (7.45) becomes

$$\dot{V}(t) \leq \frac{\prod_{h=1}^{m} D_h}{V_0(t)} \left[-\frac{\lambda_{\min}(Q)}{2}|X(t)|^2 \right.$$

$$+ \sum_{i=1}^{m} \frac{2m|Pb_i|^2}{\lambda_{\min}(Q)} w_i(0,t)^2$$

$$- g \sum_{i=1}^{m} \frac{1}{D_i} \left(w_i(0,t)^2 + \|w_i(x,t)\|^2 \right)$$

$$+ g\gamma_D N_1 \left(|X(t)|^2 + \sum_{i=1}^{m} \left(\|w_i(x,t)\|^2 + w_i(0,t)^2 \right) \right) \Bigg]$$

$$- \frac{2g}{\gamma_D} \sum_{i=1}^{m} \left(\left(\prod_{\substack{h=1 \\ h\neq i}}^{m} D_h \right) \tilde{D}_i \left(\dot{\hat{D}}_i - \gamma_D \tau_{D_i}(t) \right) \right), \tag{7.50}$$

$$\leq \frac{\prod\limits_{h=1}^{m} D_h}{V_0(t)}\left[-\left(\frac{\lambda_{\min}(Q)}{2}-g\gamma_D N_1\right)|X(t)|^2\right.$$

$$-\left(\frac{g}{\bar{D}_m}-\frac{2m\sup_i|Pb_i|^2}{\lambda_{\min}(Q)}-g\gamma_D N_1\right)\sum_{i=1}^{m} w_i(0,t)^2$$

$$\left.-\left(\frac{g}{\bar{D}_m}-g\gamma_D N_1\right)\sum_{i=1}^{m}\|w_i(x,t)\|^2\right]$$

$$-\frac{2g}{\gamma_D}\sum_{i=1}^{m}\left(\left(\prod_{\substack{h=1\\h\neq i}}^{m} D_h\right)\tilde{D}_i\left(\dot{\hat{D}}_i-\gamma_D\tau_{D_i}(t)\right)\right). \qquad (7.51)$$

Consequently, if we design the coefficients g and γ_D as well as the matrices P and Q to satisfy

$$\frac{\lambda_{\min}(Q)}{2}-g\gamma_D N_1 \geq 0, \qquad (7.52)$$

$$\frac{g}{\bar{D}_m}-\frac{2m\sup_i|Pb_i|^2}{\lambda_{\min}(Q)}-g\gamma_D N_1 \geq 0, \qquad (7.53)$$

$$\frac{g}{\bar{D}_m}-g\gamma_D N_1 \geq 0, \qquad (7.54)$$

then (7.51) becomes

$$\dot{V}(t)\leq-\frac{\prod\limits_{h=1}^{m} D_h}{V_0(t)}N_2\left[|X(t)|^2+\sum_{i=1}^{m}\left(\|w_i(x,t)\|^2+w_i(0,t)^2\right)\right], \qquad (7.55)$$

where

$$N_2=\min\left\{\frac{\lambda_{\min}(Q)}{2}-g\gamma_D N_1,\ \frac{g}{\bar{D}_m}-\frac{2m\sup_i|Pb_i|^2}{\lambda_{\min}(Q)}-g\gamma_D N_1,\ \frac{g}{\bar{D}_m}-g\gamma_D N_1\right\}.$$
$$(7.56)$$

As a result, based on the nonpositive property of the time derivative $\dot{V}(t)$, following a similar procedure in [27], we could guarantee the boundedness of states $X(t)$, $u_i(x,t)$, $U_i(t)$, $w_i(x,t)$, $\hat{D}_i(t)$, $\tilde{D}_i(t)$ for $i\in\{1,2,\cdots,m\}$. Furthermore, by applying Barbalat's lemma, the regulation, i.e., the convergence $\lim_{t\to\infty} X(t) = 0$, is also established. This completes the proof of the theorem. $\qquad\square$

7.3 Global Stabilization under Uncertain Delays and ODE State

In this section, as shown in the third row of table 7.1, we consider a more complicated case of uncertainty combination $(D_i, X(t))$.

Since θ is known, with (7.24) and (7.18) in mind, the system (7.1)–(7.2) is transformed into an ODE-PDE cascade such that

$$\dot{X}(t) = AX(t) + \sum_{i=1}^{m} b_i u_i(0,t), \tag{7.57}$$

$$Y(t) = CX(t), \tag{7.58}$$

$$D_i \partial_t u_i(x,t) = \partial_x u_i(x,t), \quad x \in [0,1], \tag{7.59}$$

$$u_i(1,t) = U_i(t), \quad i \in \{1,2,\cdots,m\}. \tag{7.60}$$

To deal with unmeasurable ODE state $X(t)$, an additional assumption is required as follows:

Assumption 7.8. *The pair (A,C) is detectable; in other words, there exists a matrix $L \in \mathbb{R}^{n \times q}$ to make $A - LC$ Hurwitz such that*

$$(A - LC)^T P_L + P_L(A - LC) = -Q_L \tag{7.61}$$

with $P_L^T = P_L > 0$ and $Q_L^T = Q_L > 0$.

Under Assumption 7.8, the ODE observer is selected as

$$\dot{\hat{X}}(t) = A\hat{X}(t) + \sum_{i=1}^{m} b_i u_i(0,t) + L(Y(t) - C\hat{X}(t)), \tag{7.62}$$

where $\hat{X}(t)$ is an estimate of unmeasurable $X(t)$, while observer error $\tilde{X}(t) = X(t) - \hat{X}(t)$ is governed by

$$\dot{\tilde{X}}(t) = (A - LC)\tilde{X}(t). \tag{7.63}$$

By replacing $X(t)$ with $\hat{X}(t)$ in (7.29)–(7.30), (7.36), we design the ODE-observer-based backstepping transformation and the boundary control law as follows:

$$w_i(x,t) = u_i(x,t) - k_i \Phi(\hat{D}_i x, \hat{D}_l^v, 0)\hat{X}(t)$$
$$- \sum_{j=1}^{m} \hat{D}_j \int_0^{\phi_j(\hat{D}_i x, \hat{D}_j)} k_i \Phi(\hat{D}_i x, \hat{D}_l^v, \hat{D}_j y)b_j u_j(y,t)dy, \tag{7.64}$$

$$u_i(x,t) = w_i(x,t) + k_i e^{A_m \hat{D}_i x}\hat{X}(t)$$
$$+ \sum_{j=1}^{m} \hat{D}_j \int_0^{\phi_j(\hat{D}_i x, \hat{D}_j)} k_i e^{A_m(\hat{D}_i x - \hat{D}_j y)}b_j w_j(y,t)dy, \tag{7.65}$$

where $\phi_j(\hat{D}_i x, \hat{D}_j)$ and $\Phi(\hat{D}_i x, \hat{D}_l^v, \hat{D}_j y)$ are exactly identical with (7.31)–(7.34) which are repeated as follows in order to help readers avoid reading back and forth:

$$\phi_j(\hat{D}_i x, \hat{D}_j) = \begin{cases} \frac{\hat{D}_i}{\hat{D}_j} x, & x \in \left[0, \frac{\hat{D}_j}{\hat{D}_i}\right] \\ 1, & x \in \left[\frac{\hat{D}_j}{\hat{D}_i}, \frac{\hat{D}_m}{\hat{D}_i}\right] \end{cases} \tag{7.66}$$

and $\hat{D}_l^v = (\hat{D}_{v-1}, \hat{D}_{v-2}, \cdots, \hat{D}_{l+1}, \hat{D}_l)$, for $\frac{\hat{D}_{v-1}}{\hat{D}_j} \leq y \leq \frac{\hat{D}_i}{\hat{D}_j} x \leq \frac{\hat{D}_v}{\hat{D}_j}$, $i, j \in \{1, 2, \cdots, m\}$, $v \in \{1, 2, \cdots, i\}$,

$$\Phi(\hat{D}_i x, \hat{D}_l^v, \hat{D}_j y) = e^{A_{v-1}(\hat{D}_i x - \hat{D}_j y)}, \tag{7.67}$$

for $\frac{\hat{D}_{v-1}}{\hat{D}_i} \leq x \leq \frac{\hat{D}_v}{\hat{D}_i}$, $\frac{\hat{D}_{l-1}}{\hat{D}_j} \leq y \leq \frac{\hat{D}_l}{\hat{D}_j}$, $i, j \in \{1, 2, \cdots, m\}$, $v, l \in \{1, 2, \cdots, i\}$, $l < v$,

$$\Phi(\hat{D}_i x, \hat{D}_l^v, \hat{D}_j y) = e^{A_{v-1}(\hat{D}_i x - \hat{D}_{v-1})} e^{A_{v-2}(\hat{D}_{v-1} - \hat{D}_{v-2})}$$
$$\cdots e^{A_l(\hat{D}_{l+1} - \hat{D}_l)} e^{A_{l-1}(\hat{D}_l - \hat{D}_j y)}, \tag{7.68}$$

and specifically for (7.68), if $l = v - 1$,

$$\Phi(\hat{D}_i x, \hat{D}_l^v, \hat{D}_j y) = e^{A_{v-1}(\hat{D}_i x - \hat{D}_{v-1})} e^{A_{v-2}(\hat{D}_{v-1} - \hat{D}_j y)} \tag{7.69}$$

with

$$\hat{D}_0 = 0, \quad \begin{cases} A_0 = A, \\ A_i = A_{i-1} + b_i k_i, i \in \{1, 2, \cdots, m\} \\ A_m = A + BK = A + \sum_{i=1}^m b_i k_i, \end{cases} \tag{7.70}$$

and the boundary control law

$$U_i(t) = u_i(1, t) = k_i \Phi(\hat{D}_i, \hat{D}_l^v, 0) \hat{X}(t)$$
$$+ \sum_{j=1}^m \hat{D}_j \int_0^{\phi_j(\hat{D}_i, \hat{D}_j)} k_i \Phi(\hat{D}_i, \hat{D}_l^v, \hat{D}_j y) b_j u_j(y, t) dy \tag{7.71}$$

is chosen to meet boundary condition $w_i(1, t) = 0$.

Employing (7.62), (7.64)–(7.71), the system (7.57)–(7.60) is transformed into the closed-loop system such that

$$\dot{\hat{X}}(t) = (A + BK)\hat{X}(t) + \sum_{i=1}^m b_i w_i(0, t) + LC\tilde{X}(t), \tag{7.72}$$

$$\dot{\tilde{X}}(t) = (A - LC)\tilde{X}(t), \tag{7.73}$$

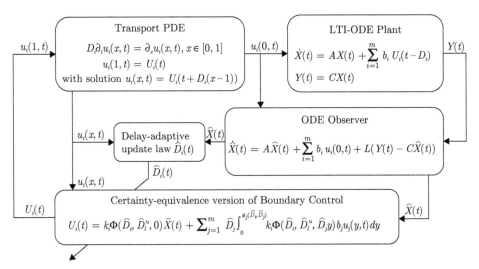

Figure 7.4. Adaptive control under uncertain delays and ODE state

$$D_i\partial_t w_i(x,t) = \partial_x w_i(x,t) - \sum_{h=1}^{m}\frac{D_i}{D_h}\tilde{D}_h\bar{p}_{ih}(x,t)$$

$$- D_i\bar{q}_i(x,t,\dot{\hat{D}}) - D_i k_i \Phi(\hat{D}_i x, \hat{D}_l^v, 0) LC\tilde{X}(t)$$

$$x \in [0,1], \tag{7.74}$$

$$w_i(1,t) = 0, \quad i \in \{1, 2, \cdots, m\}, \tag{7.75}$$

where $\bar{p}_{ih}(x,t)$ and $\bar{q}_i(x,t,\dot{\hat{D}})$ are shown as (7.137) and (7.146) in section 7.6, respectively.

Based on (7.74), the following delay-adaptive update law is selected to cancel the delay estimation error term $\tilde{D}_i$ in the Lyapunov-based analysis:

$$\dot{\hat{D}}_i(t) = \gamma_D \mathrm{Proj}_{[\underline{D}_{i-1,i},\overline{D}_{i,i+1}]}\tau_{D_i}(t), \quad \hat{D}_i(0) \in [\underline{D}_{i-1,i},\overline{D}_{i,i+1}], \tag{7.76}$$

$$\tau_{D_i}(t) = -\frac{\sum_{h=1}^{m}\int_0^1 (1+x)w_h(x,t)\bar{p}_{hi}(x,t)dx}{1+\hat{X}(t)^T P\hat{X}(t) + g\sum_{i=1}^{m}\int_0^1 (1+x)w_i(x,t)^2 dx}. \tag{7.77}$$

Then the adaptive controller in this section is shown in figure 7.4 and the results are summarized as a main theorem below.

Theorem 7.9. *Under Assumptions 7.2, 7.5, and 7.8 and assuming that the PDE states $u_i(x,t)$ of (7.59)–(7.60) are measurable, consider the closed-loop system consisting of the ODE-PDE cascade (7.57)–(7.60), the ODE-state observer (7.62), the invertible backstepping transformation (7.64)–(7.65), the*

PDE boundary control (7.71), and the delay-adaptive update law (7.76)–(7.77). All the signals of the closed-loop system are globally bounded and regulation is achieved, i.e., $\lim_{t\to\infty} X(t) = 0$.

Proof. We build a Lyapunov candidate encompassing all states of the closed-loop system (7.72)–(7.75):

$$V(t) = \left(\prod_{h=1}^{m} D_h \right) \log V_0(t) + \frac{g}{\gamma_D} \sum_{i=1}^{m} \left(\left(\prod_{\substack{h=1 \\ h \neq i}}^{m} D_h \right) \tilde{D}_i(t)^2 \right)$$

$$+ \left(\prod_{h=1}^{m} D_h \right) \tilde{X}(t)^T P_L \tilde{X}(t), \tag{7.78}$$

where $V_0(t)$ is the denominator of (7.77) such that

$$V_0(t) = 1 + \hat{X}(t)^T P \hat{X}(t) + g \sum_{i=1}^{m} \int_0^1 (1+x) w_i(x,t)^2 dx. \tag{7.79}$$

If we take the time derivative of $V(t)$ along (7.72)–(7.75), following a similar procedure as in the proof of Theorem 7.7, it is not hard to show the nonincreasing property of $V(t)$ and consequently the proof of Theorem 7.9 is completed. □

7.4   Global Stabilization under Uncertain Delays and Parameters

In this section, as shown in the fourth row of table 7.1, we further consider another case of uncertainty combination (D_i, θ), which is more challenging than the cases considered in previous sections.

To deal with unknown parameter vector θ, besides Assumptions 7.2 and 7.5, an additional assumption is made below.

Assumption 7.10. *The unknown parameter vector satisfies*

$$\Theta = \{\theta \in \mathbb{R}^p \,|\, \mathscr{P}(\theta) \leq 0\}, \tag{7.80}$$

where $\mathscr{P}(\cdot): \mathbb{R}^p \to \mathbb{R}$ is a known, convex, and smooth function, and Θ is a convex set with a smooth boundary $\partial\Theta$.

As in [80] and [27], Assumption 7.10 is unrestrictive and allows a large uncertainty within the set Θ. Since θ is unknown in this section, we focus on the ODE-PDE cascade (7.19)–(7.22).

By the certainty-equivalence principle, use the parameter estimate $\hat{\theta}(t)$ to replace the unknown θ with the estimation error $\tilde{\theta}(t) = \theta - \hat{\theta}(t)$. Recalling (7.3), we have

$$A(\hat{\theta}) = A_0 + \sum_{h=1}^{p} \hat{\theta}_h A_h, \quad b_i(\hat{\theta}) = b_{i0} + \sum_{h=1}^{p} \hat{\theta}_h b_{ih}, \tag{7.81}$$

$$A(\tilde{\theta}) = \sum_{h=1}^{p} \tilde{\theta}_h A_h, \quad b_i(\tilde{\theta}) = \sum_{h=1}^{p} \tilde{\theta}_h b_{ih}. \tag{7.82}$$

The backstepping transformation has the form below:

$$w_i(x,t) = u_i(x,t) - k_i(\hat{\theta})\Phi(\hat{D}_i x, \hat{D}_l^v, 0, \hat{\theta})X(t)$$

$$- \sum_{j=1}^{m} \hat{D}_j \int_0^{\phi_j(\hat{D}_i x, \hat{D}_j)} k_i(\hat{\theta})\Phi(\hat{D}_i x, \hat{D}_l^v, \hat{D}_j y, \hat{\theta})b_j(\hat{\theta})u_j(y,t)dy, \tag{7.83}$$

$$u_i(x,t) = w_i(x,t) + k_i(\hat{\theta})e^{A_m(\hat{\theta})\hat{D}_i x}X(t)$$

$$+ \sum_{j=1}^{m} \hat{D}_j \int_0^{\phi_j(\hat{D}_i x, \hat{D}_j)} k_i(\hat{\theta})e^{A_m(\hat{\theta})(\hat{D}_i x - \hat{D}_j y)}b_j(\hat{\theta})w_j(y,t)dy, \tag{7.84}$$

where for $j \in \{1,2,\cdots,m\}$, $\phi_j(\hat{D}_i x, \hat{D}_j)$ is exactly identical with (7.31), while $\hat{D}_l^v = (\hat{D}_{v-1}, \hat{D}_{v-2}, \cdots, \hat{D}_{l+1}, \hat{D}_l)$, for $\frac{\hat{D}_{v-1}}{\hat{D}_j} \le y \le \frac{\hat{D}_i}{\hat{D}_j} x \le \frac{\hat{D}_v}{\hat{D}_j}$, $i,j \in \{1,2,\cdots,m\}$, $v \in \{1,2,\cdots,i\}$,

$$\Phi(\hat{D}_i x, \hat{D}_l^v, \hat{D}_j y, \hat{\theta}) = e^{A_{v-1}(\hat{\theta})(\hat{D}_i x - \hat{D}_j y)}, \tag{7.85}$$

for $\frac{\hat{D}_{v-1}}{\hat{D}_i} \le x \le \frac{\hat{D}_v}{\hat{D}_i}$, $\frac{\hat{D}_{l-1}}{\hat{D}_j} \le y \le \frac{\hat{D}_l}{\hat{D}_j}$, $i,j \in \{1,2,\cdots,m\}$, $v,l \in \{1,2,\cdots,i\}$, $l < v$,

$$\Phi(\hat{D}_i x, \hat{D}_l^v, \hat{D}_j y, \hat{\theta}) = e^{A_{v-1}(\hat{\theta})(\hat{D}_i x - \hat{D}_{v-1})}e^{A_{v-2}(\hat{\theta})(\hat{D}_{v-1} - \hat{D}_{v-2})}$$

$$\cdots e^{A_l(\hat{\theta})(\hat{D}_{l+1} - \hat{D}_l)}e^{A_{l-1}(\hat{\theta})(\hat{D}_l - \hat{D}_j y)}, \tag{7.86}$$

and specifically for (7.86), if $l = v - 1$,

$$\Phi(\hat{D}_i x, \hat{D}_l^v, \hat{D}_j y, \hat{\theta}) = e^{A_{v-1}(\hat{\theta})(\hat{D}_i x - \hat{D}_{v-1})}e^{A_{v-2}(\hat{\theta})(\hat{D}_{v-1} - \hat{D}_j y)} \tag{7.87}$$

with

$$\hat{D}_0 = 0, \quad \begin{cases} A_0(\hat{\theta}) = A(\hat{\theta}), \\ A_i(\hat{\theta}) = A_{i-1}(\hat{\theta}) + b_i(\hat{\theta})k_i(\hat{\theta}), i \in \{1,2,\cdots,m\} \\ A_m(\hat{\theta}) = (A + BK)(\hat{\theta}) = A(\hat{\theta}) + \sum_{i=1}^{m} b_i(\hat{\theta})k_i(\hat{\theta}). \end{cases} \tag{7.88}$$

Subsequently, for $i=1,2,\cdots,m$, if we substitute $x=1$ into (7.83) to meet boundary condition $w_i(1,t)=0$, the prediction-based PDE boundary control in the delay-parameter-adaptive feedback scheme has the expression such that

$$U_i(t)=u_i(1,t)=k_i(\hat\theta)\Phi(\hat D_i,\hat D_l^v,0,\hat\theta)X(t)$$

$$+\sum_{j=1}^m \hat D_j\int_0^{\phi_j(\hat D_i,\hat D_j)}k_i(\hat\theta)\Phi(\hat D_i,\hat D_l^v,\hat D_j y,\hat\theta)b_j(\hat\theta)u_j(y,t)dy. \qquad (7.89)$$

Then the ODE-PDE cascade (7.19)–(7.22) could be converted into the following closed-loop system:

$$\dot X(t)=(A+BK)(\hat\theta)X(t)+\sum_{i=1}^m b_i(\hat\theta)w_i(0,t)$$

$$+\sum_{h=1}^p \tilde\theta_h\left(A_h X(t)+\sum_{i=1}^m b_{ih}u_i(0,t)\right), \qquad (7.90)$$

$$D_i\partial_t w_i(x,t)=\partial_x w_i(x,t)-\sum_{h=1}^m \frac{D_i}{D_h}\tilde D_h\bar{\bar p}_{ih}(x,t)$$

$$-D_i\bar{\bar q}_i(x,t,\dot D)-D_i\sum_{h=1}^p \tilde\theta_h r_{ih}(x,t)$$

$$-D_i\sum_{h=1}^p \dot{\hat\theta}_h s_{ih}(x,t),\quad x\in[0,1], \qquad (7.91)$$

$$w_i(1,t)=0,\quad i\in\{1,2,\cdots,m\}. \qquad (7.92)$$

where $\bar{\bar p}_{ih}(x,t)$, $\bar{\bar q}_i(x,t,\dot D)$, $r_{ih}(x,t)$, $s_{ih}(x,t)$ are shown as (7.138), (7.147), (7.155), and (7.156) of section 7.6, respectively.

Based on (7.90) and (7.91), the delay-parameter-adaptive update laws are selected to cancel the delay and parameter estimation error terms $\tilde D_i$ and $\tilde\theta_i$ in the Lyapunov-based analysis, as follows:

$i=1,2,\cdots,m,$

$$\dot{\hat D}_i(t)=\gamma_D\text{Proj}_{[\underline D_{i-1,i},\overline D_{i,i+1}]}\tau_{D_i}(t),\quad \hat D_i(0)\in[\underline D_{i-1,i},\overline D_{i,i+1}], \qquad (7.93)$$

$$\tau_{D_i}(t)=-\frac{\sum_{h=1}^m\int_0^1(1+x)w_h(x,t)\bar{\bar p}_{hi}(x,t)dx}{1+X(t)^T P(\hat\theta)X(t)+g\sum_{i=1}^m\int_0^1(1+x)w_i(x,t)^2dx}, \qquad (7.94)$$

$i=1,2,\cdots,p,$

$$\dot{\hat\theta}_i(t)=\gamma_\theta\text{Proj}_\Theta\tau_{\theta_i}(t),\quad \hat\theta_i(0)\in\Theta,\quad \gamma_\theta>0, \qquad (7.95)$$

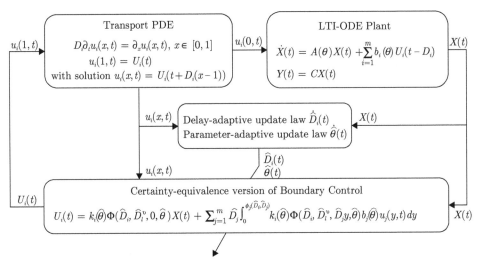

Figure 7.5. Adaptive control under uncertain delays and parameters

$$
\tau_{\theta_i}(t) = \frac{2X(t)^T P(\hat{\theta})\big(A_i X(t) + \sum_{h=1}^{m} b_{hi} u_h(0,t)\big)/g}{1 + X(t)^T P(\hat{\theta})X(t) + g \sum_{i=1}^{m} \int_0^1 (1+x) w_i(x,t)^2 dx}
$$

$$
\frac{- \sum_{h=1}^{m} \int_0^1 (1+x) w_h(x,t) r_{hi}(x,t) dx}{1 + X(t)^T P(\hat{\theta})X(t) + g \sum_{i=1}^{m} \int_0^1 (1+x) w_i(x,t)^2 dx}.
\tag{7.96}
$$

Then the adaptive controller in this section is shown in figure 7.5 and the results are summarized as a main theorem below.

Theorem 7.11. *Under Assumptions 7.2, 7.5, and 7.10 and assuming that the PDE states $u_i(x,t)$ in (7.21)–(7.22) are measurable, consider the closed-loop system consisting of the ODE-PDE cascade (7.19)–(7.22), the invertible backstepping transformation (7.83)–(7.84), the PDE boundary control (7.89), and the delay-parameter-adaptive update laws (7.93)–(7.96). All the signals of the closed-loop system are globally bounded and regulation is achieved, i.e., $\lim_{t\to\infty} X(t) = 0$.*

Proof. We build a Lyapunov candidate encompassing all states of the closed-loop system (7.90)–(7.92):

$$
V(t) = \left(\prod_{h=1}^{m} D_h\right) \log V_0(t) + \frac{g}{\gamma_D} \sum_{i=1}^{m} \left(\left(\prod_{\substack{h=1 \\ h\neq i}}^{m} D_h\right) \tilde{D}_i(t)^2\right)
$$

$$
+ \frac{g}{\gamma_\theta}\left(\prod_{h=1}^{m} D_h\right) \tilde{\theta}(t)^T \tilde{\theta}(t),
\tag{7.97}
$$

where $V_0(t)$ is the denominator of (7.94) and (7.96) such that

$$V_0(t) = 1 + X(t)^T P(\hat{\theta}) X(t) + g \sum_{i=1}^{m} \int_0^1 (1+x) w_i(x,t)^2 dx. \tag{7.98}$$

If we take the time derivative of $V(t)$ along (7.90)–(7.92), following a similar procedure of the proof of Theorem 7.7, it is not hard to show the nonincreasing property of $V(t)$ and consequently the proof of Theorem 7.11 is obtained. We do not go into details. □

7.5 Simulation

In this section, to help readers better understand the proposed control scheme, we provide a concrete example of an LTI system with two input delays. Since Theorem 7.7 in section 7.2 is the basis of subsequent results, we give a detailed computation for (7.29)–(7.41) of section 7.2.

Consider a class of LTI systems with a couple of input delays as follows:

$$\dot{X}(t) = AX(t) + b_1 U_1(t - D_1) + b_2 U_2(t - D_2), \tag{7.99}$$

where matrices A, b_1, and b_2 are known, whereas input delays D_1 and D_2 are unknown and satisfy $D_1 \le D_2$. Employing the measurable transport PDE (7.18) such that $u_1(x,t) = U_1(t + D_1(x-1))$ and $u_2(x,t) = U_2(t + D_2(x-1))$, $x \in [0,1]$, the ODE dynamics (7.99) is converted into the ODE-PDE cascade such that

$$\dot{X}(t) = AX(t) + b_1 u_1(0,t) + b_2 u_2(0,t), \tag{7.100}$$
$$D_1 \partial_t u_1(x,t) = \partial_x u_1(x,t), \quad x \in [0,1], \tag{7.101}$$
$$u_1(1,t) = U_1(t), \tag{7.102}$$
$$D_2 \partial_t u_2(x,t) = \partial_x u_2(x,t), \quad x \in [0,1], \tag{7.103}$$
$$u_2(1,t) = U_2(t). \tag{7.104}$$

Based on (7.31)–(7.35), the backstepping transformation (7.29)–(7.30) for the two-input systems (7.100)–(7.104) have concrete forms as follows:

• $i=1, x \in [0,1]$

$$w_1(x,t) = u_1(x,t) - k_1 e^{A\hat{D}_1 x} X(t)$$
$$- \hat{D}_1 \int_0^x k_1 e^{A(\hat{D}_1 x - \hat{D}_1 y)} b_1 u_1(y,t) dy$$
$$- \hat{D}_2 \int_0^{\frac{\hat{D}_1}{\hat{D}_2} x} k_1 e^{A(\hat{D}_1 x - \hat{D}_2 y)} b_2 u_2(y,t) dy, \tag{7.105}$$

$$u_1(x,t) = w_1(x,t) + k_1 e^{A_2 \hat{D}_1 x} X(t)$$

$$+ \hat{D}_1 \int_0^x k_1 e^{A_2(\hat{D}_1 x - \hat{D}_1 y)} b_1 w_1(y,t) dy$$

$$+ \hat{D}_2 \int_0^{\frac{\hat{D}_1}{\hat{D}_2} x} k_1 e^{A_2(\hat{D}_1 x - \hat{D}_2 y)} b_2 w_2(y,t) dy, \qquad (7.106)$$

- $i = 2, x \in \left[0, \frac{\hat{D}_1}{\hat{D}_2}\right]$

$$w_2(x,t) = u_2(x,t) - k_2 e^{A \hat{D}_2 x} X(t)$$

$$- \hat{D}_1 \int_0^{\frac{\hat{D}_2}{\hat{D}_1} x} k_2 e^{A(\hat{D}_2 x - \hat{D}_1 y)} b_1 u_1(y,t) dy$$

$$- \hat{D}_2 \int_0^x k_2 e^{A(\hat{D}_2 x - \hat{D}_2 y)} b_2 u_2(y,t) dy, \qquad (7.107)$$

$$u_2(x,t) = w_2(x,t) + k_2 e^{A_2 \hat{D}_2 x} X(t)$$

$$+ \hat{D}_1 \int_0^{\frac{\hat{D}_2}{\hat{D}_1} x} k_2 e^{A_2(\hat{D}_2 x - \hat{D}_1 y)} b_1 w_1(y,t) dy,$$

$$+ \hat{D}_2 \int_0^x k_2 e^{A_2(\hat{D}_2 x - \hat{D}_2 y)} b_2 w_2(y,t) dy, \qquad (7.108)$$

- $i = 2, x \in \left[\frac{\hat{D}_1}{\hat{D}_2}, 1\right]$

$$w_2(x,t) = u_2(x,t) - k_2 e^{A_1(\hat{D}_2 x - \hat{D}_1)} e^{A \hat{D}_1} X(t)$$

$$- \hat{D}_1 \int_0^1 k_2 e^{A_1(\hat{D}_2 x - \hat{D}_1)} e^{A(\hat{D}_1 - \hat{D}_1 y)} b_1 u_1(y,t) dy$$

$$- \hat{D}_2 \int_0^{\frac{\hat{D}_1}{\hat{D}_2}} k_2 e^{A_1(\hat{D}_2 x - \hat{D}_1)} e^{A(\hat{D}_1 - \hat{D}_2 y)} b_2 u_2(y,t) dy$$

$$- \hat{D}_2 \int_{\frac{\hat{D}_1}{\hat{D}_2}}^x k_2 e^{A_1(\hat{D}_2 x - \hat{D}_2 y)} b_2 u_2(y,t) dy, \qquad (7.109)$$

$$u_2(x,t) = w_2(x,t) + k_2 e^{A_2 \hat{D}_2 x} X(t)$$

$$+ \hat{D}_1 \int_0^1 k_2 e^{A_2(\hat{D}_2 x - \hat{D}_1 y)} b_1 w_1(y,t) dy$$

$$+ \hat{D}_2 \int_0^x k_2 e^{A_2(\hat{D}_2 x - \hat{D}_2 y)} b_2 w_2(y,t) dy, \qquad (7.110)$$

where $A_2 = A_1 + b_2 k_2$, $A_1 = A + b_1 k_1$.

Obtaining $p_{ih}(x,t)$ and $q_i(x,t,\hat{D})$ in (7.38) of section 7.2 involves a lengthy but straightforward calculation. Let us consider the LTI system with two distinct input delays and illustrate how to get $p_{ih}(x,t)$ and $q_i(x,t,\dot{D})$ for the first input channel $i=1$ step by step as follows. Taking the partial derivative with respect to x on both sides of (7.105), we have

$$\partial_x w_1(x,t) = \partial_x u_1(x,t) - k_1 e^{A\hat{D}_1 x} A\hat{D}_1 X(t) - \hat{D}_1 k_1 b_1 u_1(x,t)$$
$$- \hat{D}_1 k_1 b_2 u_2(\hat{D}_1 x/\hat{D}_2, t) - \hat{D}_1 \int_0^x k_1 A\hat{D}_1 e^{A(\hat{D}_1 x - \hat{D}_1 y)} b_1 u_1(y,t) dy$$
$$- \hat{D}_2 \int_0^{\frac{\hat{D}_1}{\hat{D}_2} x} k_1 A\hat{D}_1 e^{A(\hat{D}_1 x - \hat{D}_2 y)} b_2 u_2(y,t) dy. \tag{7.111}$$

Taking the partial derivative with respect to t on both sides of (7.105), we have

$$\partial_t w_1(x,t) = \partial_t u_1(x,t) - k_1 e^{A\hat{D}_1 x} A\dot{\hat{D}}_1 x X(t)$$
$$- k_1 e^{A\hat{D}_1 x} (AX(t) + b_1 u_1(0,t) + b_2 u_2(0,t))$$
$$- \dot{\hat{D}}_1 \int_0^x k_1 e^{A(\hat{D}_1 x - \hat{D}_1 y)} b_1 u_1(y,t) dy$$
$$- \hat{D}_1 \int_0^x k_1 A(\dot{\hat{D}}_1 x - \dot{\hat{D}}_1 y) e^{A(\hat{D}_1 x - \hat{D}_1 y)} b_1 u_1(y,t) dy$$
$$- \hat{D}_1 \int_0^x k_1 e^{A(\hat{D}_1 x - \hat{D}_1 y)} b_1 \partial_t u_1(y,t) dy$$
$$- \dot{\hat{D}}_2 \int_0^{\frac{\hat{D}_1}{\hat{D}_2} x} k_1 e^{A(\hat{D}_1 x - \hat{D}_2 y)} b_2 u_2(y,t) dy$$
$$- \frac{\dot{\hat{D}}_1 \hat{D}_2 - \hat{D}_1 \dot{\hat{D}}_2}{\hat{D}_2} x k_1 b_2 u_2(\hat{D}_1 x/\hat{D}_2, t)$$
$$- \hat{D}_2 \int_0^{\frac{\hat{D}_1}{\hat{D}_2} x} k_1 A(\dot{\hat{D}}_1 x - \dot{\hat{D}}_2 y) e^{A(\hat{D}_1 x - \hat{D}_2 y)} b_2 u_2(y,t) dy$$
$$- \hat{D}_2 \int_0^{\frac{\hat{D}_1}{\hat{D}_2} x} k_1 e^{A(\hat{D}_1 x - \hat{D}_2 y)} b_2 \partial_t u_2(y,t) dy. \tag{7.112}$$

By utilizing (7.101) and (7.103) and integration by parts, we have

- $-\hat{D}_1 \int_0^x k_1 e^{A(\hat{D}_1 x - \hat{D}_1 y)} b_1 \partial_t u_1(y,t) dy$

$$= -\frac{\hat{D}_1}{\hat{D}_1} \int_0^x k_1 e^{A(\hat{D}_1 x - \hat{D}_1 y)} b_1 \partial_y u_1(y,t) dy$$
$$= -\frac{\hat{D}_1}{\hat{D}_1} k_1 b_1 u_1(x,t) + \frac{\hat{D}_1}{\hat{D}_1} k_1 e^{A\hat{D}_1 x} b_1 u_1(0,t)$$
$$- \frac{\hat{D}_1}{\hat{D}_1} \int_0^x k_1 A\hat{D}_1 e^{A(\hat{D}_1 x - \hat{D}_1 y)} b_1 u_1(y,t) dy, \tag{7.113}$$

- $-\hat{D}_2 \int_0^{\frac{\hat{D}_1}{\hat{D}_2}x} k_1 e^{A(\hat{D}_1 x - \hat{D}_2 y)} b_2 \partial_t u_2(y,t)dy$

$$= -\frac{\hat{D}_2}{D_2} \int_0^{\frac{\hat{D}_1}{\hat{D}_2}x} k_1 e^{A(\hat{D}_1 x - \hat{D}_2 y)} b_2 \partial_y u_2(y,t)dy$$

$$= -\frac{\hat{D}_2}{D_2} k_1 b_2 u_2(\hat{D}_1 x/\hat{D}_2, t) + \frac{\hat{D}_2}{D_2} k_1 e^{A\hat{D}_1 x} b_2 u_2(0,t)$$

$$- \frac{\hat{D}_2}{D_2} \int_0^{\frac{\hat{D}_1}{\hat{D}_2}x} k_1 A\hat{D}_2 e^{A(\hat{D}_1 x - \hat{D}_2 y)} b_2 u_2(y,t)dy. \qquad (7.114)$$

Substituting (7.113)–(7.114) into (7.112), we have

$$\partial_t w_1(x,t) = \partial_t u_1(x,t) - k_1 e^{A\hat{D}_1 x}\left(AX(t) + \frac{\tilde{D}_1}{D_1} b_1 u_1(0,t) + \frac{\tilde{D}_2}{D_2} b_2 u_2(0,t)\right)$$

$$- \frac{\hat{D}_1}{D_1} k_1 b_1 u_1(x,t) - \frac{\hat{D}_2}{D_2} k_1 b_2 u_2(\hat{D}_1 x/\hat{D}_2, t)$$

$$- \frac{\hat{D}_1}{D_1} \int_0^x k_1 A\hat{D}_1 e^{A(\hat{D}_1 x - \hat{D}_1 y)} b_1 u_1(y,t)dy$$

$$- \frac{\hat{D}_2}{D_2} \int_0^{\frac{\hat{D}_1}{\hat{D}_2}x} k_1 A\hat{D}_2 e^{A(\hat{D}_1 x - \hat{D}_2 y)} b_2 u_2(y,t)dy$$

$$- k_1 e^{A\hat{D}_1 x} A\dot{\hat{D}}_1 xX(t) - \frac{\dot{\hat{D}}_1 \hat{D}_2 - \hat{D}_1 \dot{\hat{D}}_2}{\hat{D}_2} xk_1 b_2 u_2(\hat{D}_1 x/\hat{D}_2, t)$$

$$- \hat{D}_1 \int_0^x k_1 \left(\frac{\dot{\hat{D}}_1}{\hat{D}_1}I + A(\dot{\hat{D}}_1 x - \dot{\hat{D}}_1 y)\right) e^{A(\hat{D}_1 x - \hat{D}_1 y)} b_1 u_1(y,t)dy$$

$$- \hat{D}_2 \int_0^{\frac{\hat{D}_1}{\hat{D}_2}x} k_1 \left(\frac{\dot{\hat{D}}_2}{\hat{D}_2}I + A(\dot{\hat{D}}_1 x - \dot{\hat{D}}_2 y)\right) e^{A(\hat{D}_1 x - \hat{D}_2 y)} b_2 u_2(y,t)dy.$$

$$(7.115)$$

Bearing $\tilde{D}_1 = D_1 - \hat{D}_1$ and $\tilde{D}_2 = D_2 - \hat{D}_2$ in mind, calculating $D_1 \cdot (7.115)$–(7.111), we have

$$\partial_t w_1(x,t) = \partial_x w_1(x,t) - \tilde{D}_1\left(k_1 e^{A\hat{D}_1 x} AX(t) + k_1 e^{A\hat{D}_1 x} b_1 u_1(0,t)\right.$$

$$\left. + k_1 b_2 u_2(\hat{D}_1 x/\hat{D}_2, t) + \hat{D}_2 \int_0^{\frac{\hat{D}_1}{\hat{D}_2}x} k_1 A e^{A(\hat{D}_1 x - \hat{D}_2 y)} b_2 u_2(y,t)dy\right)$$

$$- \frac{D_1}{D_2} \tilde{D}_2 \left(k_1 e^{A\hat{D}_1 x} b_2 u_2(0,t) - k_1 b_2 u_2(\hat{D}_1 x / \hat{D}_2, t) \right.$$

$$\left. - \hat{D}_2 \int_0^{\frac{\hat{D}_1}{\hat{D}_2} x} k_1 A e^{A(\hat{D}_1 x - \hat{D}_2 y)} b_2 u_2(y,t) dy \right)$$

$$- D_1 \left(k_1 e^{A\hat{D}_1 x} A\dot{\hat{D}}_1 x X(t) + \frac{\dot{\hat{D}}_1 \hat{D}_2 - \hat{D}_1 \dot{\hat{D}}_2}{\hat{D}_2} x k_1 b_2 u_2(\hat{D}_1 x / \hat{D}_2, t) \right.$$

$$+ \hat{D}_1 \int_0^x k_1 \left(\frac{\dot{\hat{D}}_1}{\hat{D}_1} I + A(\dot{\hat{D}}_1 x - \dot{\hat{D}}_1 y) \right) e^{A(\hat{D}_1 x - \hat{D}_1 y)} b_1 u_1(y,t) dy$$

$$\left. + \hat{D}_2 \int_0^{\frac{\hat{D}_1}{\hat{D}_2} x} k_1 \left(\frac{\dot{\hat{D}}_2}{\hat{D}_2} I + A(\dot{\hat{D}}_1 x - \dot{\hat{D}}_2 y) \right) e^{A(\hat{D}_1 x - \hat{D}_2 y)} b_2 u_2(y,t) dy \right). \quad (7.116)$$

Thus $p_{ih}(x,t)$ and $q_i(x,t,\hat{D})$ in (7.38) of section 7.2 are obtained. Following the similar procedure (7.111)–(7.116), we can get similar results for the second input channel $i=2$, and can further extend this to a general m-input case.

Taking partial derivatives $\partial_t w_1(x,t)$, $\partial_x w_1(x,t)$, $\partial_t w_2(x,t)$, and $\partial_x w_2(x,t)$ on both sides of (7.105)–(7.109), the closed-loop system of (7.37)–(7.39) has the concrete form such that

$$\dot{X}(t) = (A + BK)X(t) + b_1 w_1(0,t) + b_2 w_2(0,t), \quad (7.117)$$

$$D_1 \partial_t w_1(x,t) = \partial_x w_1(x,t) - \tilde{D}_1 p_{11}(x,t) - \frac{D_1}{D_2} \tilde{D}_2 p_{12}(x,t)$$

$$- D_1 q_1(x,t,\hat{D}), \quad x \in [0,1], \quad (7.118)$$

$$w_1(1,t) = 0, \quad (7.119)$$

$$D_2 \partial_t w_2(x,t) = \partial_x w_2(x,t) - \frac{D_2}{D_1} \tilde{D}_1 p_{21}(x,t) - \tilde{D}_2 p_{22}(x,t)$$

$$- D_2 q_2(x,t,\hat{D}), \quad x \in [0,1], \quad (7.120)$$

$$w_2(1,t) = 0, \quad (7.121)$$

where $p_{11}(x,t)$, $p_{12}(x,t)$, $p_{21}(x,t)$, and $p_{22}(x,t)$ have the forms as (7.136) in section 7.6 such that

$$p_{11}(x,t) = k_1 e^{A\hat{D}_1 x}(AX(t) + b_1 u_1(0,t)) + k_1 b_2 u_2(\hat{D}_1 x / \hat{D}_2, t)$$

$$+ \hat{D}_2 \int_0^{\frac{\hat{D}_1}{\hat{D}_2} x} k_1 A e^{A(\hat{D}_1 x - \hat{D}_2 y)} b_2 u_2(y,t) dy, \quad x \in [0,1], \quad (7.122)$$

$$p_{12}(x,t) = k_1 e^{A\hat{D}_1 x} b_2 u_2(0,t) - k_1 b_2 u_2(\hat{D}_1 x / \hat{D}_2, t)$$

$$- \hat{D}_2 \int_0^{\frac{\hat{D}_1}{\hat{D}_2} x} k_1 A e^{A(\hat{D}_1 x - \hat{D}_2 y)} b_2 u_2(y,t) dy, \quad x \in [0,1], \quad (7.123)$$

$$p_{21}(x,t) = k_2 e^{A\hat{D}_2 x} b_1 u_1(0,t) - k_2 b_1 u_1(\hat{D}_2 x/\hat{D}_1, t)$$

$$- \hat{D}_1 \int_0^{\frac{\hat{D}_2}{\hat{D}_1} x} k_2 A e^{A(\hat{D}_2 x - \hat{D}_1 y)} b_1 u_1(y,t) dy, \quad x \in \left[0, \frac{\hat{D}_1}{\hat{D}_2}\right], \quad (7.124)$$

$$p_{22}(x,t) = k_2 e^{A\hat{D}_2 x} (AX(t) + b_2 u_2(0,t)) + k_2 b_1 u_1(\hat{D}_2 x/\hat{D}_1, t)$$

$$+ \hat{D}_1 \int_0^{\frac{\hat{D}_2}{\hat{D}_1} x} k_2 A e^{A(\hat{D}_2 x - \hat{D}_1 y)} b_1 u_1(y,t) dy, \quad x \in \left[0, \frac{\hat{D}_1}{\hat{D}_2}\right], \quad (7.125)$$

$$p_{21}(x,t) = k_2 e^{A_1(\hat{D}_2 x - \hat{D}_1)} (A - A_1) e^{A\hat{D}_1} X(t)$$

$$+ k_2 e^{A_1(\hat{D}_2 x - \hat{D}_1)} e^{A\hat{D}_1} b_1 u_1(0,t)$$

$$- \hat{D}_1 \int_0^1 k_2 e^{A_1(\hat{D}_2 x - \hat{D}_1)} A_1 e^{A(\hat{D}_1 - \hat{D}_1 y)} b_1 u_1(y,t) dy$$

$$+ \hat{D}_2 \int_0^{\frac{\hat{D}_1}{\hat{D}_2}} k_2 e^{A_1(\hat{D}_2 x - \hat{D}_1)} (A - A_1) e^{A(\hat{D}_1 - \hat{D}_2 y)} b_2 u_2(y,t) dy,$$

$$x \in \left[\frac{\hat{D}_1}{\hat{D}_2}, 1\right], \quad (7.126)$$

$$p_{22}(x,t) = k_2 e^{A_1(\hat{D}_2 x - \hat{D}_1)} A_1 e^{A\hat{D}_1} X(t)$$

$$+ k_2 e^{A_1(\hat{D}_2 x - \hat{D}_1)} e^{A\hat{D}_1} b_2 u_2(0,t)$$

$$+ \hat{D}_1 \int_0^1 k_2 e^{A_1(\hat{D}_2 x - \hat{D}_1)} A_1 e^{A(\hat{D}_1 - \hat{D}_1 y)} b_1 u_1(y,t) dy$$

$$+ \hat{D}_2 \int_0^{\frac{\hat{D}_1}{\hat{D}_2}} k_2 e^{A_1(\hat{D}_2 x - \hat{D}_1)} (A_1 - A) e^{A(\hat{D}_1 - \hat{D}_2 y)} b_2 u_2(y,t) dy,$$

$$x \in \left[\frac{\hat{D}_1}{\hat{D}_2}, 1\right]. \quad (7.127)$$

Substituting $x = 1$ into (7.105) and (7.109) to make (7.119) and (7.121) hold, we obtain the boundary control law (7.36) expressions below:

$$U_1(t) = u_1(1,t) = k_1 e^{A\hat{D}_1} X(t)$$

$$- \hat{D}_1 \int_0^1 k_1 e^{A(\hat{D}_1 - \hat{D}_1 y)} b_1 u_1(y,t) dy$$

$$- \hat{D}_2 \int_0^{\frac{\hat{D}_1}{\hat{D}_2}} k_1 e^{A(\hat{D}_1 - \hat{D}_2 y)} b_2 u_2(y,t) dy, \quad (7.128)$$

$$U_2(t) = u_2(1,t) = k_2 e^{A_1(\hat{D}_2 - \hat{D}_1)} e^{A\hat{D}_1} X(t)$$

$$- \hat{D}_1 \int_0^1 k_2 e^{A_1(\hat{D}_2 - \hat{D}_1)} e^{A(\hat{D}_1 - \hat{D}_1 y)} b_1 u_1(y,t) dy$$

$$- \hat{D}_2 \int_0^{\frac{\hat{D}_1}{\hat{D}_2}} k_2 e^{A_1(\hat{D}_2 - \hat{D}_1)} e^{A(\hat{D}_1 - \hat{D}_2 y)} b_2 u_2(y,t) dy$$

$$- \hat{D}_2 \int_{\frac{\hat{D}_1}{\hat{D}_2}}^1 k_2 e^{A_1(\hat{D}_2 - \hat{D}_2 y)} b_2 u_2(y,t) dy. \qquad (7.129)$$

Build a Lyapunov candidate (7.42)–(7.43) as

$$V(t) = D_1 D_2 \log V_0(t) + \left(D_2 \tilde{D}_1(t)^2 + D_1 \tilde{D}_2(t)^2 \right), \qquad (7.130)$$

where $V_0(t)$ is

$$V_0(t) = 1 + X(t)^T P X(t) + g \int_0^1 (1+x)\left(w_1(x,t)^2 + w_2(x,t)^2\right) dx. \qquad (7.131)$$

Taking the time derivative of (7.130) along the closed-loop system (7.117)–(7.121), we design the parameter update laws (7.40)–(7.41) to remove the estimation error terms $\tilde{D}_1$ and $\tilde{D}_2$ in $\dot{V}(t)$ such that

$$\dot{\hat{D}}_1(t) = \gamma_D \mathrm{Proj}_{[\underline{D}_1, \overline{D}_{1,2}]} \tau_{D_1}(t), \quad \hat{D}_1(0) \in [\underline{D}_1, \overline{D}_{1,2}], \qquad (7.132)$$

$$\tau_{D_1}(t) = \frac{- \int_0^1 (1+x)\left(w_1(x,t)p_{11}(x,t) + w_2(x,t)p_{21}(x,t)\right) dx}{1 + X(t)^T P X(t) + g \int_0^1 (1+x)\left(w_1(x,t)^2 + w_2(x,t)^2\right) dx}, \qquad (7.133)$$

$$\dot{\hat{D}}_2(t) = \gamma_D \mathrm{Proj}_{[\underline{D}_{1,2}, \overline{D}_2]} \tau_{D_2}(t), \quad \hat{D}_2(0) \in [\underline{D}_{1,2}, \overline{D}_2], \qquad (7.134)$$

$$\tau_{D_2}(t) = \frac{- \int_0^1 (1+x)\left(w_1(x,t)p_{12}(x,t) + w_2(x,t)p_{22}(x,t)\right) dx}{1 + X(t)^T P X(t) + g \int_0^1 (1+x)\left(w_1(x,t)^2 + w_2(x,t)^2\right) dx}. \qquad (7.135)$$

For simulation, we consider the unstable system with $A = 1$, $b_1 = 1$, $b_2 = 1$, $D_1 = 0.5$, and $D_2 = 1$. The stabilizing gains are selected as $k_1 = -1$, $k_2 = -0.5$, and the simulation results are shown in figures 7.6 and 7.7. It is apparent that the stabilization of the time-delay system is achieved and the convergence of delay estimates towards actual values is established.

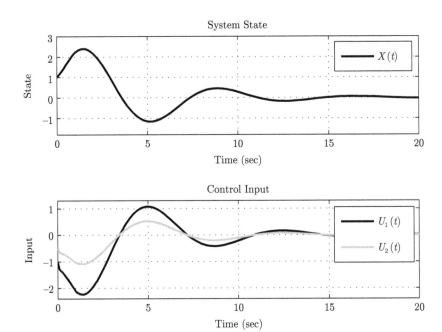

Figure 7.6. State and input

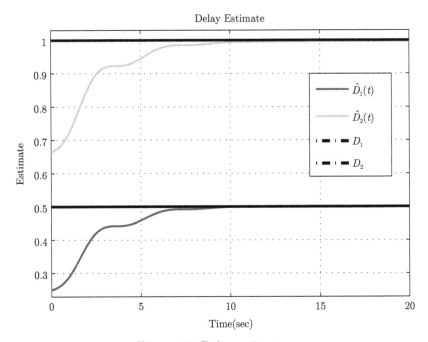

Figure 7.7. Delay estimates

7.6   Auxiliary Calculations for Sections 7.2–7.4

- The explicit forms of $p_{ih}(x,t)$, $\bar{p}_{ih}(x,t)$, and $\bar{\bar{p}}_{ih}(x,t)$ in (7.38), (7.74), and (7.91) are listed as follows:

$$p_{ih}(x,t) = k_i P_{\Phi(\hat{D}_i x, \hat{D}_l^v, 0)} X(t) + k_i \Phi(\hat{D}_i x, \hat{D}_l^v, 0) b_h u_h(0,t)$$

$$+ \sum_{j=1}^{m} \hat{D}_j \int_0^{\phi_j(\hat{D}_i x, \hat{D}_j)} k_i P_{\Phi(\hat{D}_i x, \hat{D}_l^v, \hat{D}_j y)} b_j u_j(y,t) dy$$

$$+ \sum_{j=1}^{m} \rho_j k_i b_j u_j(\phi_j(\hat{D}_i x, \hat{D}_j), t), \tag{7.136}$$

$$\bar{p}_{ih}(x,t) = k_i P_{\Phi(\hat{D}_i x, \hat{D}_l^v, 0)} \dot{X}(t) + k_i \Phi(\hat{D}_i x, \hat{D}_l^v, 0) b_h u_h(0,t)$$

$$+ \sum_{j=1}^{m} \hat{D}_j \int_0^{\phi_j(\hat{D}_i x, \hat{D}_j)} k_i P_{\Phi(\hat{D}_i x, \hat{D}_l^v, \hat{D}_j y)} b_j u_j(y,t) dy$$

$$+ \sum_{j=1}^{m} \rho_j k_i b_j u_j(\phi_j(\hat{D}_i x, \hat{D}_j), t), \tag{7.137}$$

$$\bar{\bar{p}}_{ih}(x,t) = k_i(\hat{\theta}) P_{\Phi(\hat{D}_i x, \hat{D}_l^v, 0, \hat{\theta})} X(t) + k_i(\hat{\theta}) \Phi(\hat{D}_i x, \hat{D}_l^v, 0, \hat{\theta}) b_h(\hat{\theta}) u_h(0,t)$$

$$+ \sum_{j=1}^{m} \hat{D}_j \int_0^{\phi_j(\hat{D}_i x, \hat{D}_j)} k_i(\hat{\theta}) P_{\Phi(\hat{D}_i x, \hat{D}_l^v, \hat{D}_j y, \hat{\theta})} b_j(\hat{\theta}) u_j(y,t) dy$$

$$+ \sum_{j=1}^{m} \rho_j k_i(\hat{\theta}) b_j(\hat{\theta}) u_j(\phi_j(\hat{D}_i x, \hat{D}_j), t), \tag{7.138}$$

where

$$\rho_j = \begin{cases} 1, & \text{if } h = i, \phi_j(\hat{D}_i x, \hat{D}_j) = \frac{\hat{D}_i}{\hat{D}_j} x, i \neq j \\ 0, & \text{if } h \neq i, j, \text{ or } \phi_j(\hat{D}_i x, \hat{D}_j) = \frac{\hat{D}_i}{\hat{D}_j} x, i = j, \text{ or } \phi_j(\hat{D}_i x, \hat{D}_j) = 1 \\ -1, & \text{if } h = j, \phi_j(\hat{D}_i x, \hat{D}_j) = \frac{\hat{D}_i}{\hat{D}_j} x, i \neq j, \end{cases}$$

$$\tag{7.139}$$

$$P_{\Phi(\hat{D}_i x, \hat{D}_l^v, 0)} = \begin{cases} \frac{\partial \Phi(\hat{D}_i x, \hat{D}_l^v, 0)}{\partial(\hat{D}_i x)}, & h = i \\ \frac{\partial \Phi(\hat{D}_i x, \hat{D}_l^v, 0)}{\partial \hat{D}_h}, & h \neq i \end{cases},$$

$$P_{\Phi(\hat{D}_i x, \hat{D}_l^v, 0, \hat{\theta})} = \begin{cases} \frac{\partial \Phi(\hat{D}_i x, \hat{D}_l^v, 0, \hat{\theta})}{\partial(\hat{D}_i x)}, & h = i \\ \frac{\partial \Phi(\hat{D}_i x, \hat{D}_l^v, 0, \hat{\theta})}{\partial \hat{D}_h}, & h \neq i, \end{cases} \tag{7.140}$$

and $P_{\Phi(\hat{D}_i x, \hat{D}_l^v, \hat{D}_j y)}$ has exactly the same structure as $P_{\Phi(\hat{D}_i x, \hat{D}_l^v, \hat{D}_j y, \hat{\theta})}$ by taking the place of $\hat{\theta}$ with θ, while $P_{\Phi(\hat{D}_i x, \hat{D}_l^v, \hat{D}_j y, \hat{\theta})}$ has the form corresponding to $\Phi(\hat{D}_i x, \hat{D}_l^v, \hat{D}_j y, \hat{\theta})$ of (7.85)–(7.87) such that:

For $\Phi(\hat{D}_i x, \hat{D}_l^v, \hat{D}_j y, \hat{\theta})$ of (7.85)

$$
P_{\Phi(\hat{D}_i x, \hat{D}_l^v, \hat{D}_j y, \hat{\theta})} =
\begin{cases}
\dfrac{\partial \Phi(\hat{D}_i x, \hat{D}_l^v, \hat{D}_j y, \hat{\theta})}{\partial(\hat{D}_i x)}, & h = i, h \neq j \\[2ex]
\dfrac{\partial \Phi(\hat{D}_i x, \hat{D}_l^v, \hat{D}_j y, \hat{\theta})}{\partial \hat{D}_h} = 0, & h \neq i, j \\[2ex]
0, & h = i = j \\[2ex]
\dfrac{\partial \Phi(\hat{D}_i x, \hat{D}_l^v, \hat{D}_j y, \hat{\theta})}{\partial(\hat{D}_j y)}, & h = j, h \neq i.
\end{cases}
\tag{7.141}
$$

For $\Phi(\hat{D}_i x, \hat{D}_l^v, \hat{D}_j y, \hat{\theta})$ of (7.86)–(7.87)

$$
\begin{aligned}
&P_{\Phi(\hat{D}_i x, \hat{D}_l^v, \hat{D}_j y, \hat{\theta})} \\
&=
\begin{cases}
A_{v-1}(\hat{\theta}) e^{A_{v-1}(\hat{\theta})\hat{D}_i x} \Gamma(\hat{D}_l^v, \hat{\theta}) e^{-A_{l-1}(\hat{\theta})\hat{D}_j y} \\
\quad + e^{A_{v-1}(\hat{\theta})\hat{D}_i x} \dfrac{\partial \Gamma(\hat{D}_l^v, \hat{\theta})}{\partial \hat{D}_h} e^{-A_{l-1}(\hat{\theta})\hat{D}_j y}, & h = i, h \neq j \\[2ex]
e^{A_{v-1}(\hat{\theta})\hat{D}_i x} \dfrac{\partial \Gamma(\hat{D}_l^v, \hat{\theta})}{\partial \hat{D}_h} e^{-A_{l-1}(\hat{\theta})\hat{D}_j y}, & h \neq i, j \\[2ex]
A_{v-1}(\hat{\theta}) e^{A_{v-1}(\hat{\theta})\hat{D}_i x} \Gamma(\hat{D}_l^v, \hat{\theta}) e^{-A_{l-1}(\hat{\theta})\hat{D}_j y} \\
\quad + e^{A_{v-1}(\hat{\theta})\hat{D}_i x} \dfrac{\partial \Gamma(\hat{D}_l^v, \hat{\theta})}{\partial \hat{D}_h} e^{-A_{l-1}(\hat{\theta})\hat{D}_j y} \\
\quad + e^{A_{v-1}(\hat{\theta})\hat{D}_i x} \Gamma(\hat{D}_l^v, \hat{\theta}) e^{-A_{l-1}(\hat{\theta})\hat{D}_j y}(-A_{l-1}(\hat{\theta})), & h = i = j \\[2ex]
e^{A_{v-1}(\hat{\theta})\hat{D}_i x} \dfrac{\partial \Gamma(\hat{D}_l^v, \hat{\theta})}{\partial \hat{D}_h} e^{-A_{l-1}(\hat{\theta})\hat{D}_j y} \\
\quad + e^{A_{v-1}(\hat{\theta})\hat{D}_i x} \Gamma(\hat{D}_l^v, \hat{\theta}) e^{-A_{l-1}(\hat{\theta})\hat{D}_j y}(-A_{l-1}(\hat{\theta})), & h = j, h \neq i,
\end{cases}
\end{aligned}
\tag{7.142}
$$

with $\Gamma(\hat{D}_l^v, \hat{\theta})$ coming from (7.86)–(7.87) such that

$$
\begin{aligned}
\Phi(\hat{D}_i x, \hat{D}_l^v, \hat{D}_j y, \hat{\theta}) &= e^{A_{v-1}(\hat{\theta})(\hat{D}_i x - \hat{D}_{v-1})} e^{A_{v-2}(\hat{\theta})(\hat{D}_{v-1} - \hat{D}_{v-2})} \cdots \\
&\quad \times e^{A_l(\hat{\theta})(\hat{D}_{l+1} - \hat{D}_l)} e^{A_{l-1}(\hat{\theta})(\hat{D}_l - \hat{D}_j y)} \\
&= e^{A_{v-1}(\hat{\theta})\hat{D}_i x}\left(e^{-A_{v-1}(\hat{\theta})\hat{D}_{v-1}} e^{A_{v-2}(\hat{\theta})(\hat{D}_{v-1} - \hat{D}_{v-2})} \cdots \right. \\
&\quad \left. \times e^{A_l(\hat{\theta})(\hat{D}_{l+1} - \hat{D}_l)} e^{A_{l-1}(\hat{\theta})\hat{D}_l} \right) e^{-A_{l-1}(\hat{\theta})\hat{D}_j y} \\
&= e^{A_{v-1}(\hat{\theta})\hat{D}_i x} \Gamma(\hat{D}_l^v, \hat{\theta}) e^{-A_{l-1}(\hat{\theta})\hat{D}_j y},
\end{aligned}
\tag{7.143}
$$

and specifically, if $l = v - 1$,

$$\Phi(\hat{D}_i x, \hat{D}_l^v, \hat{D}_j y, \hat{\theta}) = e^{A_{v-1}(\hat{\theta})(\hat{D}_i x - \hat{D}_{v-1})} e^{A_{v-2}(\hat{\theta})(\hat{D}_{v-1} - \hat{D}_j y)}$$

$$= e^{A_{v-1}(\hat{\theta})\hat{D}_i x} \left(e^{-A_{v-1}(\hat{\theta})\hat{D}_{v-1}} e^{A_{v-2}(\hat{\theta})\hat{D}_{v-1}} \right) e^{-A_{v-2}(\hat{\theta})\hat{D}_j y}$$

$$= e^{A_{v-1}(\hat{\theta})\hat{D}_i x} \Gamma(\hat{D}_l^v, \hat{\theta}) e^{-A_{v-2}(\hat{\theta})\hat{D}_j y}. \tag{7.144}$$

- The explicit forms of $q_{ih}(x,t)$, $\bar{q}_{ih}(x,t)$, and $\bar{\bar{q}}_{ih}(x,t)$ in (7.38), (7.74), and (7.91) are listed as follows:

$$q_i(x,t,\dot{\hat{D}}) = k_i \sum_h \frac{\partial \Phi(\hat{D}_i x, \hat{D}_l^v, 0)}{\partial \hat{D}_h} \dot{\hat{D}}_h X(t)$$

$$+ \sum_{j=1}^m \hat{D}_j \int_0^{\phi_j(\hat{D}_i x, \hat{D}_j)} k_i \left(\frac{\dot{\hat{D}}_j}{\hat{D}_j} \Phi(\hat{D}_i x, \hat{D}_l^v, \hat{D}_j y) \right.$$

$$+ \sum_h \frac{\partial \Phi(\hat{D}_i x, \hat{D}_l^v, \hat{D}_j y)}{\partial \hat{D}_h} \dot{\hat{D}}_h \Bigg) b_j u_j(y,t) dy$$

$$+ \sum_{j=1}^m \epsilon_j \frac{\dot{\hat{D}}_i \hat{D}_j - \hat{D}_i \dot{\hat{D}}_j}{\hat{D}_j} x k_i b_j u_j(\phi_j(\hat{D}_i x, \hat{D}_j), t), \tag{7.145}$$

$$\bar{q}_i(x,t,\dot{\hat{D}}) = k_i \sum_h \frac{\partial \Phi(\hat{D}_i x, \hat{D}_l^v, 0)}{\partial \hat{D}_h} \dot{\hat{D}}_h \hat{X}(t)$$

$$+ \sum_{j=1}^m \hat{D}_j \int_0^{\phi_j(\hat{D}_i x, \hat{D}_j)} k_i \left(\frac{\dot{\hat{D}}_j}{\hat{D}_j} \Phi(\hat{D}_i x, \hat{D}_l^v, \hat{D}_j y) \right.$$

$$+ \sum_h \frac{\partial \Phi(\hat{D}_i x, \hat{D}_l^v, \hat{D}_j y)}{\partial \hat{D}_h} \dot{\hat{D}}_h \Bigg) b_j u_j(y,t) dy$$

$$+ \sum_{j=1}^m \epsilon_j \frac{\dot{\hat{D}}_i \hat{D}_j - \hat{D}_i \dot{\hat{D}}_j}{\hat{D}_j} x k_i b_j u_j(\phi_j(\hat{D}_i x, \hat{D}_j), t), \tag{7.146}$$

$$\bar{\bar{q}}_i(x,t,\dot{\hat{D}}) = k_i(\hat{\theta}) \sum_h \frac{\partial \Phi(\hat{D}_i x, \hat{D}_l^v, 0, \hat{\theta})}{\partial \hat{D}_h} \dot{\hat{D}}_h X(t)$$

$$+ \sum_{j=1}^m \hat{D}_j \int_0^{\phi_j(\hat{D}_i x, \hat{D}_j)} k_i(\hat{\theta}) \left(\frac{\dot{\hat{D}}_j}{\hat{D}_j} \Phi(\hat{D}_i x, \hat{D}_l^v, \hat{D}_j y, \hat{\theta}) \right.$$

$$+ \sum_h \frac{\partial \Phi(\hat{D}_i x, \hat{D}_l^v, \hat{D}_j y, \hat{\theta})}{\partial \hat{D}_h} \dot{\hat{D}}_h \Bigg) b_j(\hat{\theta}) u_j(y,t) dy$$

$$+ \sum_{j=1}^m \epsilon_j \frac{\dot{\hat{D}}_i \hat{D}_j - \hat{D}_i \dot{\hat{D}}_j}{\hat{D}_j} x k_i(\hat{\theta}) b_j(\hat{\theta}) u_j(\phi_j(\hat{D}_i x, \hat{D}_j), t), \tag{7.147}$$

where

$$\epsilon_j = \begin{cases} 1, & \phi_j(\hat{D}_i x, \hat{D}_j) = \frac{\hat{D}_i}{\hat{D}_j} x \\ 0, & \phi_j(\hat{D}_i x, \hat{D}_j) = 1, \end{cases} \tag{7.148}$$

and $\sum_h \frac{\partial \Phi(\hat{D}_i x, \hat{D}_l^v, \hat{D}_j y)}{\partial \hat{D}_h} \dot{D}_h$ has the form corresponding to $\Phi(\hat{D}_i x, \hat{D}_l^v, \hat{D}_j y)$ of (7.32)–(7.34), while $\sum_h \frac{\partial \Phi(\hat{D}_i x, \hat{D}_l^v, \hat{D}_j y, \hat{\theta})}{\partial \hat{D}_h} \dot{D}_h$ has the form corresponding to $\Phi(\hat{D}_i x, \hat{D}_l^v, \hat{D}_j y, \hat{\theta})$ of (7.85)–(7.87) such that:

For $\Phi(\hat{D}_i x, \hat{D}_l^v, \hat{D}_j y, \hat{\theta})$ of (7.85)

$$\sum_h \frac{\partial \Phi(\hat{D}_i x, \hat{D}_l^v, \hat{D}_j y, \hat{\theta})}{\partial \hat{D}_h} \dot{D}_h = A_{v-1}(\hat{\theta})(\dot{\hat{D}}_i x - \dot{\hat{D}}_j y) e^{A_{v-1}(\hat{\theta})(\hat{D}_i x - \hat{D}_j y)}$$

$$\tag{7.149}$$

$$\sum_h \frac{\partial \Phi(\hat{D}_i x, \hat{D}_l^v, 0, \hat{\theta})}{\partial \hat{D}_h} \dot{D}_h = A_{v-1}(\hat{\theta}) \dot{\hat{D}}_i x \, e^{A_{v-1}(\hat{\theta}) \hat{D}_i x}. \tag{7.150}$$

For $\Phi(\hat{D}_i x, \hat{D}_l^v, \hat{D}_j y, \hat{\theta})$ of (7.86)

$$\sum_h \frac{\partial \Phi(\hat{D}_i x, \hat{D}_l^v, \hat{D}_j y, \hat{\theta})}{\partial \hat{D}_h} \dot{D}_h$$

$$= A_{v-1}(\hat{\theta})(\dot{\hat{D}}_i x - \dot{\hat{D}}_{v-1}) e^{A_{v-1}(\hat{\theta})(\hat{D}_i x - \hat{D}_{v-1})} e^{A_{v-2}(\hat{\theta})(\hat{D}_{v-1} - \hat{D}_{v-2})} \dots$$

$$\times e^{A_l(\hat{\theta})(\hat{D}_{l+1} - \hat{D}_l)} e^{A_{l-1}(\hat{\theta})(\hat{D}_l - \hat{D}_j y)}$$

$$+ e^{A_{v-1}(\hat{\theta})(\hat{D}_i x - \hat{D}_{v-1})} A_{v-2}(\hat{\theta})(\dot{\hat{D}}_{v-1} - \dot{\hat{D}}_{v-2}) e^{A_{v-2}(\hat{\theta})(\hat{D}_{v-1} - \hat{D}_{v-2})} \dots$$

$$\times e^{A_l(\hat{\theta})(\hat{D}_{l+1} - \hat{D}_l)} e^{A_{l-1}(\hat{\theta})(\hat{D}_l - \hat{D}_j y)}$$

$$+ \quad \dots \quad \dots$$

$$+ e^{A_{v-1}(\hat{\theta})(\hat{D}_i x - \hat{D}_{v-1})} e^{A_{v-2}(\hat{\theta})(\hat{D}_{v-1} - \hat{D}_{v-2})} \dots$$

$$\times A_l(\hat{\theta})(\dot{\hat{D}}_{l+1} - \dot{\hat{D}}_l) e^{A_l(\hat{\theta})(\hat{D}_{l+1} - \hat{D}_l)} e^{A_{l-1}(\hat{\theta})(\hat{D}_l - \hat{D}_j y)}$$

$$+ e^{A_{v-1}(\hat{D}_i x - \hat{D}_{v-1})} e^{A_{v-2}(\hat{D}_{v-1} - \hat{D}_{v-2})} \dots$$

$$\times e^{A_l(\hat{\theta})(\hat{D}_{l+1} - \hat{D}_l)} e^{A_{l-1}(\hat{\theta})(\hat{D}_l - \hat{D}_j y)} A_{l-1}(\hat{\theta})(\dot{\hat{D}}_l - \dot{\hat{D}}_j y), \tag{7.151}$$

$$\sum_h \frac{\partial \Phi(\hat{D}_i x, \hat{D}_l^v, 0, \hat{\theta})}{\partial \hat{D}_h} \dot{D}_h$$

$$= A_{v-1}(\hat{\theta})(\dot{\hat{D}}_i x - \dot{\hat{D}}_{v-1}) e^{A_{v-1}(\hat{\theta})(\hat{D}_i x - \hat{D}_{v-1})} e^{A_{v-2}(\hat{\theta})(\hat{D}_{v-1} - \hat{D}_{v-2})} \dots$$

$$\times e^{A_l(\hat{\theta})(\hat{D}_{l+1} - \hat{D}_l)} e^{A_{l-1}(\hat{\theta}) \hat{D}_l}$$

$$+ e^{A_{v-1}(\hat{\theta})(\hat{D}_i x - \hat{D}_{v-1})} A_{v-2}(\hat{\theta})(\dot{\hat{D}}_{v-1} - \dot{\hat{D}}_{v-2}) e^{A_{v-2}(\hat{\theta})(\hat{D}_{v-1} - \hat{D}_{v-2})} \cdots$$

$$\times\, e^{A_l(\hat{\theta})(\hat{D}_{l+1} - \hat{D}_l)} e^{A_{l-1}(\hat{\theta})\hat{D}_l}$$

$$+\quad \cdots \quad \cdots$$

$$+ e^{A_{v-1}(\hat{\theta})(\hat{D}_i x - \hat{D}_{v-1})} e^{A_{v-2}(\hat{\theta})(\hat{D}_{v-1} - \hat{D}_{v-2})} \cdots$$

$$\times\, A_l(\hat{\theta})(\dot{\hat{D}}_{l+1} - \dot{\hat{D}}_l) e^{A_l(\hat{\theta})(\hat{D}_{l+1} - \hat{D}_l)} e^{A_{l-1}(\hat{\theta})\hat{D}_l}$$

$$+ e^{A_{v-1}(\hat{\theta})(\hat{D}_i x - \hat{D}_{v-1})} e^{A_{v-2}(\hat{\theta})(\hat{D}_{v-1} - \hat{D}_{v-2})} \cdots$$

$$\times\, e^{A_l(\hat{\theta})(\hat{D}_{l+1} - \hat{D}_l)} e^{A_{l-1}(\hat{\theta})\hat{D}_l} A_{l-1}(\hat{\theta})\dot{\hat{D}}_l. \tag{7.152}$$

For $\Phi(\hat{D}_i x, \hat{D}_l^v, \hat{D}_j y, \hat{\theta})$ of (7.87)

$$\sum_h \frac{\partial \Phi(\hat{D}_i x, \hat{D}_l^v, \hat{D}_j y, \hat{\theta})}{\partial \hat{D}_h} \dot{\hat{D}}_h = e^{A_{v-1}(\hat{\theta})(\hat{D}_i x - \hat{D}_{v-1})} \left(A_{v-1}(\hat{\theta})(\dot{\hat{D}}_i x - \dot{\hat{D}}_{v-1}) \right.$$

$$\left. + A_{v-2}(\hat{\theta})(\dot{\hat{D}}_{v-1} - \dot{\hat{D}}_j y) \right) e^{A_{v-2}(\hat{\theta})(\hat{D}_{v-1} - \hat{D}_j y)}, \tag{7.153}$$

$$\sum_h \frac{\partial \Phi(\hat{D}_i x, \hat{D}_l^v, 0, \hat{\theta})}{\partial \hat{D}_h} \dot{\hat{D}}_h = e^{A_{v-1}(\hat{\theta})(\hat{D}_i x - \hat{D}_{v-1})} \left(A_{v-1}(\hat{\theta})(\dot{\hat{D}}_i x - \dot{\hat{D}}_{v-1}) \right.$$

$$\left. + A_{v-2}(\hat{\theta})\dot{\hat{D}}_{v-1} \right) e^{A_{v-2}(\hat{\theta})\hat{D}_{v-1}}. \tag{7.154}$$

- The explicit forms of $r_{ih}(x, t)$ and $s_{ih}(x, t)$ in (7.91) are listed as follows:

$$r_{ih}(x, t) = k_i(\hat{\theta})\Phi(\hat{D}_i x, \hat{D}_l^v, 0, \hat{\theta}) \left(A_h X(t) + \sum_{j=1}^m b_{jh} u_j(0, t) \right), \tag{7.155}$$

$$s_{ih}(x, t) = \frac{\partial k_i(\hat{\theta})}{\partial \hat{\theta}_h} \Phi(\hat{D}_i x, \hat{D}_l^v, 0, \hat{\theta}) X(t) + k_i(\hat{\theta}) \frac{\partial \Phi(\hat{D}_i x, \hat{D}_l^v, 0, \hat{\theta})}{\partial \hat{\theta}_h} X(t)$$

$$+ \sum_{j=1}^m \hat{D}_j \int_0^{\phi_j(\hat{D}_i x, \hat{D}_j)} \left(\frac{\partial k_i(\hat{\theta})}{\partial \hat{\theta}_h} \Phi(\hat{D}_i x, \hat{D}_l^v, \hat{D}_j y, \hat{\theta}) b_j(\hat{\theta}) \right.$$

$$+ k_i(\hat{\theta}) \frac{\partial \Phi(\hat{D}_i x, \hat{D}_l^v, \hat{D}_j y, \hat{\theta})}{\partial \hat{\theta}_h} b_j(\hat{\theta})$$

$$\left. + k_i(\hat{\theta}) \Phi(\hat{D}_i x, \hat{D}_l^v, \hat{D}_j y, \hat{\theta}) \frac{\partial b_j(\hat{\theta})}{\partial \hat{\theta}_h} \right) u_j(y, t) dy, \tag{7.156}$$

and $\frac{\partial \Phi(\hat{D}_i x, \hat{D}_l^v, \hat{D}_j y, \hat{\theta})}{\partial \hat{\theta}_h}$ has the form corresponding to $\Phi(\hat{D}_i x, \hat{D}_l^v, \hat{D}_j y, \hat{\theta})$ of (7.85)–(7.87) such that:

For $\Phi(\hat{D}_i x, \hat{D}_l^v, \hat{D}_j y, \hat{\theta})$ of (7.85)

$$\frac{\partial \Phi(\hat{D}_i x, \hat{D}_l^v, \hat{D}_j y, \hat{\theta})}{\partial \hat{\theta}_h} = \frac{\partial A_{v-1}(\hat{\theta})}{\partial \hat{\theta}_h}(\hat{D}_i x - \hat{D}_j y)e^{A_{v-1}(\hat{\theta})(\hat{D}_i x - \hat{D}_j y)}, \quad (7.157)$$

$$\frac{\partial \Phi(\hat{D}_i x, \hat{D}_l^v, 0, \hat{\theta})}{\partial \hat{\theta}_h} = \frac{\partial A_{v-1}(\hat{\theta})}{\partial \hat{\theta}_h}\hat{D}_i x e^{A_{v-1}(\hat{\theta})\hat{D}_i x}. \quad (7.158)$$

For $\Phi(\hat{D}_i x, \hat{D}_l^v, \hat{D}_j y, \hat{\theta})$ of (7.86)

$$\frac{\partial \Phi(\hat{D}_i x, \hat{D}_l^v, \hat{D}_j y, \hat{\theta})}{\partial \hat{\theta}_h}$$

$$= \frac{\partial A_{v-1}(\hat{\theta})}{\partial \hat{\theta}_h}(\hat{D}_i x - \hat{D}_{v-1})e^{A_{v-1}(\hat{\theta})(\hat{D}_i x - \hat{D}_{v-1})}e^{A_{v-2}(\hat{\theta})(\hat{D}_{v-1} - \hat{D}_{v-2})}\cdots$$

$$\times e^{A_l(\hat{\theta})(\hat{D}_{l+1} - \hat{D}_l)}e^{A_{l-1}(\hat{\theta})(\hat{D}_l - \hat{D}_j y)}$$

$$+ e^{A_{v-1}(\hat{\theta})(\hat{D}_i x - \hat{D}_{v-1})}\frac{\partial A_{v-2}(\hat{\theta})}{\partial \hat{\theta}_h}(\hat{D}_{v-1} - \hat{D}_{v-2})e^{A_{v-2}(\hat{\theta})(\hat{D}_{v-1} - \hat{D}_{v-2})}\cdots$$

$$\times e^{A_l(\hat{\theta})(\hat{D}_{l+1} - \hat{D}_l)}e^{A_{l-1}(\hat{\theta})(\hat{D}_l - \hat{D}_j y)}$$

$$+ \quad \cdots \quad \cdots$$

$$+ e^{A_{v-1}(\hat{\theta})(\hat{D}_i x - \hat{D}_{v-1})}e^{A_{v-2}(\hat{\theta})(\hat{D}_{v-1} - \hat{D}_{v-2})}\cdots$$

$$\times \frac{\partial A_l(\hat{\theta})}{\partial \hat{\theta}_h}(\hat{D}_{l+1} - \hat{D}_l)e^{A_l(\hat{\theta})(\hat{D}_{l+1} - \hat{D}_l)}e^{A_{l-1}(\hat{\theta})(\hat{D}_l - \hat{D}_j y)}$$

$$+ e^{A_{v-1}(\hat{D}_i x - \hat{D}_{v-1})}e^{A_{v-2}(\hat{D}_{v-1} - \hat{D}_{v-2})}\cdots$$

$$\times e^{A_l(\hat{\theta})(\hat{D}_{l+1} - \hat{D}_l)}e^{A_{l-1}(\hat{\theta})(\hat{D}_l - \hat{D}_j y)}\frac{\partial A_{l-1}(\hat{\theta})}{\partial \hat{\theta}_h}(\hat{D}_l - \hat{D}_j y), \quad (7.159)$$

$$\frac{\partial \Phi(\hat{D}_i x, \hat{D}_l^v, 0, \hat{\theta})}{\partial \hat{\theta}_h}$$

$$= \frac{\partial A_{v-1}(\hat{\theta})}{\partial \hat{\theta}_h}(\hat{D}_i x - \hat{D}_{v-1})e^{A_{v-1}(\hat{\theta})(\hat{D}_i x - \hat{D}_{v-1})}e^{A_{v-2}(\hat{\theta})(\hat{D}_{v-1} - \hat{D}_{v-2})}\cdots$$

$$\times e^{A_l(\hat{\theta})(\hat{D}_{l+1} - \hat{D}_l)}e^{A_{l-1}(\hat{\theta})\hat{D}_l}$$

$$+ e^{A_{v-1}(\hat{\theta})(\hat{D}_i x - \hat{D}_{v-1})}\frac{\partial A_{v-2}(\hat{\theta})}{\partial \hat{\theta}_h}(\hat{D}_{v-1} - \hat{D}_{v-2})e^{A_{v-2}(\hat{\theta})(\hat{D}_{v-1} - \hat{D}_{v-2})}\cdots$$

$$\times e^{A_l(\hat{\theta})(\hat{D}_{l+1} - \hat{D}_l)}e^{A_{l-1}(\hat{\theta})\hat{D}_l}$$

$$+ \quad \cdots \quad \cdots$$

$$+ e^{A_{v-1}(\hat{\theta})(\hat{D}_i x - \hat{D}_{v-1})} e^{A_{v-2}(\hat{\theta})(\hat{D}_{v-1} - \hat{D}_{v-2})} \cdots$$

$$\times \frac{\partial A_l(\hat{\theta})}{\partial \hat{\theta}_h} (\hat{D}_{l+1} - \hat{D}_l) e^{A_l(\hat{\theta})(\hat{D}_{l+1} - \hat{D}_l)} e^{A_{l-1}(\hat{\theta})\hat{D}_l}$$

$$+ e^{A_{v-1}(\hat{\theta})(\hat{D}_i x - \hat{D}_{v-1})} e^{A_{v-2}(\hat{\theta})(\hat{D}_{v-1} - \hat{D}_{v-2})} \cdots$$

$$\times e^{A_l(\hat{\theta})(\hat{D}_{l+1} - \hat{D}_l)} e^{A_{l-1}(\hat{\theta})\hat{D}_l} \frac{\partial A_{l-1}(\hat{\theta})}{\partial \hat{\theta}_h} \hat{D}_l. \tag{7.160}$$

For $\Phi(\hat{D}_i x, \hat{D}_l^v, \hat{D}_j y, \hat{\theta})$ of (7.87)

$$\frac{\partial \Phi(\hat{D}_i x, \hat{D}_l^v, \hat{D}_j y, \hat{\theta})}{\partial \hat{\theta}_h} = e^{A_{v-1}(\hat{\theta})(\hat{D}_i x - \hat{D}_{v-1})} \left(\frac{\partial A_{v-1}(\hat{\theta})}{\partial \hat{\theta}_h} (\hat{D}_i x - \hat{D}_{v-1}) \right.$$

$$\left. + \frac{\partial A_{v-2}(\hat{\theta})}{\partial \hat{\theta}_h} (\hat{D}_{v-1} - \hat{D}_j y) \right) e^{A_{v-2}(\hat{\theta})(\hat{D}_{v-1} - \hat{D}_j y)},$$

$$\tag{7.161}$$

$$\frac{\partial \Phi(\hat{D}_i x, \hat{D}_l^v, 0, \hat{\theta})}{\partial \hat{\theta}_h} = e^{A_{v-1}(\hat{\theta})(\hat{D}_i x - \hat{D}_{v-1})} \left(\frac{\partial A_{v-1}(\hat{\theta})}{\partial \hat{\theta}_h} (\hat{D}_i x - \hat{D}_{v-1}) \right.$$

$$\left. + \frac{\partial A_{v-2}(\hat{\theta})}{\partial \hat{\theta}_h} \hat{D}_{v-1} \right) e^{A_{v-2}(\hat{\theta})\hat{D}_{v-1}}. \tag{7.162}$$

Chapter Eight

Output Feedback of Uncertain Multi-Input Systems

IN CHAPTER 7 we solved the problem of adaptive stabilization in the presence of distinct discrete multiple actuator delays that are long and unknown, under the assumption that the actuator state in each input channel is available for measurement. In this chapter this assumption of measurable actuator state is removed, as shown in table 7.1.

Similar to the case of single-input delay, the result of multi-input delays that we obtain in this chapter is not global, as we do not believe the problem where the actuator state is not measurable and the delay value is unknown at the same time is solvable globally, since the problem is not linearly parameterized. We want to state up front that, in a practical sense, the stability result we prove in this chapter is not a highly satisfactory result since it is local both in the initial state and in the initial parameter error. This means that the initial delay estimate needs to be sufficiently close to the true delay. (The delay can be long, but it needs to be known quite closely.) Under such an assumption, we would argue, one might as well use a linear controller and rely on robustness of the feedback law to small errors in the assumed delay value (just as what we presented in section 3.4). Nevertheless, we present the local result here as it highlights quite clearly why a global result is not obtainable when both the delay value and the delay state are unavailable. Hence, this result amplifies the significance of the global result in chapter 7 and highlights the importance of employing the measurement of the delay state, when available, such as when the delay is the result of a physical transport process.

8.1 Model Depiction

Consider the general ODE-PDE cascade (7.25)–(7.27) in section 7.2, which is copied as follows for readers' convenience:

$$\dot{X}(t) = AX(t) + \sum_{i=1}^{m} b_i u_i(0, t), \tag{8.1}$$

$$D_i \partial_t u_i(x, t) = \partial_x u_i(x, t), \quad x \in [0, 1], \tag{8.2}$$

$$u_i(1, t) = U_i(t), \quad i \in \{1, 2, \cdots, m\}, \tag{8.3}$$

where the solution of transport PDE (8.2)–(8.3) representing the delayed input has the following form:

$$u_i(x,t) = U_i(t + D_i(x-1)), x \in [0,1]. \tag{8.4}$$

The ODE-PDE cascade (8.1)–(8.3) represents the finite-dimensional LTI system with distinct multi-input delays whose dynamics satisfies the ODE as follows:

$$\dot{X}(t) = AX(t) + \sum_{i=1}^{m} b_i U_i(t - D_i), \tag{8.5}$$

where the matrices A and b_i satisfy Assumption 6.1 in section 6.1, and the delays D_i satisfy Assumption 7.5 in section 7.2.

8.2   Local Stabilization under Uncertain Delays and PDE States

In chapter 7, we dealt with the stabilization problem for the case of unknown delays under the assumption that the actuator state $u_i(x,t)$ is measurable. In this chapter, as shown in the fifth row of table 7.1, we deal with the more challenging case where neither is the delay D_i known nor is the transport PDE $u_i(x,t)$ of (8.4) measurable, which is more common in practice. Since the transport PDE $u_i(x,t)$ is unmeasurable, it cannot be utilized directly. To solve this technical difficulty, we introduce the PDE observer to estimate the unmeasured $u_i(x,t)$ such that, for $i = 1, 2, \cdots, m$,

$$\hat{u}_i(x,t) = U_i(t + \hat{D}_i(t)(x-1)), \quad x \in [0,1], \tag{8.6}$$

which is governed by the PDE observer dynamics

$$\hat{D}_i(t)\partial_t \hat{u}_i(x,t) = \partial_x \hat{u}_i(x,t) + \dot{\hat{D}}_i(t)(x-1)\partial_x \hat{u}_i(x,t), \tag{8.7}$$

$$\hat{u}_i(1,t) = U_i(t), \tag{8.8}$$

with the PDE state estimation error being defined as

$$\tilde{u}_i(x,t) = u_i(x,t) - \hat{u}_i(x,t), \quad x \in [0,1], \tag{8.9}$$

$$= U_i(t + D_i(x-1)) - U_i(t + \hat{D}_i(t)(x-1)), \tag{8.10}$$

where $\hat{D}_i(t)$ is an estimate of the unknown delay D_i with the estimation error $\tilde{D}_i(t) = D_i - \hat{D}_i(t)$.

As a consequence, for $i \in \{1, 2, \cdots, m\}$ and on the interval $x \in [0,1]$, the ODE-PDE cascade (8.1)–(8.3) could be expressed as the following system:

$$\dot{X}(t) = AX(t) + \sum_{i=1}^{m} b_i \big(\hat{u}_i(0,t) + \tilde{u}_i(0,t) \big), \tag{8.11}$$

$$D_i \partial_t \tilde{u}_i(x,t) = \partial_x \tilde{u}_i(x,t) - \tilde{D}_i \frac{1}{\hat{D}_i} \partial_x \hat{u}_i(x,t)$$

$$- \dot{\hat{D}}_i D_i(x-1) \frac{1}{\hat{D}_i} \partial_x \hat{u}_i(x,t), \tag{8.12}$$

$$\tilde{u}_i(1,t) = 0, \tag{8.13}$$

$$\hat{D}_i \partial_t \hat{u}_i(x,t) = \partial_x \hat{u}_i(x,t) + \dot{\hat{D}}_i(x-1) \partial_x \hat{u}_i(x,t), \tag{8.14}$$

$$\hat{u}_i(1,t) = U_i(t). \tag{8.15}$$

Remark 8.1. Now we give an explanation for the unity-interval rescaling in chapter 6.2. By unity-interval rescaling the system from the interval $\sigma \in [0, D_i]$ into the interval $x \in [0,1]$ and applying the certainty-equivalence principle to identify the unknown delay, we satisfy the boundary condition (8.13) such that $\tilde{u}_i(1,t) = u_i(1,t) - \hat{u}_i(1,t) = U_i(t + D_i(1-1)) - U_i(t + \hat{D}_i(1-1)) = U_i(t) - U_i(t) = 0$. For the Lyapunov-based stability analysis later, this stable boundary condition (8.13) plays an important role in establishing the boundedness of the PDE state estimation error $\tilde{u}_i(0,t) = u_i(0,t) - \hat{u}_i(0,t) = U_i(t - D_i) - U_i(t - \hat{D}_i)$. If we do not employ the unity-interval rescaling and apply the certainty-equivalence principle directly to (6.3) such that $\hat{u}(\sigma,t) = U(t + \sigma - \hat{D}_i)$, $\sigma \in [0, \hat{D}_i]$, then the boundary condition becomes $\tilde{u}_i(D_i, t) = u_i(D_i, t) - \hat{u}_i(D_i, t) = U_i(t + D_i - D_i) - U_i(t + D_i - \hat{D}_i) \neq 0$ or $\tilde{u}_i(\hat{D}_i, t) = u_i(\hat{D}_i, t) - \hat{u}_i(\hat{D}_i, t) = U_i(t + \hat{D}_i - D_i) - U_i(t + \hat{D}_i - \hat{D}_i) \neq 0$; thus the boundedness of the PDE state estimation error $\tilde{u}_i(0,t) = u_i(0,t) - \hat{u}_i(0,t) = U_i(t - D_i) - U_i(t - \hat{D}_i)$ could not be guaranteed.

Similarly, as D_i and $u_i(x,t)$ are uncertain, based on the certainty-equivalence principle, by replacing D_i with $\hat{D}_i$ and $u_i(x,t)$ with $\hat{u}_i(x,t)$ in (6.58) of section 6.2, we present the invertible backstepping transformation $(X(t), \hat{u}_i(x,t)) \rightarrow (X(t), \hat{w}_i(x,t))$ in the adaptive control scheme as follows:

$$\hat{w}_i(x,t) = \hat{u}_i(x,t) - k_i \Phi(\hat{D}_i x, \hat{D}_l^v, 0) X(t)$$

$$- \sum_{j=1}^{m} \hat{D}_j \int_0^{\phi_j(\hat{D}_i x, \hat{D}_j)} k_i \Phi(\hat{D}_i x, \hat{D}_l^v, \hat{D}_j y) b_j \hat{u}_j(y,t) dy, \tag{8.16}$$

$$\hat{u}_i(x,t) = \hat{w}_i(x,t) + k_i e^{A_m \hat{D}_i x} X(t)$$

$$+ \sum_{j=1}^{m} \hat{D}_j \int_0^{\phi_j(\hat{D}_i x, \hat{D}_j)} k_i e^{A_m(\hat{D}_i x - \hat{D}_j y)} b_j \hat{w}_j(y,t) dy, \tag{8.17}$$

where for $j \in \{1, 2, \cdots, m\}$,

$$\phi_j(\hat{D}_i x, \hat{D}_j) = \begin{cases} \frac{\hat{D}_i}{\hat{D}_j} x, & x \in \left[0, \frac{\hat{D}_j}{\hat{D}_i}\right] \\ 1, & x \in \left[\frac{\hat{D}_j}{\hat{D}_i}, \frac{\hat{D}_m}{\hat{D}_i}\right] \end{cases} \tag{8.18}$$

and $\quad \hat{D}_l^v = (\hat{D}_{v-1}, \hat{D}_{v-2}, \cdots, \hat{D}_{l+1}, \hat{D}_l), \quad$ for $\quad \frac{\hat{D}_{v-1}}{\hat{D}_j} \leq y \leq \frac{\hat{D}_v}{\hat{D}_j}, x \leq \frac{\hat{D}_v}{\hat{D}_j}, \quad i, j \in \{1, 2, \cdots, m\}, v \in \{1, 2, \cdots, i\}$,

$$\Phi(\hat{D}_i x, \hat{D}_l^v, \hat{D}_j y) = e^{A_{v-1}(\hat{D}_i x - \hat{D}_j y)}, \tag{8.19}$$

for $\frac{\hat{D}_{v-1}}{\hat{D}_i} \leq x \leq \frac{\hat{D}_v}{\hat{D}_i}, \frac{\hat{D}_{l-1}}{\hat{D}_j} \leq y \leq \frac{\hat{D}_l}{\hat{D}_j}, i, j \in \{1, 2, \cdots, m\}, v, l \in \{1, 2, \cdots, i\}, l < v$,

$$\Phi(\hat{D}_i x, \hat{D}_l^v, \hat{D}_j y) = e^{A_{v-1}(\hat{D}_i x - \hat{D}_{v-1})} e^{A_{v-2}(\hat{D}_{v-1} - \hat{D}_{v-2})}$$
$$\cdots e^{A_l(\hat{D}_{l+1} - \hat{D}_l)} e^{A_{l-1}(\hat{D}_l - \hat{D}_j y)}, \tag{8.20}$$

and specifically for (8.20), if $l = v - 1$,

$$\Phi(\hat{D}_i x, \hat{D}_l^v, \hat{D}_j y) = e^{A_{v-1}(\hat{D}_i x - \hat{D}_{v-1})} e^{A_{v-2}(\hat{D}_{v-1} - \hat{D}_j y)}, \tag{8.21}$$

with $\hat{D}_0 = 0$ and A_i for $i = 0, 1, \cdots, m$ defined the same as those in (7.35) such that

$$\hat{D}_0 = 0, \quad \begin{cases} A_0 = A, \\ A_i = A_{i-1} + b_i k_i, i \in \{1, 2, \cdots, m\} \\ A_m = A + BK = A + \sum_{i=1}^m b_i k_i. \end{cases} \tag{8.22}$$

Remark 8.2. Here we would like to draw the reader's attention to the fact that the spatial domain of the transport PDE $u_i(x, t)$ in (8.4) and $\hat{u}_i(x, t)$ in (8.6) is $x \in [0, 1]$, and for $\hat{w}_i(x, t)$ in (8.16). At the boundary $x = 0$ and $x = 1$, the delayed input and the actual input are represented, respectively. If Assumption 7.5 in section 7.2 were absent, then the order of the delay estimate $\hat{D}_1 \leq \hat{D}_2 \leq \cdots \leq \hat{D}_m$ cannot be guaranteed by the later projection operator. Consequently, $\frac{\hat{D}_j}{\hat{D}_i} > 1$ may happen to the limit (8.18) of the integration of backstepping transformation (8.16)–(8.17), which makes $\hat{u}_j(y, t)$ and $\hat{w}_j(y, t)$ potentially go beyond their domain $y \in [0, 1]$. Thus the backstepping transformation, the subsequent control laws and update laws are no longer meaningfully defined. To address this problem, a switched scheme by an LTI system with two inputs is offered in a simulation shown later in this chapter.

Subsequently, for $i = 1, 2, \cdots, m$, if we substitute $x = 1$ into (8.16) to make $\hat{w}_i(1, t) = 0$, the prediction-based PDE boundary control in the adaptive feedback scheme is

$$U_i(t) = \hat{u}_i(1, t) = k_i \Phi(\hat{D}_i, \hat{D}_l^v, 0) X(t)$$
$$+ \sum_{j=1}^m \hat{D}_j \int_0^{\phi_j(\hat{D}_i, \hat{D}_j)} k_i \Phi(\hat{D}_i, \hat{D}_l^v, \hat{D}_j y) b_j \hat{u}_j(y, t) dy. \tag{8.23}$$

Then, if we follow a similar design choice in [28], the delay update law is selected as follows: for $i = \{1, 2, \cdots, m\}$,

$$\dot{\hat{D}}_i(t) = \gamma_D \mathrm{Proj}_{[\underline{D}_{i-1,i}, \overline{D}_{i,i+1}]} \tau_{D_i}(t), \quad \hat{D}_i(0) \in [\underline{D}_{i-1,i}, \overline{D}_{i,i+1}], \tag{8.24}$$

$$\tau_{D_i}(t) = -\sum_{h=1}^{m} \int_0^1 (1+x)\hat{w}_h(x,t)\hat{p}_{hi}(x,t)dx, \tag{8.25}$$

where $\gamma_D > 0$ is a design coefficient, and $\mathrm{Proj}_{[\underline{D}_{i-1,i}, \overline{D}_{i,i+1}]}(\cdot)$ is a standard projection operator defined on the interval $[\underline{D}_{i-1,i}, \overline{D}_{i,i+1}]$ (whose explicit form could be found in appendix E of [80, 27, 28]). It is used here to guarantee the length sequence of delay estimates such that $0 < \hat{D}_1 \leq \hat{D}_2 \leq \cdots \leq \hat{D}_m$, and to keep the denominator in (8.18) away from zero. How to get (8.25) is discussed in Remark 8.4 later and $\hat{p}_{hi}(x,y)$ has the form

$$\hat{p}_{ih}(x,t) = k_i P_{\Phi(\hat{D}_i x, \hat{D}_l^v, 0)} X(t) + k_i \Phi(\hat{D}_i x, \hat{D}_l^v, 0) b_h \hat{u}_h(0,t)$$

$$+ \sum_{j=1}^{m} \hat{D}_j \int_0^{\phi_j(\hat{D}_i x, \hat{D}_j)} k_i P_{\Phi(\hat{D}_i x, \hat{D}_l^v, \hat{D}_j y)} b_j \hat{u}_j(y,t)dy$$

$$+ \sum_{j=1}^{m} \rho_j k_i b_j \hat{u}_j(\phi_j(\hat{D}_i x, \hat{D}_j), t), \tag{8.26}$$

where

$$\rho_j = \begin{cases} 1, & \text{if } h = i, \phi_j(\hat{D}_i x, \hat{D}_j) = \frac{\hat{D}_i}{\hat{D}_j} x, i \neq j \\ 0, & \text{if } h \neq i, j, \quad \text{or} \quad \phi_j(\hat{D}_i x, \hat{D}_j) = \frac{\hat{D}_i}{\hat{D}_j} x, i = j, \\ & \text{or} \quad \phi_j(\hat{D}_i x, \hat{D}_j) = 1 \\ -1, & \text{if } h = j, \phi_j(\hat{D}_i x, \hat{D}_j) = \frac{\hat{D}_i}{\hat{D}_j} x, i \neq j, \end{cases} \tag{8.27}$$

$$P_{\Phi(\hat{D}_i x, \hat{D}_l^v, 0)} = \begin{cases} \frac{\partial \Phi(\hat{D}_i x, \hat{D}_l^v, 0)}{\partial(\hat{D}_i x)}, & h = i \\ \frac{\partial \Phi(\hat{D}_i x, \hat{D}_l^v, 0)}{\partial \hat{D}_h}, & h \neq i, \end{cases} \tag{8.28}$$

and $P_{\Phi(\hat{D}_i x, \hat{D}_l^v, \hat{D}_j y)}$ has the form corresponding to $\Phi(\hat{D}_i x, \hat{D}_l^v, \hat{D}_j y)$ of (8.19)–(8.21), as given next.

For $\Phi(\hat{D}_i x, \hat{D}_l^v, \hat{D}_j y)$ of (8.19),

$$P_{\Phi(\hat{D}_i x, \hat{D}_l^v, \hat{D}_j y)} = \begin{cases} \frac{\partial \Phi(\hat{D}_i x, \hat{D}_l^v, \hat{D}_j y)}{\partial(\hat{D}_i x)}, & h = i, h \neq j \\ \frac{\partial \Phi(\hat{D}_i x, \hat{D}_l^v, \hat{D}_j y)}{\partial \hat{D}_h} = 0, & h \neq i, j \\ 0, & h = i = j \\ \frac{\partial \Phi(\hat{D}_i x, \hat{D}_l^v, \hat{D}_j y)}{\partial(\hat{D}_j y)}, & h = j, h \neq i. \end{cases} \tag{8.29}$$

For $\Phi(\hat{D}_i x, \hat{D}_l^v, \hat{D}_j y)$ of (8.20)–(8.21)

$$
P_{\Phi(\hat{D}_i x, \hat{D}_l^v, \hat{D}_j y)} =
\begin{cases}
A_{v-1} e^{A_{v-1}\hat{D}_i x} \Gamma_l^v(\hat{D}) e^{-A_{l-1}\hat{D}_j y} \\[4pt]
\quad + e^{A_{v-1}\hat{D}_i x} \dfrac{\partial \Gamma_l^v(\hat{D})}{\partial \hat{D}_h} e^{-A_{l-1}\hat{D}_j y}, & h=i, h\neq j \\[10pt]
e^{A_{v-1}\hat{D}_i x} \dfrac{\partial \Gamma_l^v(\hat{D})}{\partial \hat{D}_h} e^{-A_{l-1}\hat{D}_j y}, & h\neq i, j \\[10pt]
A_{v-1} e^{A_{v-1}\hat{D}_i x} \Gamma_l^v(\hat{D}) e^{-A_{l-1}\hat{D}_j y} \\[4pt]
\quad + e^{A_{v-1}\hat{D}_i x} \dfrac{\partial \Gamma_l^v(\hat{D})}{\partial \hat{D}_h} e^{-A_{l-1}\hat{D}_j y} \\[4pt]
\quad + e^{A_{v-1}\hat{D}_i x} \Gamma_l^v(\hat{D}) e^{-A_{l-1}\hat{D}_j y}(-A_{l-1}), & h=i=j \\[10pt]
e^{A_{v-1}\hat{D}_i x} \dfrac{\partial \Gamma_l^v(\hat{D})}{\partial \hat{D}_h} e^{-A_{l-1}\hat{D}_j y} \\[4pt]
\quad + e^{A_{v-1}\hat{D}_i x} \Gamma_l^v(\hat{D}) e^{-A_{l-1}\hat{D}_j y}(-A_{l-1}), & h=j, h\neq i,
\end{cases}
$$

with $\Gamma_l^v(\hat{D})$ coming from (8.20)–(8.21) such that

$$
\begin{aligned}
\Phi(\hat{D}_i x, \hat{D}_l^v, \hat{D}_j y) &= e^{A_{v-1}(\hat{D}_i x - \hat{D}_{v-1})} e^{A_{v-2}(\hat{D}_{v-1}-\hat{D}_{v-2})} \cdots \\
&\quad \times e^{A_l(\hat{D}_{l+1}-\hat{D}_l)} e^{A_{l-1}(\hat{D}_l - \hat{D}_j y)} \\
&= e^{A_{v-1}\hat{D}_i x} \left(e^{-A_{v-1}\hat{D}_{v-1}} e^{A_{v-2}(\hat{D}_{v-1}-\hat{D}_{v-2})} \cdots \right. \\
&\quad \left. \times \ e^{A_l(\hat{D}_{l+1}-\hat{D}_l)} e^{A_{l-1}\hat{D}_l} \right) e^{-A_{l-1}\hat{D}_j y} \\
&= e^{A_{v-1}\hat{D}_i x} \Gamma_l^v(\hat{D}) e^{-A_{l-1}\hat{D}_j y}, \tag{8.30}
\end{aligned}
$$

and specifically, if $l = v - 1$,

$$
\begin{aligned}
\Phi(\hat{D}_i x, \hat{D}_l^v, \hat{D}_j y) &= e^{A_{v-1}(\hat{D}_i x - \hat{D}_{v-1})} e^{A_{v-2}(\hat{D}_{v-1}-\hat{D}_j y)} \\
&= e^{A_{v-1}\hat{D}_i x} \left(e^{-A_{v-1}\hat{D}_{v-1}} e^{A_{v-2}\hat{D}_{v-1}} \right) e^{-A_{v-2}\hat{D}_j y} \\
&= e^{A_{v-1}\hat{D}_i x} \Gamma_l^v(\hat{D}) e^{-A_{l-1}\hat{D}_j y}. \tag{8.31}
\end{aligned}
$$

Remark 8.3. Referring to the statement of [25], we list a couple of potentially necessary conditions to protect the stability of the closed-loop system, for the delay update law design to satisfy, as follows:

Condition 1: There exists a constant $M_D > 0$ such that

$$
|\tau_{D_i}(t)| \leq M_D \bar{V}, \tag{8.32}
$$

$$
\bar{V} = |X(t)|^2 + \sum_{i=1}^{m} \left(\|u_i(x,t)\|^2 + \|\hat{u}_i(x,t)\|^2 + \|\partial_x \hat{u}_i(x,t)\|^2 \right). \tag{8.33}
$$

Condition 2: There exists a constant $M_D > 0$ such that

$$\forall t \geq 0, \quad \tau_{D_i}(t)\tilde{D}_i(t) \geq 0 \quad \text{and} \quad |\tau_{D_i}(t)| \leq M_D. \tag{8.34}$$

Obviously our update law (8.25) meets Condition 1, and as shown in later Lyapunov-based analysis, this is useful for the establishment of the stability of the closed-loop system. Similar to the claim in [25], other adaptive update laws being a product of a regressor and a regulation error and satisfying one of the above conditions could be a possible choice.

Under the above backstepping transformation (8.16)–(8.17), the PDE boundary control (8.23), and the delay update law (8.24)–(8.25), after a lengthy and sophisticated calculation, the ODE-PDE cascade (8.11)–(8.15) could be converted into the target system as follows:

$$\dot{X}(t) = (A + BK)X(t)$$
$$+ \sum_{i=1}^{m} b_i\big(\hat{w}_i(0,t) + \tilde{u}_i(0,t)\big), \tag{8.35}$$

$$D_i \partial_t \tilde{u}_i(x,t) = \partial_x \tilde{u}_i(x,t) - \tilde{D}_i r_i(x,t)$$
$$- \dot{\hat{D}}_i D_i(x-1)r_i(x,t), \tag{8.36}$$

$$\tilde{u}_i(1,t) = 0, \tag{8.37}$$

$$\hat{D}_i \partial_t \hat{w}_i(x,t) = \partial_x \hat{w}_i(x,t) - \hat{D}_i s_i(x,t)$$
$$- \hat{D}_i k_i \Phi(\hat{D}_i x, \hat{D}_l^v, 0) \sum_{j=1}^{m} b_j \tilde{u}_j(0,t), \tag{8.38}$$

$$\hat{w}_i(1,t) = 0, \tag{8.39}$$

$$\hat{D}_i \partial_{xt} \hat{w}_i(x,t) = \partial_{xx} \hat{w}_i(x,t) - \hat{D}_i \partial_x s_i(x,t)$$
$$- \hat{D}_i k_i \partial_x \Phi(\hat{D}_i x, \hat{D}_l^v, 0) \sum_{j=1}^{m} b_j \tilde{u}_j(0,t), \tag{8.40}$$

$$\partial_x \hat{w}_i(1,t) = \hat{D}_i s_i(1,t)$$
$$+ \hat{D}_i k_i \Phi(\hat{D}_i, \hat{D}_l^v, 0) \sum_{j=1}^{m} b_j \tilde{u}_j(0,t), \tag{8.41}$$

where, for $i \in \{1, 2, \cdots, m\}$,

$$r_i(x,t) = \frac{1}{\hat{D}_i} \partial_x \hat{u}_i(x,t)$$

$$= \frac{1}{\hat{D}_i} \partial_x \hat{w}_i(x,t) + k_i A_m e^{A_m \hat{D}_i x} X(t)$$

$$+ \frac{1}{\hat{D}_i} \sum_{j=1}^{m} \hat{D}_j \partial_x \phi_j(\hat{D}_i x, \hat{D}_j) k_i b_j \hat{w}_j(\phi_j(\hat{D}_i x, \hat{D}_j), t)$$

$$+ \sum_{j=1}^{m} \hat{D}_j \int_0^{\phi_j(\hat{D}_i x, \hat{D}_j)} k_i A_m e^{A_m(\hat{D}_i x - \hat{D}_j y)} b_j \hat{w}_j(y,t) dy,$$

$$(8.42)$$

$$s_i(x,t) = \dot{\hat{D}}_i \frac{(1-x)}{\hat{D}_i} \partial_x \hat{w}_i(x,t)$$

$$+ k_i \Phi(\hat{D}_i x, \hat{D}_l^v, 0) \sum_{j=1}^{m} \dot{\hat{D}}_j b_j \hat{u}_j(0,t)$$

$$+ k_i H_{\Phi(\hat{D}_i x, \hat{D}_l^v, 0)}(\dot{\hat{D}}) X(t)$$

$$+ \sum_{\phi_j(\hat{D}_i x, \hat{D}_j)} (\dot{\hat{D}}_i - \dot{\hat{D}}_j) k_i b_j \hat{u}_j \left(\frac{\hat{D}_i}{\hat{D}_j} x, t \right)$$

$$+ \sum_{j=1}^{m} \hat{D}_j \int_0^{\phi_j(\hat{D}_i x, \hat{D}_j)} k_i H_{\Phi(\hat{D}_i x, \hat{D}_l^v, \hat{D}_j y)}(\dot{\hat{D}})$$

$$\times b_j \hat{u}_j(y,t) dy,$$

$$(8.43)$$

where $H_{\Phi(\hat{D}_i x, \hat{D}_l^v, \hat{D}_j y)}(\dot{\hat{D}})$ has the form corresponding to $\Phi(\hat{D}_i x, \hat{D}_l^v, \hat{D}_j y)$ of (8.19)–(8.21), its elaborate form is given in (8.44)–(8.48) below, and the term containing $\hat{u}_j \left(\frac{\hat{D}_i}{\hat{D}_j} x, t \right)$ in (8.43) exists only when $\phi_j(\hat{D}_i x, \hat{D}_j) = \frac{\hat{D}_i}{\hat{D}_j} x$.

For $\Phi(\hat{D}_i x, \hat{D}_l^v, \hat{D}_j y)$ of (8.19),

$$H_{\Phi(\hat{D}_i x, \hat{D}_l^v, \hat{D}_j y)}(\dot{\hat{D}}) = A_{v-1}(\dot{\hat{D}}_i - \dot{\hat{D}}_j) e^{A_{v-1}(\hat{D}_i x - \hat{D}_j y)}, \qquad (8.44)$$

$$H_{\Phi(\hat{D}_i x, \hat{D}_l^v, 0)}(\dot{\hat{D}}) = A_{v-1} \dot{\hat{D}}_i e^{A_{v-1} \hat{D}_i x}. \qquad (8.45)$$

For $\Phi(\hat{D}_i x, \hat{D}_l^v, \hat{D}_j y)$ of (8.20),

$$H_{\Phi(\hat{D}_i x, \hat{D}_l^v, \hat{D}_j y)}(\dot{\hat{D}})$$

$$= A_{v-1}(\dot{\hat{D}}_i - \dot{\hat{D}}_{v-1}) e^{A_{v-1}(\hat{D}_i x - \hat{D}_{v-1})} e^{A_{v-2}(\hat{D}_{v-1} - \hat{D}_{v-2})} \cdots$$

$$\times e^{A_l(\hat{D}_{l+1} - \hat{D}_l)} e^{A_{l-1}(\hat{D}_l - \hat{D}_j y)}$$

$$+ e^{A_{v-1}(\hat{D}_i x - \hat{D}_{v-1})} A_{v-2}(\dot{\hat{D}}_{v-1} - \dot{\hat{D}}_{v-2}) e^{A_{v-2}(\hat{D}_{v-1} - \hat{D}_{v-2})} \cdots$$

$$\times e^{A_l(\hat{D}_{l+1} - \hat{D}_l)} e^{A_{l-1}(\hat{D}_l - \hat{D}_j y)}$$

$$+ \quad \cdots \quad \cdots$$

$$+ e^{A_{v-1}(\hat{D}_i x - \hat{D}_{v-1})} e^{A_{v-2}(\hat{D}_{v-1} - \hat{D}_{v-2})} \cdots$$

$$\times A_l(\dot{\hat{D}}_{l+1} - \dot{\hat{D}}_l) e^{A_l(\hat{D}_{l+1} - \hat{D}_l)} e^{A_{l-1}(\hat{D}_l - \hat{D}_j y)}$$

$$+ e^{A_{v-1}(\hat{D}_i x - \hat{D}_{v-1})} e^{A_{v-2}(\hat{D}_{v-1} - \hat{D}_{v-2})} \cdots$$

$$\times e^{A_l(\hat{D}_{l+1} - \hat{D}_l)} e^{A_{l-1}(\hat{D}_l - \hat{D}_j y)} A_{l-1}(\dot{\hat{D}}_l - \dot{\hat{D}}_j), \tag{8.46}$$

$$H_{\Phi(\hat{D}_i x, \hat{D}_l^v, 0)}(\dot{\hat{D}})$$

$$= A_{v-1}(\dot{\hat{D}}_i - \dot{\hat{D}}_{v-1}) e^{A_{v-1}(\hat{D}_i x - \hat{D}_{v-1})} e^{A_{v-2}(\hat{D}_{v-1} - \hat{D}_{v-2})} \cdots$$

$$\times e^{A_l(\hat{D}_{l+1} - \hat{D}_l)} e^{A_{l-1}\hat{D}_l}$$

$$+ e^{A_{v-1}(\hat{D}_i x - \hat{D}_{v-1})} A_{v-2}(\dot{\hat{D}}_{v-1} - \dot{\hat{D}}_{v-2}) e^{A_{v-2}(\hat{D}_{v-1} - \hat{D}_{v-2})} \cdots$$

$$\times e^{A_l(\hat{D}_{l+1} - \hat{D}_l)} e^{A_{l-1}\hat{D}_l}$$

$$+ \quad \cdots \quad \cdots$$

$$+ e^{A_{v-1}(\hat{D}_i x - \hat{D}_{v-1})} e^{A_{v-2}(\hat{D}_{v-1} - \hat{D}_{v-2})} \cdots$$

$$\times A_l(\dot{\hat{D}}_{l+1} - \dot{\hat{D}}_l) e^{A_l(\hat{D}_{l+1} - \hat{D}_l)} e^{A_{l-1}\hat{D}_l}$$

$$+ e^{A_{v-1}(\hat{D}_i x - \hat{D}_{v-1})} e^{A_{v-2}(\hat{D}_{v-1} - \hat{D}_{v-2})} \cdots$$

$$\times e^{A_l(\hat{D}_{l+1} - \hat{D}_l)} e^{A_{l-1}\hat{D}_l} A_{l-1}\dot{\hat{D}}_l. \tag{8.47}$$

For $\Phi(\hat{D}_i x, \hat{D}_l^v, \hat{D}_j y)$ of (8.21),

$$H_{\Phi(\hat{D}_i x, \hat{D}_l^v, \hat{D}_j y)}(\dot{\hat{D}}) = e^{A_{v-1}(\hat{D}_i x - \hat{D}_{v-1})} \left(A_{v-1}(\dot{\hat{D}}_i - \dot{\hat{D}}_{v-1}) \right.$$

$$\left. + A_{v-2}(\dot{\hat{D}}_{v-1} - \dot{\hat{D}}_j) \right) e^{A_{v-2}(\hat{D}_{v-1} - \hat{D}_j y)}, \tag{8.48}$$

$$H_{\Phi(\hat{D}_i x, \hat{D}_l^v, 0)}(\dot{\hat{D}}) = e^{A_{v-1}(\hat{D}_i x - \hat{D}_{v-1})} \left(A_{v-1}(\dot{\hat{D}}_i - \dot{\hat{D}}_{v-1}) \right.$$

$$\left. + A_{v-2}\dot{\hat{D}}_{v-1} \right) e^{A_{v-2}\hat{D}_{v-1}}. \tag{8.49}$$

Remark 8.4. Here we give a more elaborate explanation of the update law (8.25). The present chapter considers the case where the delay D_i is unknown and the actuator state $u_i(x,t)$ is unmeasured. The basic idea to design the update law (8.25) follows the certainty-equivalence idea from the case where the delay D_i is unknown but the actuator state $u_i(x,t)$ is measured in section 7.2. In that situation of the measurable actuator $u_i(x,t)$, the closed-loop system (8.35)–(8.39) becomes

$$D_i \partial_t w_i(x,t) = \partial_x w_i(x,t) - \sum_{h=1}^m \frac{D_i}{D_h} \tilde{D}_h p_{ih}(x,t)$$

$$- D_i q_i(x,t,\dot{\hat{D}}), \tag{8.50}$$

$$w_i(1,t) = 0, \quad x \in [0,1], \quad i \in \{1, 2, \cdots, m\}, \tag{8.51}$$

where $p_{ih}(x,t) = f(X(t), u_i(x,t))$ is a function with respect to the ODE state $X(t)$ and the measured actuator $u_i(x,t)$, whose structure is exactly the same as (8.25) with $u_i(x,t)$ taking the place of $\hat{u}_i(x,t)$. To eliminate the estimation error $\tilde{D}_i$ in the Lyapunov-based analysis, the update law under the measurable actuator has the form below, for $i \in \{1, 2, \cdots, m\}$,

$$\dot{\hat{D}}_i(t) = \gamma_D \mathrm{Proj}_{[\underline{D}_{i-1,i}, \overline{D}_{i,i+1}]} \tau_{D_i}(t), \quad \hat{D}_i(0) \in [\underline{D}_{i-1,i}, \overline{D}_{i,i+1}], \tag{8.52}$$

$$\tau_{D_i}(t) = -\sum_{h=1}^{m} \int_0^1 (1+x) w_h(x,t) p_{hi}(x,t) dx. \tag{8.53}$$

In this chapter, as the actuator state $u_i(x,t)$ is unmeasurable, we use $\hat{w}_h(x,t)$ and $\hat{p}_{hi}(x,t) = f(X(t), \hat{u}_i(x,t))$ to replace $w_h(x,t)$ and $p_{hi}(x,t) = f(X(t), u_i(x,t))$ in (8.53). Thus (8.25) is acquired. In addition, comparing the delay update law in this chapter with that in section 7.2, the reader might have noted in (8.25) that we do not employ normalization in the update law in this chapter. This is because the purpose of the update law normalization in section 7.2 was to achieve a global result. When both the delay state and the delay value are unmeasurable, a global result is not achievable independently of whether or not we employ update law normalization. So, for simplicity, in this chapter we have forgone the normalization.

8.3   Stability Analysis

The adaptive controller in chapter 8 is shown in figure 8.1. We summarize the results we obtain as a main theorem below and subsequently provide its proof.

Theorem 8.5. *Consider the closed-loop system consisting of the ODE-PDE cascade (8.1)–(8.3), the PDE-state observer (8.7)–(8.8), the backstepping transformation (8.16)–(8.17), the adaptive PDE boundary control (8.23), and the delay update law (8.24)–(8.25). There exists a constant $M^* > 0$ such that if the initial states satisfy the condition*

$$|X(0)|^2 + \sum_{i=1}^{m} \left(\|u_i(x,0)\|^2 + \|\hat{u}_i(x,0)\|^2 \right.$$

$$\left. + \|\partial_x \hat{u}_i(x,0)\|^2 + |\tilde{D}_i(0)|^2 \right) \le M^*, \tag{8.54}$$

then all the signals of the closed-loop system are bounded and the regulation for the ODE in (8.1) is achieved, i.e., $\lim_{t \to \infty} X(t) = 0$.

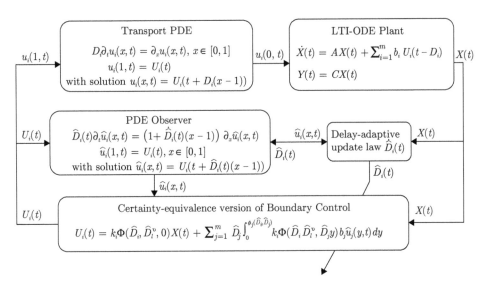

Figure 8.1. Adaptive control under uncertain delays and PDE states

Proof. We build a Lyapunov candidate encompassing all states of the closed-loop system (8.35)–(8.41):

$$V(t) = X(t)^T P X(t) + g_u \sum_{i=1}^{m} D_i \int_0^1 (1+x)\tilde{u}_i(x,t)^2 dx$$

$$+ g_w \sum_{i=1}^{m} \hat{D}_i(t) \left(\int_0^1 (1+x)\hat{w}_i(x,t)^2 dx \right.$$

$$\left. + \int_0^1 (1+x)(\partial_x \hat{w}_i(x,t))^2 dx \right) + \sum_{i=1}^{m} \frac{1}{2}\tilde{D}_i(t)^2, \qquad (8.55)$$

where P is a symmetric and positive matrix defined in (6.2), while g_u and g_w are positive coefficients. Taking the time derivative of $V(t)$ along the dynamics of the target closed-loop system (8.35)–(8.41), we have

$$\dot{V} = -X(t)^T Q X(t) + 2X(t)^T P \sum_{i=1}^{m} \left(b_i \hat{w}_i(0,t) + b_i \tilde{u}_i(0,t) \right)$$

$$+ g_u \sum_{i=1}^{m} \left(-\tilde{u}_i(0,t)^2 - \|\tilde{u}_i(x,t)\|^2 \right.$$

$$- \tilde{D}_i \int_0^1 2(1+x)\tilde{u}_i(x,t)r_i(x,t)dx$$

$$\left. - \dot{\hat{D}}_i \int_0^1 2D_i(x^2-1)\tilde{u}_i(x,t)r_i(x,t)dx \right)$$

$$+g_w \sum_{i=1}^{m} \left(-\hat{w}_i(0,t)^2 - \|\hat{w}_i(x,t)\|^2 \right.$$

$$-\int_0^1 2\hat{D}_i(1+x)\hat{w}_i(x,t)s_i(x,t)dx$$

$$\left. -\int_0^1 2\hat{D}_i(1+x)\hat{w}_i(x,t)k_i\Phi(x,0)\sum_{j=1}^{m} b_j\tilde{u}_j(0,t)dx \right)$$

$$+g_w \sum_{i=1}^{m} \left(2(\partial_x\hat{w}_i(1,t))^2 - (\partial_x\hat{w}_i(0,t))^2 - \|\partial_x\hat{w}_i(x,t)\|^2 \right.$$

$$-\int_0^1 2\hat{D}_i(1+x)\partial_x\hat{w}_i(x,t)\partial_x s_i(x,t)dx$$

$$\left. -\int_0^1 2\hat{D}_i(1+x)\partial_x\hat{w}_i(x,t)k_i\partial_x\Phi(x,0)\sum_{j=1}^{m} b_j\tilde{u}_j(0,t)dx \right)$$

$$+g_w \sum_{i=1}^{m} \left(\dot{\hat{D}}_i \int_0^1 (1+x)\big(\hat{w}_i(x,t)^2 + (\partial_x\hat{w}_i(x,t))^2\big)dx \right)$$

$$-\sum_{i=1}^{m} \dot{\hat{D}}_i\tilde{D}_i. \tag{8.56}$$

Recalling (8.16)–(8.17), according to Young's and Cauchy-Schwartz inequalities, for $i \in \{1,2,\cdots,m\}$, we have the following inequalities:

$$\|\hat{u}_i(x,t)\|^2 \le r_1 \left(|X(t)|^2 + \sum_{i=1}^{m} \|\hat{w}_i(x,t)\|^2 \right), \tag{8.57}$$

$$\|\partial_x\hat{u}_i(x,t)\|^2 \le r_2 \left(|X(t)|^2 + \sum_{i=1}^{m} \|\hat{w}_i(x,t)\|^2 + \sum_{i=1}^{m} \|\partial_x\hat{w}_i(x,t)\|^2 \right), \tag{8.58}$$

where $r_1 > 0$ and $r_2 > 0$ are constants independent of initial conditions. Furthermore, with the help of (8.39), Poincaré and Agmon's inequalities imply that $\hat{w}_i(0,t)^2 \le 4\|\partial_x\hat{w}_i(x,t)\|^2$. By the backstepping transformation, Condition 1 could be converted as $\bar{V} \le V^*$, where

$$V^* = |X(t)|^2 + \sum_{i=1}^{m} \big(\|\tilde{u}_i(x,t)\|^2 + \|\hat{w}_i(x,t)\|^2 + \|\partial_x\hat{w}_i(x,t)\|^2\big).$$

In parallel with [24–28], according to Young's and Cauchy-Schwartz inequalities, we have the following inequalities:

$$-X(t)^T Q X(t) + 2X(t)^T P \sum_{i=1}^{m} \left(b_i \hat{w}_i(0,t) + b_i \tilde{u}_i(0,t) \right)$$

$$\leq -\frac{\lambda_{\min}(Q)}{2} |X(t)|^2$$

$$-\frac{\lambda_{\min}(Q)}{4m} \sum_{i=1}^{m} \left(X(t) - \frac{4mPb_i}{\lambda_{\min}(Q)} \hat{w}_i(0,t) \right)^2$$

$$-\frac{\lambda_{\min}(Q)}{4m} \sum_{i=1}^{m} \left(X(t) - \frac{4mPb_i}{\lambda_{\min}(Q)} \tilde{u}_i(0,t) \right)^2$$

$$+\sum_{i=1}^{m} \frac{4m|Pb_i|^2}{\lambda_{\min}(Q)} \hat{w}_i(0,t)^2 + \sum_{i=1}^{m} \frac{4m|Pb_i|^2}{\lambda_{\min}(Q)} \tilde{u}_i(0,t)^2$$

$$\leq -\frac{\lambda_{\min}(Q)}{2} |X(t)|^2$$

$$+\sum_{i=1}^{m} \frac{4m|Pb_i|^2}{\lambda_{\min}(Q)} \hat{w}_i(0,t)^2 + \sum_{i=1}^{m} \frac{4m|Pb_i|^2}{\lambda_{\min}(Q)} \tilde{u}_i(0,t)^2, \qquad (8.59)$$

$$g_u \sum_{i=1}^{m} \left(-\tilde{D}_i \int_0^1 2(1+x)\tilde{u}_i(x,t) r_i(x,t) dx \right)$$

$$\leq g_u \sum_{i=1}^{m} |\tilde{D}_i| M_1' \left(|X(t)|^2 + \|\tilde{u}_i(x,t)\|^2 \right.$$

$$\left. + \|\partial_x \hat{w}_i(x,t)\|^2 + \sum_{j=1}^{m} \|\hat{w}_j(x,t)\|^2 \right)$$

$$\leq g_u M_1 V^* \sum_{i=1}^{m} |\tilde{D}_i|, \qquad (8.60)$$

$$g_u \sum_{i=1}^{m} \left(-\dot{\tilde{D}}_i \int_0^1 2D_i(x^2 - 1)\tilde{u}_i(x,t) r_i(x,t) dx \right)$$

$$\leq g_u \sum_{i=1}^{m} |\dot{\tilde{D}}_i| M_2' \left(|X(t)|^2 + \|\tilde{u}_i(x,t)\|^2 \right.$$

$$\left. + \|\partial_x \hat{w}_i(x,t)\|^2 + \sum_{j=1}^{m} \|\hat{w}_j(x,t)\|^2 \right)$$

$$\leq g_u M_2 V^* \sum_{i=1}^{m} |\dot{\tilde{D}}_i|, \qquad (8.61)$$

$$g_w \sum_{i=1}^{m} \left(- \int_0^1 2\hat{D}_i (1+x)\hat{w}_i(x,t)s_i(x,t)dx \right)$$

$$\le g_w \sum_{i=1}^{m} |\dot{\hat{D}}_i| M_3' \left(|X(t)|^2 + \sum_{j=1}^{m} \|\hat{w}_j(x,t)\|^2 \right.$$

$$+ \|\partial_x \hat{w}_i(x,t)\|^2 + \sum_{j=1}^{m} \|\hat{w}_j(0,t)\|^2 \Bigg)$$

$$\le g_w M_3 V^* \sum_{i=1}^{m} |\dot{\hat{D}}_i|, \tag{8.62}$$

$$g_w \sum_{i=1}^{m} \left(- \int_0^1 2\hat{D}_i (1+x)\hat{w}_i(x,t)k_i \Phi(x,0) \sum_{j=1}^{m} b_j \tilde{u}_j(0,t)dx \right)$$

$$\le \frac{1}{2} g_w \sum_{i=1}^{m} \|\hat{w}_i(x,t)\|^2 + g_w M_4 \sum_{i=1}^{m} \tilde{u}_i(0,t)^2, \tag{8.63}$$

$$g_w \sum_{i=1}^{m} \left(- \int_0^1 2\hat{D}_i (1+x)\partial_x \hat{w}_i(x,t)\partial_x s_i(x,t)dx \right)$$

$$\le g_w \sum_{i=1}^{m} |\dot{\hat{D}}_i| M_5' \left(|X(t)|^2 + (\partial_x \hat{w}_i(0,t))^2 + \sum_{j=1}^{m} \|\hat{w}_j(x,t)\|^2 \right.$$

$$+ \sum_{j=1}^{m} \|\partial_x \hat{w}_j(x,t)\|^2 + \sum_{j=1}^{m} \|\hat{w}_j(0,t)\|^2 \Bigg)$$

$$\le g_w M_5 V^* \sum_{i=1}^{m} |\dot{\hat{D}}_i| + g_w M_5 \sum_{i=1}^{m} |\dot{\hat{D}}_i| \sum_{i=1}^{m} (\partial_x \hat{w}_i(0,t))^2, \tag{8.64}$$

$$g_w \sum_{i=1}^{m} \left(- \int_0^1 2\hat{D}_i (1+x)\partial_x \hat{w}_i(x,t)k_i \partial_x \Phi(x,0) \sum_{j=1}^{m} b_j \tilde{u}_j(0,t)dx \right)$$

$$\le \frac{1}{2} g_w \sum_{i=1}^{m} \|\partial_x \hat{w}_i(x,t)\|^2 + g_w M_6 \sum_{i=1}^{m} \tilde{u}_i(0,t)^2, \tag{8.65}$$

$$g_w \sum_{i=1}^{m} \left(\dot{\hat{D}}_i \int_0^1 (1+x)\left(\hat{w}_i(x,t)^2 + (\partial_x \hat{w}_i(x,t))^2 \right)dx \right)$$

$$\le g_w M_7 V^* \sum_{i=1}^{m} |\dot{\hat{D}}_i|, \tag{8.66}$$

$$g_w \sum_{i=1}^{m} 2(\partial_x \hat{w}_i(1,t))^2$$

$$\leq g_w M_8' \sum_{i=1}^{m} |\dot{\hat{D}}_i|^2 \left(|X(t)|^2 + \sum_{j=1}^{m} \|\hat{w}_j(x,t)\|^2 + \sum_{j=1}^{m} \hat{w}_j(0,t)^2 \right)$$

$$+ g_w M_8' \sum_{i=1}^{m} \tilde{u}_i(0,t)^2$$

$$\leq g_w M_8 V^* \sum_{i=1}^{m} |\dot{\hat{D}}_i|^2 + g_w M_8 \sum_{i=1}^{m} \tilde{u}_i(0,t)^2, \tag{8.67}$$

where M_1', M_1, M_2', M_2, M_3', M_3, M_4, M_5', M_5, M_6, M_7, M_8', M_8 are all positive constants independent of initial conditions. As a consequence, substituting the inequalities (8.59)–(8.67) into (8.56), we have

$$\dot{V} \leq -\frac{\lambda_{\min}(Q)}{2}|X(t)|^2$$

$$+ \sum_{i=1}^{m} \frac{4m|Pb_i|^2}{\lambda_{\min}(Q)}\hat{w}_i(0,t)^2 + \sum_{i=1}^{m} \frac{4m|Pb_i|^2}{\lambda_{\min}(Q)}\tilde{u}_i(0,t)^2$$

$$- g_u \sum_{i=1}^{m} \tilde{u}_i(0,t)^2 - g_u \sum_{i=1}^{m} \|\tilde{u}_i(x,t)\|^2$$

$$+ g_u M_1 V^* \sum_{i=1}^{m} |\tilde{D}_i| + g_u M_2 V^* \sum_{i=1}^{m} |\dot{\hat{D}}_i|$$

$$- g_w \sum_{i=1}^{m} \hat{w}_i(0,t)^2 - g_w \sum_{i=1}^{m} \|\hat{w}_i(x,t)\|^2$$

$$+ g_w M_3 V^* \sum_{i=1}^{m} |\dot{\hat{D}}_i| + \frac{1}{2}g_w \sum_{i=1}^{m} \|\hat{w}_i(x,t)\|^2$$

$$+ g_w M_4 \sum_{i=1}^{m} \tilde{u}_i(0,t)^2$$

$$- g_w \sum_{i=1}^{m} (\partial_x \hat{w}_i(0,t))^2 - g_w \sum_{i=1}^{m} \|\partial_x \hat{w}_i(x,t)\|^2$$

$$+ g_w M_5 V^* \sum_{i=1}^{m} |\dot{\hat{D}}_i| + g_w M_5 \sum_{i=1}^{m} |\dot{\hat{D}}_i| \sum_{i=1}^{m} (\partial_x \hat{w}_i(0,t))^2$$

$$+ \frac{1}{2}g_w \sum_{i=1}^{m} \|\partial_x \hat{w}_i(x,t)\|^2 + g_w M_6 \sum_{i=1}^{m} \tilde{u}_i(0,t)^2$$

$$+ g_w M_7 V^* \sum_{i=1}^{m} |\dot{\hat{D}}_i| + g_w M_8 V^* \sum_{i=1}^{m} |\dot{\hat{D}}_i|^2$$

$$+g_w M_8 \sum_{i=1}^{m} \tilde{u}_i(0,t)^2 - \sum_{i=1}^{m} \dot{\tilde{D}}_i \tilde{D}_i. \tag{8.68}$$

If we unite similar terms, (8.68) becomes

$$\dot{V} \leq -\sum_{i=1}^{m} \left(g_w - \frac{4m|Pb_i|^2}{\lambda_{\min}(Q)} \right) \hat{w}_i(0,t)^2$$

$$-\sum_{i=1}^{m} \left(g_u - \frac{4m|Pb_i|^2}{\lambda_{\min}(Q)} - g_w(M_4 + M_6 + M_8) \right) \tilde{u}_i(0,t)^2$$

$$-\frac{\lambda_{\min}(Q)}{2}|X(t)|^2 - g_u \sum_{i=1}^{m} \|\tilde{u}_i(x,t)\|^2$$

$$-\frac{1}{2}g_w \sum_{i=1}^{m} \|\hat{w}_i(x,t)\|^2 - \frac{1}{2}g_w \sum_{i=1}^{m} \|\partial_x \hat{w}_i(x,t)\|^2$$

$$+\left(g_u M_1 \sum_{i=1}^{m} |\tilde{D}_i| + g_w M_8 \sum_{i=1}^{m} |\dot{\tilde{D}}_i|^2 \right.$$

$$\left. +g_w(M_2 + M_3 + M_5 + M_7) \sum_{i=1}^{m} |\dot{\tilde{D}}_i| \right) V^*$$

$$-g_w(1 - M_5 \sum_{i=1}^{m} |\dot{\tilde{D}}_i|) \sum_{i=1}^{m} (\partial_x \hat{w}_i(0,t))^2 - \sum_{i=1}^{m} \dot{\tilde{D}}_i \tilde{D}_i. \tag{8.69}$$

To make the terms $\hat{w}_i(0,t)^2$ and $\tilde{u}_i(0,t)^2$ vanish, one can choose the coefficients Q, g_u, g_w to satisfy $g_w \geq \sup_i \frac{4m|Pb_i|^2}{\lambda_{\min}(Q)}$ and $g_u \geq \sup_i \frac{4m|Pb_i|^2}{\lambda_{\min}(Q)} + g_w(M_4 + M_6 + M_8)$; thus (8.69) becomes

$$\dot{V} \leq -g_v V^* - g_w(1 - M_5 \sum_{i=1}^{m} |\dot{\tilde{D}}_i|) \sum_{i=1}^{m} (\partial_x \hat{w}_i(0,t))^2$$

$$+\left(g_u M_1 \sum_{i=1}^{m} |\tilde{D}_i| + g_w M_8 \sum_{i=1}^{m} |\dot{\tilde{D}}_i|^2 + g_w M_9 \sum_{i=1}^{m} |\dot{\tilde{D}}_i| \right) V^*$$

$$-\sum_{i=1}^{m} \dot{\tilde{D}}_i \tilde{D}_i, \tag{8.70}$$

where $g_v = \min \left\{ \frac{\lambda_{\min}(Q)}{2}, g_u, \frac{1}{2}g_w \right\}$, $M_9 = M_2 + M_3 + M_5 + M_7$.

The techniques to treat the remaining nonnegative terms are slightly dependent on Condition 1 or Condition 2. Referring to [24–28], we now resume the proof. We first consider Condition 1.

From (8.24)–(8.25), the update law meets $|\dot{\tilde{D}}_i| \leq \gamma_D M_D V^*$, and then (8.70) becomes

$$\dot{V} \leq - \left(g_v - g_u M_1 \sum_{i=1}^{m} |\tilde{D}_i| - \gamma_D M_D \sum_{i=1}^{m} |\tilde{D}_i| \right) V^*$$

$$+ \gamma_D M_D g_w m M_9 V^{*2} + \gamma_D^2 M_D^2 g_w m M_8 V^{*3}$$

$$- g_w (1 - \gamma_D M_D m M_5 V^*) \sum_{i=1}^{m} (\partial_x \hat{w}_i(0, t))^2. \tag{8.71}$$

Furthermore, there exist constants $\varepsilon > 0$, $M_{10} > 0$, and $\eta > 0$ to make the following inequality hold:

$$\sum_{i=1}^{m} |\tilde{D}_i| \leq \frac{\varepsilon}{2} + \frac{M_{10}}{2\varepsilon} \sum_{i=1}^{m} |\tilde{D}_i|^2 \tag{8.72}$$

$$\leq \frac{\varepsilon}{2} + \frac{M_{10}}{2\varepsilon} (V - \eta V^*). \tag{8.73}$$

Substituting (8.73) into (8.71), we obtain

$$\dot{V} \leq - \left(g_v - (g_u M_1 + \gamma_D M_D) \left(\frac{\varepsilon}{2} + \frac{M_{10}}{2\varepsilon} V \right) \right) V^*$$

$$- \left((g_u M_1 + \gamma_D M_D) \frac{M_{10} \eta}{2\varepsilon} \right.$$

$$\left. - \gamma_D M_D g_w m M_9 - \gamma_D^2 M_D^2 g_w m M_8 V^* \right) V^{*2}$$

$$- g_w (1 - \gamma_D M_D M_5 m V^*) \sum_{i=1}^{m} (\partial_x \hat{w}_i(0, t))^2. \tag{8.74}$$

As a result, if the coefficients are chosen to satisfy

$$g_v \geq (g_u M_1 + \gamma_D M_D) \frac{\varepsilon}{2}, \tag{8.75}$$

$$(g_u M_1 + \gamma_D M_D) \frac{M_{10} \eta}{2\varepsilon} \geq \gamma_D M_D g_w m M_9, \tag{8.76}$$

and the initial states satisfy

$$V(0) \leq \min \left\{ \left(\frac{g_v}{g_u M_1 + \gamma_D M_D} - \frac{\varepsilon}{2} \right) \frac{2\varepsilon}{M_{10}}, \frac{\eta}{\gamma_D M_D M_5 m}, \right.$$

$$\left. \frac{\eta \left((g_u M_1 + \gamma_D M_D) \frac{M_{10} \eta}{2\varepsilon} - \gamma_D M_D g_w m M_9 \right)}{\gamma_D^2 M_D^2 g_w m M_8} \right\}, \tag{8.77}$$

we conclude that there exist nonnegative functions $\mu_1(t)$ and $\mu_2(t)$ such that

$$\dot{V} \leq -\mu_1(t)V^* - \mu_2(t)V^{*2}, \tag{8.78}$$

and thus $\forall t \in [0, +\infty), V(t) \leq V(0)$.

Next we consider Condition 2. From (8.34), the update law meets

$$|\dot{\tilde{D}}_i| \leq \gamma_D M_D, \quad \dot{\tilde{D}}_i \tilde{D}_i \geq 0, \tag{8.79}$$

Then (8.70) becomes

$$\dot{V} \leq - \left(g_v - g_u M_1 \sum_{i=1}^{m} |\tilde{D}_i| - \gamma_D^2 M_D^2 g_w M_8 - \gamma_D M_D g_w M_9 \right) V^*$$

$$- g_w (1 - \gamma_D M_D M_5) \sum_{i=1}^{m} (\partial_x \hat{w}_i(0, t))^2. \tag{8.80}$$

Based on inequalities (8.72)–(8.73), we obtain

$$\dot{V} \leq - \left(g_v - g_u M_1 \frac{\varepsilon}{2} - g_u M_1 \frac{M_{10}}{2\varepsilon} V \right.$$

$$\left. - \gamma_D^2 M_D^2 g_w M_8 - \gamma_D M_D g_w M_9 \right) V^*$$

$$- g_w (1 - \gamma_D M_D M_5) \sum_{i=1}^{m} (\partial_x \hat{w}_i(0, t))^2. \tag{8.81}$$

Thus if the coefficients are chosen to satisfy

$$g_v \geq g_u M_1 \frac{\varepsilon}{2} + \gamma_D^2 M_D^2 g_w M_8 + \gamma_D M_D g_w M_9, \tag{8.82}$$

$$1 \geq \gamma_D M_D M_5, \tag{8.83}$$

and the initial following condition is satisfied such that

$$V(0) \leq \frac{g_v - g_u M_1 \frac{\varepsilon}{2} - \gamma_D^2 M_D^2 g_w M_8 - \gamma_D M_D g_w M_9}{g_u M_1 \frac{M_{10}}{2\varepsilon}}, \tag{8.84}$$

one can finally obtain

$$\dot{V} \leq -\mu(t)V^*, \tag{8.85}$$

where $\mu(t)$ is a nonnegative function and consequently

$$\forall t \in [0, +\infty), V(t) \leq V(0). \tag{8.86}$$

As a result, based on the nonincreasing property of the time derivative $\dot{V}$ for Condition 1 or 2, following a similar procedure in [24–28], the boundedness of states $X(t)$, $\tilde{u}_i(x,t)$, $\hat{u}_i(x,t)$, $U_i(t)$, $\hat{w}_i(x,t)$, $\partial_x w_i(x,t)$ for $i \in \{1, 2, \cdots, m\}$ could be guaranteed. Furthermore, by applying Barbalat's lemma, the convergence $\lim_{t \to \infty} X(t) = 0$ is also established. This completes the proof.   □

8.4 Simulation

Consider a simple one-dimensional LTI systems with two inputs as follows:

$$\dot{X}(t) = AX(t) + b_1 U_1(t - D_1) + b_2 U_2(t - D_2), \tag{8.87}$$

where $D_1 \leq D_2$ is assumed without loss of generality.

By transport PDE (8.4), we have

$$u_1(x, t) = U_1(t + D_1(x - 1)), \quad x \in [0, 1], \tag{8.88}$$
$$u_2(x, t) = U_2(t + D_2(x - 1)), \quad x \in [0, 1]. \tag{8.89}$$

As delays D_1, D_2 are unknown and actuator states $u_1(x,t)$, $u_2(x,t)$ are unmeasured, the PDE observers (8.6) are brought in as follows:

$$\hat{u}_1(x, t) = U_1(t + \hat{D}_1(t)(x - 1)), \quad x \in [0, 1], \tag{8.90}$$
$$\hat{u}_2(x, t) = U_2(t + \hat{D}_2(t)(x - 1)), \quad x \in [0, 1], \tag{8.91}$$

with the estimation error (8.9)–(8.10) such that

$$\tilde{u}_1(x, t) = u_1(x, t) - \hat{u}_1(x, t), \quad x \in [0, 1], \tag{8.92}$$
$$= U_1(t + D_1(x - 1)) - U_1(t + \hat{D}_1(t)(x - 1)), \tag{8.93}$$
$$\tilde{u}_2(x, t) = u_2(x, t) - \hat{u}_2(x, t), \quad x \in [0, 1], \tag{8.94}$$
$$= U_2(t + D_2(x - 1)) - U_2(t + \hat{D}_2(t)(x - 1)). \tag{8.95}$$

With (8.88)–(8.89), ODE dynamics (8.87) is transformed as the ODE-PDE cascade below:

$$\dot{X}(t) = AX(t) + b_1 u_1(0, t) + b_2 u_2(0, t), \tag{8.96}$$
$$D_1 \partial_t u_1(x, t) = \partial_x u_1(x, t), \quad x \in [0, 1], \tag{8.97}$$
$$u_1(1, t) = U_1(t), \tag{8.98}$$
$$D_2 \partial_t u_2(x, t) = \partial_x u_2(x, t), \quad x \in [0, 1], \tag{8.99}$$
$$u_2(1, t) = U_2(t). \tag{8.100}$$

By (8.88)–(8.89) and (8.90)–(8.91), the ODE-PDE cascade (8.96)–(8.100) is further converted into

$$\dot{X}(t) = AX(t) + b_1\hat{u}_1(0,t) + b_2\hat{u}_2(0,t) + b_1\tilde{u}_1(0,t) + b_2\tilde{u}_2(0,t), \quad (8.101)$$

$$D_1\partial_t\tilde{u}_1(x,t) = \partial_x\tilde{u}_1(x,t) - \tilde{D}_1\frac{1}{\tilde{D}_1}\partial_x\hat{u}_1(x,t) - \dot{\tilde{D}}_1 D_1(x-1)\frac{1}{\tilde{D}_1}\partial_x\hat{u}_1(x,t),$$
$$\qquad\qquad\qquad\qquad\qquad\qquad\qquad\qquad (8.102)$$

$$\tilde{u}_1(1,t) = 0, \qquad\qquad\qquad\qquad\qquad\qquad\qquad (8.103)$$

$$D_2\partial_t\tilde{u}_2(x,t) = \partial_x\tilde{u}_2(x,t) - \tilde{D}_2\frac{1}{\tilde{D}_2}\partial_x\hat{u}_2(x,t) - \dot{\tilde{D}}_2 D_2(x-1)\frac{1}{\tilde{D}_2}\partial_x\hat{u}_2(x,t),$$
$$\qquad\qquad\qquad\qquad\qquad\qquad\qquad\qquad (8.104)$$

$$\tilde{u}_2(1,t) = 0, \qquad\qquad\qquad\qquad\qquad\qquad\qquad (8.105)$$

$$\hat{D}_1\partial_t\hat{u}_1(x,t) = \partial_x\hat{u}_1(x,t) + \dot{\hat{D}}_1(x-1)\partial_x\hat{u}_1(x,t), \qquad (8.106)$$
$$\hat{u}_1(1,t) = U_1(t), \qquad\qquad\qquad\qquad\qquad (8.107)$$

$$\hat{D}_2\partial_t\hat{u}_2(x,t) = \partial_x\hat{u}_2(x,t) + \dot{\hat{D}}_2(x-1)\partial_x\hat{u}_2(x,t), \qquad (8.108)$$
$$\hat{u}_2(1,t) = U_2(t). \qquad\qquad\qquad\qquad\qquad (8.109)$$

Then the invertible backstepping transformations (8.16)–(8.17) have the concrete forms as follows:

- $i = 1, x \in [0,1]$

$$\hat{w}_1(x,t) = \hat{u}_1(x,t) - k_1 e^{A\hat{D}_1 x}X(t) - \hat{D}_1\int_0^x k_1 e^{A(\hat{D}_1 x - \hat{D}_1 y)}b_1\hat{u}_1(y,t)dy$$
$$- \hat{D}_2\int_0^{\frac{\hat{D}_1}{\hat{D}_2}x} k_1 e^{A(\hat{D}_1 x - \hat{D}_2 y)}b_2\hat{u}_2(y,t)dy, \qquad (8.110)$$

$$\hat{u}_1(x,t) = \hat{w}_1(x,t) + k_1 e^{A_2\hat{D}_1 x}X(t) + \hat{D}_1\int_0^x k_1 e^{A_2(\hat{D}_1 x - \hat{D}_1 y)}b_1\hat{w}_1(y,t)dy$$
$$+ \hat{D}_2\int_0^{\frac{\hat{D}_1}{\hat{D}_2}x} k_1 e^{A_2(\hat{D}_1 x - \hat{D}_2 y)}b_2\hat{w}_2(y,t)dy. \qquad (8.111)$$

- $i = 2, x \in \left[0, \frac{\hat{D}_1}{\hat{D}_2}\right]$

$$\hat{w}_2(x,t) = \hat{u}_2(x,t) - k_2 e^{A\hat{D}_2 x}X(t) - \hat{D}_1\int_0^{\frac{\hat{D}_2}{\hat{D}_1}x} k_2 e^{A(\hat{D}_2 x - \hat{D}_1 y)}b_1\hat{u}_1(y,t)dy$$
$$- \hat{D}_2\int_0^x k_2 e^{A(\hat{D}_2 x - \hat{D}_2 y)}b_2\hat{u}_2(y,t)dy, \qquad (8.112)$$

$$\hat{u}_2(x,t) = \hat{w}_2(x,t) + k_2 e^{A_2\hat{D}_2 x}X(t) + \hat{D}_1\int_0^{\frac{\hat{D}_2}{\hat{D}_1}x} k_2 e^{A_2(\hat{D}_2 x - \hat{D}_1 y)}b_1\hat{w}_1(y,t)dy$$

$$+ \hat{D}_2 \int_0^x k_2 e^{A_2(\hat{D}_2 x - \hat{D}_2 y)} b_2 \hat{w}_2(y,t) dy. \tag{8.113}$$

- $i = 2, x \in \left[\frac{\hat{D}_1}{\hat{D}_2}, 1 \right]$

$$\hat{w}_2(x,t) = \hat{u}_2(x,t) - k_2 e^{A_1(\hat{D}_2 x - \hat{D}_1)} e^{A\hat{D}_1} X(t)$$

$$- \hat{D}_1 \int_0^1 k_2 e^{A_1(\hat{D}_2 x - \hat{D}_1)} e^{A(\hat{D}_1 - \hat{D}_1 y)} b_1 \hat{u}_1(y,t) dy$$

$$- \hat{D}_2 \int_0^{\frac{\hat{D}_1}{\hat{D}_2}} k_2 e^{A_1(\hat{D}_2 x - \hat{D}_1)} e^{A(\hat{D}_1 - \hat{D}_2 y)} b_2 \hat{u}_2(y,t) dy$$

$$- \hat{D}_2 \int_{\frac{\hat{D}_1}{\hat{D}_2}}^x k_2 e^{A_1(\hat{D}_2 x - \hat{D}_2 y)} b_2 \hat{u}_2(y,t) dy, \tag{8.114}$$

$$\hat{u}_2(x,t) = \hat{w}_2(x,t) + k_2 e^{A_2 \hat{D}_2 x} X(t) + \hat{D}_1 \int_0^1 k_2 e^{A_2(\hat{D}_2 x - \hat{D}_1 y)} b_1 \hat{w}_1(y,t) dy$$

$$+ \hat{D}_2 \int_0^x k_2 e^{A_2(\hat{D}_2 x - \hat{D}_2 y)} b_2 \hat{w}_2(y,t) dy, \tag{8.115}$$

where $A_2 = A_1 + b_2 k_2$, $A_1 = A + b_1 k_1$.

If we substitute $x = 1$ into (8.110) and (8.114) to meet boundary conditions $\hat{w}_1(1,t) = 0$ and $\hat{w}_2(1,t) = 0$, the boundary control laws (8.23) have the concrete forms

$$U_1(t) = \hat{u}_1(1,t)$$

$$= k_1 e^{A\hat{D}_1} X(t) + \hat{D}_1 \int_0^1 k_1 e^{A(\hat{D}_1 - \hat{D}_1 y)} b_1 \hat{u}_1(y,t) dy$$

$$+ \hat{D}_2 \int_0^{\frac{\hat{D}_1}{\hat{D}_2}} k_1 e^{A(\hat{D}_1 - \hat{D}_2 y)} b_2 \hat{u}_2(y,t) dy, \tag{8.116}$$

$$U_2(t) = \hat{u}_2(1,t)$$

$$= k_2^T e^{A_1(\hat{D}_2 - \hat{D}_1)} e^{A\hat{D}_1} X(t) + \hat{D}_1 \int_0^1 k_2 e^{A_1(\hat{D}_2 - \hat{D}_1)} e^{A(\hat{D}_1 - \hat{D}_1 y)} b_1 \hat{u}_1(y,t) dy$$

$$+ \hat{D}_2 k_2 \left(\int_0^{\frac{\hat{D}_1}{\hat{D}_2}} e^{A_1(\hat{D}_2 - \hat{D}_1)} e^{A(\hat{D}_1 - \hat{D}_2 y)} b_2 \hat{u}_2(y,t) dy \right.$$

$$\left. + \int_{\frac{\hat{D}_1}{\hat{D}_2}}^1 e^{A_1(\hat{D}_2 - \hat{D}_2 y)} b_2 \hat{u}_2(y,t) dy \right). \tag{8.117}$$

How to obtain the target system (8.35)–(8.39) involves a lengthy and complicated calculation. We use the example as a step-by-step mathematical derivation. Substituting $x=0$ into (8.110) and (8.112), we have

$$\hat{u}_1(0,t)=\hat{w}_1(0,t)+k_1X(t), \tag{8.118}$$
$$\hat{u}_2(0,t)=\hat{w}_2(0,t)+k_2X(t). \tag{8.119}$$

Substituting (8.118)–(8.119) into (8.101), we could obtain the target ODE (8.35) such that

$$\dot{X}(t)=(A+b_1k_1+b_2k_2)X(t)+b_1\hat{w}_1(0,t)+b_2\hat{w}_2(0,t)+b_1\tilde{u}_1(0,t)+b_2\tilde{u}_2(0,t)$$
$$=(A+BK)X(t)+b_1\hat{w}_1(0,t)+b_2\hat{w}_2(0,t)+b_1\tilde{u}_1(0,t)+b_2\tilde{u}_2(0,t). \tag{8.120}$$

Taking the partial derivative with respect to x on both sides of (8.111), (8.113), and (8.115), and substituting $\partial_x\hat{u}_1(x,t)$ and $\partial_x\hat{u}_2(x,t)$ into (8.102) and (8.104), we could then derive (8.36)–(8.37). Next we show how to get (8.38)–(8.39) for the first input channel $i=1$. Taking the partial derivative with respect to x on both sides of (8.110), we have

$$\partial_x\hat{w}_1(x,t)=\partial_x\hat{u}_1(x,t)-k_1e^{A\hat{D}_1x}A\hat{D}_1X(t)-\hat{D}_1k_1b_1\hat{u}_1(x,t)$$
$$-\hat{D}_1k_1b_2\hat{u}_2\left(\frac{\hat{D}_1}{\hat{D}_2}x,t\right)$$
$$-\hat{D}_1\int_0^x k_1A\hat{D}_1e^{A(\hat{D}_1x-\hat{D}_1y)}b_1\hat{u}_1(y,t)dy$$
$$-\hat{D}_2\int_0^{\frac{\hat{D}_1}{\hat{D}_2}x} k_1A\hat{D}_1e^{A(\hat{D}_1x-\hat{D}_2y)}b_2\hat{u}_2(y,t)dy. \tag{8.121}$$

Taking the partial derivative with respect to t on both sides of (8.110), we have

$$\partial_t\hat{w}_1(x,t)=\partial_t\hat{u}_1(x,t)-k_1e^{A\hat{D}_1x}A\dot{\hat{D}}_1xX(t)$$
$$-k_1e^{A\hat{D}_1x}(AX(t)+b_1\hat{u}_1(0,t)+b_2\hat{u}_2(0,t)+b_1\tilde{u}_1(0,t)+b_2\tilde{u}_2(0,t))$$
$$-\dot{\hat{D}}_1\int_0^x k_1e^{A(\hat{D}_1x-\hat{D}_1y)}b_1\hat{u}_1(y,t)dy$$
$$-\hat{D}_1\int_0^x k_1A(\dot{\hat{D}}_1x-\dot{\hat{D}}_1y)e^{A(\hat{D}_1x-\hat{D}_1y)}b_1\hat{u}_1(y,t)dy$$
$$-\hat{D}_1\int_0^x k_1e^{A(\hat{D}_1x-\hat{D}_1y)}b_1\partial_t\hat{u}_1(y,t)dy$$
$$-\dot{\hat{D}}_2\int_0^{\frac{\hat{D}_1}{\hat{D}_2}x} k_1e^{A(\hat{D}_1x-\hat{D}_2y)}b_2\hat{u}_2(y,t)dy$$

$$-\frac{\dot{\hat{D}}_1\hat{D}_2 - \hat{D}_1\dot{\hat{D}}_2}{\hat{D}_2}xk_1b_2\hat{u}_2\left(\frac{\hat{D}_1}{\hat{D}_2}x, t\right)$$

$$-\hat{D}_2\int_0^{\frac{\hat{D}_1}{\hat{D}_2}x}k_1A(\dot{\hat{D}}_1x - \dot{\hat{D}}_2y)e^{A(\hat{D}_1x - \hat{D}_2y)}b_2\hat{u}_2(y, t)dy$$

$$-\hat{D}_2\int_0^{\frac{\hat{D}_1}{\hat{D}_2}x}k_1e^{A(\hat{D}_1x - \hat{D}_2y)}b_2\partial_t\hat{u}_2(y, t)dy. \tag{8.122}$$

By utilizing (8.106) and (8.108) and integration by parts, we have

$$-\hat{D}_1\int_0^x k_1e^{A(\hat{D}_1x - \hat{D}_1y)}b_1\partial_t\hat{u}_1(y, t)dy$$

$$= -\int_0^x k_1e^{A(\hat{D}_1x - \hat{D}_1y)}b_1(1 + \dot{\hat{D}}_1(y - 1))\partial_y\hat{u}_1(y, t)dy$$

$$= -k_1b_1(1 + \dot{\hat{D}}_1(x - 1))\hat{u}_1(x, t) + k_1e^{A\hat{D}_1x}b_1(1 - \dot{\hat{D}}_1)\hat{u}_1(0, t)$$

$$\quad - \int_0^x k_1A\hat{D}_1e^{A(\hat{D}_1x - \hat{D}_1y)}b_1(1 + \dot{\hat{D}}_1(y - 1))\hat{u}_1(y, t)dy$$

$$\quad + \int_0^x k_1e^{A(\hat{D}_1x - \hat{D}_1y)}b_1\dot{\hat{D}}_1\hat{u}_1(y, t)dy, \tag{8.123}$$

$$-\hat{D}_2\int_0^{\frac{\hat{D}_1}{\hat{D}_2}x}k_1e^{A(\hat{D}_1x - \hat{D}_2y)}b_2\partial_t\hat{u}_2(y, t)dy$$

$$= -\int_0^{\frac{\hat{D}_1}{\hat{D}_2}x}k_1e^{A(\hat{D}_1x - \hat{D}_2y)}b_2(1 + \dot{\hat{D}}_2(y - 1))\partial_y\hat{u}_2(y, t)dy$$

$$= -k_1b_2\left(1 + \dot{\hat{D}}_2\left(\frac{\hat{D}_1}{\hat{D}_2}x - 1\right)\right)\hat{u}_2\left(\frac{\hat{D}_1}{\hat{D}_2}x, t\right) + k_1e^{A\hat{D}_1x}b_2(1 - \dot{\hat{D}}_2)\hat{u}_2(0, t)$$

$$\quad - \int_0^{\frac{\hat{D}_1}{\hat{D}_2}x}k_1A\hat{D}_2e^{A(\hat{D}_1x - \hat{D}_2y)}b_2(1 + \dot{\hat{D}}_2(y - 1))\hat{u}_2(y, t)dy$$

$$\quad + \int_0^{\frac{\hat{D}_1}{\hat{D}_2}x}k_1e^{A(\hat{D}_1x - \hat{D}_2y)}b_2\dot{\hat{D}}_2\hat{u}_2(y, t)dy. \tag{8.124}$$

Substituting (8.123)–(8.124) into (8.122), we have

$$\partial_t\hat{w}_1(x, t) = \partial_t\hat{u}_1(x, t) - (1 + \dot{\hat{D}}_1x)k_1e^{A\hat{D}_1x}AX(t)$$

$$\quad - k_1e^{A\hat{D}_1x}\left(\dot{\hat{D}}_1b_1\hat{u}_1(0, t) + \dot{\hat{D}}_2b_2\hat{u}_2(0, t)\right)$$

$$\quad - k_1e^{A\hat{D}_1x}\left(b_1\tilde{u}_1(0, t) + b_2\tilde{u}_2(0, t)\right)$$

$$-(1+\dot{\hat{D}}_1(x-1))k_1 b_1 \hat{u}_1(x,t)-(1+\dot{\hat{D}}_1 x-\dot{\hat{D}}_2)k_1 b_2 \hat{u}_2\left(\frac{\hat{D}_1}{\hat{D}_2}x,t\right)$$

$$-\hat{D}_1 \int_0^x (1+\dot{\hat{D}}_1(x-1))k_1 A e^{A(\hat{D}_1 x-\hat{D}_1 y)} b_1 \hat{u}_1(y,t)dy$$

$$-\hat{D}_2 \int_0^{\frac{\hat{D}_1}{\hat{D}_2}x} (1+\dot{\hat{D}}_1 x-\dot{\hat{D}}_2)k_1 A e^{A(\hat{D}_1 x-\hat{D}_2 y)} b_2 \hat{u}_2(y,t)dy. \qquad (8.125)$$

Calculating $\hat{D}_1 \cdot (8.125)-(1+\dot{\hat{D}}_1(x-1)) \cdot (8.121)$, we have

$$\partial_t \hat{w}_1(x,t)=(1+\dot{\hat{D}}_1(x-1))\partial_x \hat{w}_1(x,t)-\hat{D}_1 k_1 e^{A\hat{D}_1 x} A\dot{\hat{D}}_1 X(t)$$

$$-\hat{D}_1 k_1 e^{A\hat{D}_1 x}\left(\dot{\hat{D}}_1 b_1 \hat{u}_1(0,t)+\dot{\hat{D}}_2 b_2 \hat{u}_2(0,t)\right)$$

$$-\hat{D}_1 k_1 e^{A\hat{D}_1 x}\left(b_1 \tilde{u}_1(0,t)+b_2 \tilde{u}_2(0,t)\right)$$

$$-\hat{D}_1(\dot{\hat{D}}_1-\dot{\hat{D}}_2)k_1 b_2 \hat{u}_2\left(\frac{\hat{D}_1}{\hat{D}_2}x,t\right)$$

$$-\hat{D}_1 \hat{D}_2 \int_0^{\frac{\hat{D}_1}{\hat{D}_2}x} k_1 A(\dot{\hat{D}}_1-\dot{\hat{D}}_2)e^{A(\hat{D}_1 x-\hat{D}_2 y)} b_2 \hat{u}_2(y,t)dy. \qquad (8.126)$$

Thus (8.38)–(8.39) is obtained. Taking the partial derivative with respect to x on both sides of (8.126), (8.40)–(8.41) is obtained. Following the procedure (8.121)–(8.126), we can get similar results for the second input channel $i=2$, and can further extend this to a general m-input case.

Next, we consider the numerical simulation. For the LTI system with two-input (8.87), where $A=1$, $b_1=b_2=1$, the distinct delays $D_1=0.5s$ and $D_2=1s$ are unknown. However, the sequence $D_1 \leq D_2$ is known. If there is no control or controls are delayed significantly, it is apparent that $X(t)$ diverges exponentially. The control law is designed to achieve

$$U_1(t-D_1)=-X(t), \quad U_2(t-D_2)=-0.5X(t). \qquad (8.127)$$

The initial conditions are set as $X(0)=1$, $D_1(0)=0.25$, $D_2(0)=0.5$. The results are shown in figures 8.2 and 8.3.

As stated in Remark 7.6 in section 7.2, we also hope to test the sensitivity of the algorithm to the order of the delays. We consider another case that $D_1 < D_2$ is assumed but actually $D_2 < D_1$ such that $D_1=0.8s$ and $D_2=0.5s$. Suppose we do not know the exact value and the length order of the delay about D_1 and D_2. The initial states are set as $\hat{D}_1(0)=0.4s$, $\hat{D}_2(0)=0.7s$.

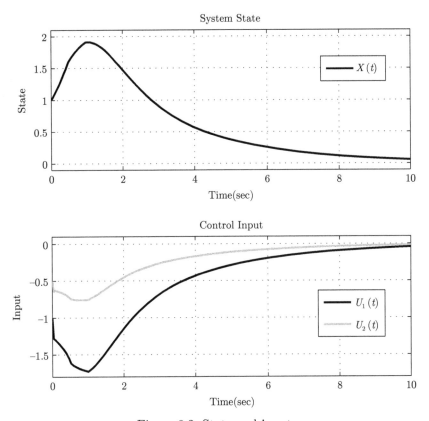

Figure 8.2. State and input

As explained in Remark 8.2, if $\frac{\hat{D}_1}{\hat{D}_2} \leq 1$ in the limit of the integration is violated, the backstepping transformation is no longer meaningfully defined. To address this difficulty, we utilize the switched control scheme as follows:

♦ If $\hat{D}_1(t) \leq \hat{D}_2(t)$, the backstepping transformation (8.16) has the form such that

- $i = 1, x \in [0, 1]$

$$\hat{w}_1(x, t) = \hat{u}_1(x, t) - k_1 e^{A\hat{D}_1 x} X(t)$$

$$- \hat{D}_1 \int_0^x k_1 e^{A(\hat{D}_1 x - \hat{D}_1 y)} b_1 \hat{u}_1(y, t) dy$$

$$- \hat{D}_2 \int_0^{\frac{\hat{D}_1}{\hat{D}_2} x} k_1 e^{A(\hat{D}_1 x - \hat{D}_2 y)} b_2 \hat{u}_2(y, t) dy, \qquad (8.128)$$

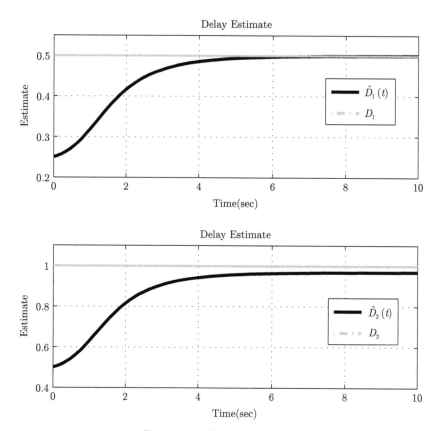

Figure 8.3. Delay estimate

$$\hat{u}_1(x,t) = \hat{w}_1(x,t) + k_1 e^{A_2 \hat{D}_1 x} X(t)$$

$$+ \hat{D}_1 \int_0^x k_1 e^{A_2(\hat{D}_1 x - \hat{D}_1 y)} b_1 \hat{w}_1(y,t) dy$$

$$+ \hat{D}_2 \int_0^{\frac{\hat{D}_1}{\hat{D}_2} x} k_1 e^{A_2(\hat{D}_1 x - \hat{D}_2 y)} b_2 \hat{w}_2(y,t) dy, \qquad (8.129)$$

- $i = 2, x \in \left[0, \frac{\hat{D}_1}{\hat{D}_2}\right]$

$$\hat{w}_2(x,t) = \hat{u}_2(x,t) - k_2 e^{A \hat{D}_2 x} X(t)$$

$$- \hat{D}_1 \int_0^{\frac{\hat{D}_2}{\hat{D}_1} x} k_2 e^{A(\hat{D}_2 x - \hat{D}_1 y)} b_1 \hat{u}_1(y,t) dy$$

$$- \hat{D}_2 \int_0^x k_2 e^{A(\hat{D}_2 x - \hat{D}_2 y)} b_2 \hat{u}_2(y,t) dy, \qquad (8.130)$$

$$\hat{u}_2(x,t) = \hat{w}_2(x,t) + k_2 e^{A_2 \hat{D}_2 x} X(t)$$

$$+ \hat{D}_1 \int_0^{\frac{\hat{D}_2}{\hat{D}_1} x} k_2 e^{A_2(\hat{D}_2 x - \hat{D}_1 y)} b_1 \hat{w}_1(y,t) dy$$

$$+ \hat{D}_2 \int_0^x k_2 e^{A_2(\hat{D}_2 x - \hat{D}_2 y)} b_2 \hat{w}_2(y,t) dy, \qquad (8.131)$$

- $i = 2, x \in \left[\frac{\hat{D}_1}{\hat{D}_2}, 1\right]$

$$\hat{w}_2(x,t) = \hat{u}_2(x,t) - k_2 e^{A_1(\hat{D}_2 x - \hat{D}_1)} e^{A \hat{D}_1} X(t)$$

$$- \hat{D}_1 \int_0^1 k_2 e^{A_1(\hat{D}_2 x - \hat{D}_1)} e^{A(\hat{D}_1 - \hat{D}_1 y)} b_1 \hat{u}_1(y,t) dy$$

$$- \hat{D}_2 \int_0^{\frac{\hat{D}_1}{\hat{D}_2}} k_2 e^{A_1(\hat{D}_2 x - \hat{D}_1)} e^{A(\hat{D}_1 - \hat{D}_2 y)} b_2 \hat{u}_2(y,t) dy$$

$$- \hat{D}_2 \int_{\frac{\hat{D}_1}{\hat{D}_2}}^x k_2 e^{A_1(\hat{D}_2 x - \hat{D}_2 y)} b_2 \hat{u}_2(y,t) dy, \qquad (8.132)$$

$$\hat{u}_2(x,t) = \hat{w}_2(x,t) + k_2 e^{A_2 \hat{D}_2 x} X(t)$$

$$+ \hat{D}_1 \int_0^1 k_2 e^{A_2(\hat{D}_2 x - \hat{D}_1 y)} b_1 \hat{w}_1(y,t) dy$$

$$+ \hat{D}_2 \int_0^x k_2 e^{A_2(\hat{D}_2 x - \hat{D}_2 y)} b_2 \hat{w}_2(y,t) dy, \qquad (8.133)$$

where $A_2 = A_1 + b_2 k_2$, $A_1 = A + b_1 k_1$.

♦ Else, i.e., $\hat{D}_1(t) > \hat{D}_2(t)$, the backstepping transformation (8.16) has the form below:

- $i = 2, x \in [0, 1]$

$$\hat{w}_2(x,t) = \hat{u}_2(x,t) - k_2 e^{A \hat{D}_2 x} X(t)$$

$$- \hat{D}_2 \int_0^x k_2 e^{A(\hat{D}_2 x - \hat{D}_2 y)} b_2 \hat{u}_2(y,t) dy$$

$$- \hat{D}_1 \int_0^{\frac{\hat{D}_2}{\hat{D}_1} x} k_2 e^{A(\hat{D}_2 x - \hat{D}_1 y)} b_1 \hat{u}_1(y,t) dy, \qquad (8.134)$$

$$\hat{u}_2(x,t) = \hat{w}_2(x,t) + k_2 e^{A_2 \hat{D}_2 x} X(t)$$

$$+ \hat{D}_2 \int_0^x k_2 e^{A_2(\hat{D}_2 x - \hat{D}_2 y)} b_2 \hat{w}_2(y,t) dy$$

$$+\hat{D}_1 \int_0^{\frac{\hat{D}_2}{\hat{D}_1}x} k_2 e^{A_2(\hat{D}_2 x - \hat{D}_1 y)} b_1 \hat{w}_1(y,t) dy, \qquad (8.135)$$

- $i = 1, x \in \left[0, \frac{\hat{D}_2}{\hat{D}_1}\right]$

$$\hat{w}_1(x,t) = \hat{u}_1(x,t) - k_1 e^{A\hat{D}_1 x} X(t)$$

$$-\hat{D}_2 \int_0^{\frac{\hat{D}_1}{\hat{D}_2}x} k_1 e^{A(\hat{D}_1 x - \hat{D}_2 y)} b_2 \hat{u}_2(y,t) dy$$

$$-\hat{D}_1 \int_0^x k_1 e^{A(\hat{D}_1 x - \hat{D}_1 y)} b_1 \hat{u}_1(y,t) dy, \qquad (8.136)$$

$$\hat{u}_1(x,t) = \hat{w}_1(x,t) + k_1 e^{A_2 \hat{D}_1 x} X(t)$$

$$+\hat{D}_2 \int_0^{\frac{\hat{D}_1}{\hat{D}_2}x} k_1 e^{A_2(\hat{D}_1 x - \hat{D}_2 y)} b_2 \hat{w}_2(y,t) dy$$

$$+\hat{D}_1 \int_0^x k_1 e^{A_2(\hat{D}_1 x - \hat{D}_1 y)} b_1 \hat{w}_1(y,t) dy, \qquad (8.137)$$

- $i = 1, x \in \left[\frac{\hat{D}_2}{\hat{D}_1}, 1\right]$

$$\hat{w}_1(x,t) = \hat{u}_1(x,t) - k_1 e^{A_1(\hat{D}_1 x - \hat{D}_2)} e^{A\hat{D}_2} X(t)$$

$$-\hat{D}_2 \int_0^1 k_1 e^{A_1(\hat{D}_1 x - \hat{D}_2)} e^{A(\hat{D}_2 - \hat{D}_2 y)} b_2 \hat{u}_2(y,t) dy$$

$$-\hat{D}_1 \int_0^{\frac{\hat{D}_2}{\hat{D}_1}} k_1 e^{A_1(\hat{D}_1 x - \hat{D}_2)} e^{A(\hat{D}_2 - \hat{D}_1 y)} b_1 \hat{u}_1(y,t) dy$$

$$-\hat{D}_1 \int_{\frac{\hat{D}_2}{\hat{D}_1}}^x k_1 e^{A_1(\hat{D}_1 x - \hat{D}_1 y)} b_1 \hat{u}_1(y,t) dy, \qquad (8.138)$$

$$\hat{u}_1(x,t) = \hat{w}_1(x,t) + k_1 e^{A_2 \hat{D}_1 x} X(t)$$

$$+\hat{D}_2 \int_0^1 k_1 e^{A_2(\hat{D}_1 x - \hat{D}_2 y)} b_2 \hat{w}_2(y,t) dy$$

$$+\hat{D}_1 \int_0^x k_1 e^{A_2(\hat{D}_1 x - \hat{D}_1 y)} b_1 \hat{w}_1(y,t) dy, \qquad (8.139)$$

where $A_2 = A_1 + b_1 k_1$, $A_1 = A + b_2 k_2$.

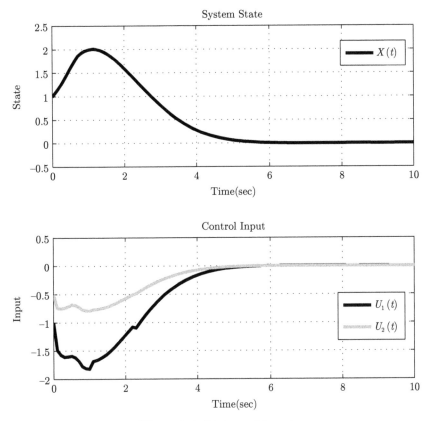

Figure 8.4. State and input

Please note that (8.128), (8.130), (8.134), and (8.136) are structurally the same at all times when $\hat{D}_1(t) = \hat{D}_2(t)$, so there is no jump in $\hat{w}_1(x,t)$ and $\hat{w}_2(x,t)$ at those times. As $\hat{w}_1(x,t)$, $\hat{w}_2(x,t)$, and $\hat{u}_1(x,t)$, $\hat{u}_2(x,t)$ are well defined by the switched scheme, the feedback law and the update law are obtained accordingly. The result is shown in figures 8.4 and 8.5.

The result of simulation for the case that $D_1 < D_2$ is assumed, but actually $D_2 < D_1$ may give readers an impression that Assumption 7.5 about the order of the delays can be relaxed by this switched control scheme. However, we would like to emphasize the fact that the order of the delays can be addressed in a simple manner when LTI systems have two inputs. When three or more inputs are taken into account, the switched control scheme gets increasingly complicated, since more comparisons among $\hat{D}_i$ with $\hat{D}_j$ have to be applied.

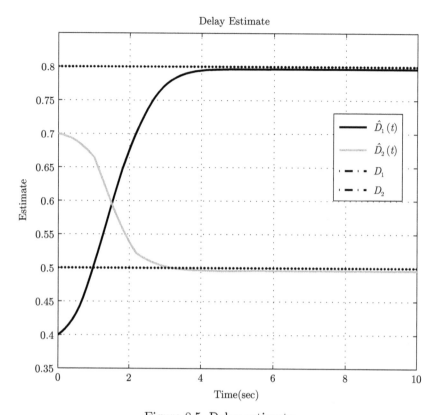

Figure 8.5. Delay estimate

Chapter Nine

Output Feedback of Systems with Uncertain Delays, Parameters, and ODE State

IN CHAPTER 8 we solved the problem of adaptive stabilization of multi-input LTI systems under the assumption that distinct discrete multiple actuator delays are long and unknown, and the actuator state in each input channel is unavailable for measurement. A local stabilization result is achieved. In this chapter, as shown in table 7.1, we consider the more challenging case where uncertainties exist not only in the input delay and the PDE actuator state, but also appear in the ODE plant state and the parameters of the system and input matrices. To be specific, the uncertainty collections $\left(D_i, u_i(x,t), X(t)\right)$ and $\left(D_i, u_i(x,t), \theta\right)$ will be addressed, respectively. Similar to the result in chapter 8, when the actuator state is not measurable and the delay value is unknown at the same time, the result of multi-input delays obtained in this chapter is also not global.

9.1 Model Depiction

Referring to (7.1)–(7.2) in section 7.1, consider a general class of LTI systems with distinct discrete multi-input delays as follows:

$$\dot{X}(t) = A(\theta)X(t) + \sum_{i=1}^{m} b_i(\theta)U_i(t - D_i), \tag{9.1}$$

$$Y(t) = CX(t), \tag{9.2}$$

where $X(t) \in \mathbb{R}^n$ is the state vector of ODE, $Y(t) \in \mathbb{R}^q$ is the output vector, and $U_i(t) \in \mathbb{R}$ is the ith scalar control input with a constant delay $D_i \in \mathbb{R}^+$ for $i \in \{1, \cdots, m\}$. The multi-input distinct delays satisfy Assumption 7.5 in section 7.2. The constant parameter vector $\theta \in \mathbb{R}^p$ represents plant parameters and control coefficients. The matrices $A(\theta) \in \mathbb{R}^{n \times n}$, $b_i(\theta) \in \mathbb{R}^n$, and $C \in \mathbb{R}^{q \times n}$ are the system matrix, the input matrix, and the output matrix, respectively. The system and input matrices $A(\theta)$ and $b_i(\theta)$ are linearly parameterized, namely,

$$A(\theta) = A_0 + \sum_{h=1}^{p} \theta_h A_h, \quad b_i(\theta) = b_{i0} + \sum_{h=1}^{p} \theta_h b_{ih} \tag{9.3}$$

and satisfy Assumption 7.2 in section 7.1. Our objective is to stabilize the potentially unstable LTI systems (9.1). The control design is to realize the exact predictor-feedback below, for $i \in \{1, \cdots, m\}$:

$$U_i(t - D_i) = k_i(\theta)X(t), \text{i.e., } U_i(t) = k_i(\theta)X(t + D_i). \tag{9.4}$$

To deal with the delayed input, a transport PDE state is introduced as follows:

$$u_i(x, t) = U_i(t + D_i(x - 1)), x \in [0, 1] \tag{9.5}$$

Thus the ODE (9.1)–(9.2) is converted to the following ODE-PDE cascade:

$$\dot{X}(t) = A(\theta)X(t) + \sum_{i=1}^{m} b_i(\theta)u_i(0, t), \tag{9.6}$$

$$Y(t) = CX(t), \tag{9.7}$$

$$D_i \partial_t u_i(x, t) = \partial_x u_i(x, t), \quad x \in [0, 1], \tag{9.8}$$

$$u_i(1, t) = U_i(t), \quad i \in \{1, \cdots, m\}. \tag{9.9}$$

For the ODE-PDE cascade (9.6)–(9.9) representing LTI systems with multi-actuator delays (9.1)–(9.2), we have dealt with uncertainty combinations $(D_i, X(t))$, (D_i, θ), $(D_i, u_i(x, t))$ in previous chapters. As described in the last two rows of table 7.1, in this chapter we deal with adaptive stabilization with more uncertainties such that $(D_i, u_i(x, t), X(t))$ and $(D_i, u_i(x, t), \theta)$.

9.2   Local Stabilization under Uncertain Delays and ODE State

In this section, as shown in the second last row in table 7.1, we consider the case of uncertainty combination $(D_i, X(t), u_i(x, t))$.

As the parameter vector θ in (9.1) and (9.6) is assumed to be known in this section, for the sake of brevity, we denote

$$A = A(\theta), \quad b_i = b_i(\theta), \quad k_i = k_i(\theta). \tag{9.10}$$

Thus (9.6)–(9.9) are rewritten briefly as

$$\dot{X}(t) = AX(t) + \sum_{i=1}^{m} b_i u_i(0, t), \tag{9.11}$$

$$Y(t) = CX(t), \tag{9.12}$$

$$D_i \partial_t u_i(x, t) = \partial_x u_i(x, t), \quad x \in [0, 1], \tag{9.13}$$

$$u_i(1,t) = U_i(t), \quad i \in \{1, \cdots, m\}. \tag{9.14}$$

The estimate $\hat{D}_i(t)$ is brought in for the unknown D_i with the estimation error $\tilde{D}_i(t) = D_i - \hat{D}_i(t)$. To solve the technical difficulty of the unmeasurable PDE state $u_i(x,t)$, we introduce the PDE observer by using the delay estimate such that, for $i \in \{1, 2, \cdots, m\}$,

$$\hat{u}_i(x,t) = U_i(t + \hat{D}_i(t)(x-1)), \quad x \in [0,1], \tag{9.15}$$

which is governed by the PDE observer dynamics

$$\hat{D}_i(t)\partial_t \hat{u}_i(x,t) = \partial_x \hat{u}_i(x,t) + \dot{\hat{D}}_i(t)(x-1)\partial_x \hat{u}_i(x,t), \tag{9.16}$$
$$\hat{u}_i(1,t) = U_i(t), \tag{9.17}$$

with the observer error being defined as

$$\tilde{u}_i(x,t) = u_i(x,t) - \hat{u}_i(x,t), \quad x \in [0,1], \tag{9.18}$$
$$= U_i(t + D_i(x-1)) - U_i(t + \hat{D}_i(t)(x-1)). \tag{9.19}$$

To deal with the unmeasurable ODE state $X(t)$, bearing Assumption 7.8 in section 7.3 in mind, the ODE observer is selected as

$$\dot{\hat{X}}(t) = A\hat{X}(t) + \sum_{i=1}^{m} b_i \hat{u}_i(0,t) + L(Y(t) - C\hat{X}(t)), \tag{9.20}$$

where $\hat{X}(t)$ is an estimate of the unmeasurable $X(t)$, with the observer error being defined as

$$\tilde{X}(t) = X(t) - \hat{X}(t). \tag{9.21}$$

With the PDE observer (9.16)–(9.17) and the ODE observer (9.20), the ODE-PDE cascade (9.11)–(9.14) is transformed into

$$\dot{\hat{X}}(t) = A\hat{X}(t) + \sum_{i=1}^{m} b_i \hat{u}_i(0,t) + LC\tilde{X}(t), \tag{9.22}$$

$$\dot{\tilde{X}}(t) = (A - LC)\tilde{X}(t) + \sum_{i=1}^{m} b_i \tilde{u}_i(0,t), \tag{9.23}$$

$$D_i \partial_t \tilde{u}_i(x,t) = \partial_x \tilde{u}_i(x,t) - \tilde{D}_i \frac{1}{\hat{D}_i}\partial_x \hat{u}_i(x,t)$$

$$\qquad\qquad - \dot{\hat{D}}_i D_i(x-1)\frac{1}{\hat{D}_i}\partial_x \hat{u}_i(x,t), \tag{9.24}$$

$$\tilde{u}_i(1,t)=0, \tag{9.25}$$

$$\hat{D}_i\partial_t\hat{u}_i(x,t)=\partial_x\hat{u}_i(x,t)+\dot{\hat{D}}_i(x-1)\partial_x\hat{u}_i(x,t), \tag{9.26}$$

$$\hat{u}_i(1,t)=U_i(t). \tag{9.27}$$

If we substitute $X(t)$ and $u_i(x,t)$ with $\hat{X}(t)$ and $\hat{u}_i(x,t)$ in (7.29)–(7.30), the backstepping transformation is designed as follows:

$$\hat{w}_i(x,t)=\hat{u}_i(x,t)-k_i\Phi(\hat{D}_ix,\hat{D}_l^v,0)\hat{X}(t)$$

$$-\sum_{j=1}^m\hat{D}_j\int_0^{\phi_j(\hat{D}_ix,\hat{D}_j)}k_i\Phi(\hat{D}_ix,\hat{D}_l^v,\hat{D}_jy)b_j\hat{u}_j(y,t)dy, \tag{9.28}$$

$$\hat{u}_i(x,t)=\hat{w}_i(x,t)+k_ie^{A_m\hat{D}_ix}\hat{X}(t)$$

$$+\sum_{j=1}^m\hat{D}_j\int_0^{\phi_j(\hat{D}_ix,\hat{D}_j)}k_ie^{A_m(\hat{D}_ix-\hat{D}_jy)}b_j\hat{w}_j(y,t)dy, \tag{9.29}$$

where $\phi_j(\hat{D}_ix,\hat{D}_j)$ and $\Phi(\hat{D}_ix,\hat{D}_l^v,\hat{D}_jy)$ are defined as follows: for $j\in\{1,2,\cdots,m\}$,

$$\phi_j(\hat{D}_ix,\hat{D}_j)=\begin{cases}\frac{\hat{D}_i}{\hat{D}_j}x, & x\in\left[0,\frac{\hat{D}_j}{\hat{D}_i}\right]\\ 1, & x\in\left[\frac{\hat{D}_j}{\hat{D}_i},\frac{\hat{D}_m}{\hat{D}_i}\right],\end{cases} \tag{9.30}$$

and for $\frac{\hat{D}_{v-1}}{\hat{D}_j}\le y\le\frac{\hat{D}_i}{\hat{D}_j}x\le\frac{\hat{D}_v}{\hat{D}_j}$, $i,j\in\{1,2,\cdots,m\}$, $v\in\{1,2,\cdots,i\}$,

$$\Phi(\hat{D}_ix,\hat{D}_l^v,\hat{D}_jy)=e^{A_{v-1}(\hat{D}_ix-\hat{D}_jy)}, \tag{9.31}$$

for $\frac{\hat{D}_{v-1}}{\hat{D}_i}\le x\le\frac{\hat{D}_v}{\hat{D}_i}$, $\frac{\hat{D}_{l-1}}{\hat{D}_j}\le y\le\frac{\hat{D}_l}{\hat{D}_j}$, $i,j\in\{1,2,\cdots,m\}$, $v,l\in\{1,2,\cdots,i\}$, $l<v$, $\hat{D}_l^v=(\hat{D}_{v-1},\hat{D}_{v-2},\cdots,\hat{D}_{l+1},\hat{D}_l)$,

$$\Phi(\hat{D}_ix,\hat{D}_l^v,\hat{D}_jy)=e^{A_{v-1}(\hat{D}_ix-\hat{D}_{v-1})}e^{A_{v-2}(\hat{D}_{v-1}-\hat{D}_{v-2})}$$

$$\cdots e^{A_l(\hat{D}_{l+1}-\hat{D}_l)}e^{A_{l-1}(\hat{D}_l-\hat{D}_jy)}, \tag{9.32}$$

and specifically for (9.32), if $l=v-1$,

$$\Phi(\hat{D}_ix,\hat{D}_l^v,\hat{D}_jy)=e^{A_{v-1}(\hat{D}_ix-\hat{D}_{v-1})}e^{A_{v-2}(\hat{D}_{v-1}-\hat{D}_jy)}, \tag{9.33}$$

with

$$\hat{D}_0 = 0, \quad \begin{cases} A_0 = A, \\ A_i = A_{i-1} + b_i k_i, \quad i \in \{1, \cdots, m\} \\ A_m = A + BK = A + \sum_{i=1}^{m} b_i k_i. \end{cases} \tag{9.34}$$

The feedback law is chosen as follows:

$$U_i(t) = \hat{u}_i(1, t) = k_i \Phi(\hat{D}_i, \hat{D}_l^v, 0)\hat{X}(t)$$

$$+ \sum_{j=1}^{m} \hat{D}_j \int_0^{\phi_j(\hat{D}_i, \hat{D}_j)} k_i \Phi(\hat{D}_i, \hat{D}_l^v, \hat{D}_j y) b_j \hat{u}_j(y, t) dy \tag{9.35}$$

to meet boundary condition $\hat{w}_i(1, t) = 0$ of the closed-loop system. Furthermore, the delay-adaptive law is designed below, for $i \in \{1, 2, \cdots, m\}$:

$$\dot{\hat{D}}_i(t) = \gamma_D \mathrm{Proj}_{[\underline{D}_{i-1,i}, \overline{D}_{i,i+1}]} \tau_{D_i}(t), \quad \hat{D}_i(0) \in [\underline{D}_{i-1,i}, \overline{D}_{i,i+1}], \tag{9.36}$$

$$\tau_{D_i}(t) = -\sum_{h=1}^{m} \int_0^1 (1+x)\hat{w}_h(x, t)\bar{p}_{hi}(x, t)dx, \tag{9.37}$$

where $\bar{p}_{hi}(x, t)$ is given in (9.71) of section 9.4.

Remark 9.1. If we compare (9.37) with (7.41) in section 7.2, through replacing unmeasurable $X(t)$ and $u_i(x, t)$ with $\hat{X}(t)$ and $\hat{u}_i(x, t)$, it is obvious that the regulation error $\hat{w}_h(x, t)$ and the regressor $\bar{p}_{hi}(x, t)$ in (9.37) are certainty-equivalence versions of $w_h(x, t)$ and $p_{hi}(x, t)$ in the numerator of (7.41). Furthermore, as stated in chapter 8 of [80], if the delay is unknown and the actuator state is unmeasured, the solution of the transport PDE is not linearly parametrizable (statically or dynamically) in the delay, so the global result is not achievable. Thus the normalization by the denominator of (7.41) is not used in (9.37).

If we employ (9.15), (9.20), and (9.28)–(9.35), the system (9.22)–(9.27) is transformed into the closed-loop system:

$$\dot{\hat{X}}(t) = (A + BK)\hat{X}(t) + \sum_{i=1}^{m} b_i \hat{w}_i(0, t) + LC\tilde{X}(t), \tag{9.38}$$

$$\dot{\tilde{X}}(t) = (A - LC)\tilde{X}(t) + \sum_{i=1}^{m} b_i \tilde{u}_i(0, t), \tag{9.39}$$

$$D_i \partial_t \tilde{u}_i(x, t) = \partial_x \tilde{u}_i(x, t) - \hat{D}_i \bar{f}_i(x, t)$$

$$- \dot{\hat{D}}_i D_i(x - 1)\bar{f}_i(x, t), \tag{9.40}$$

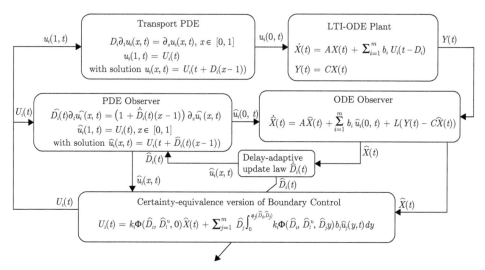

Figure 9.1. Adaptive control under uncertain delays and ODE and PDE states

$$\tilde{u}_i(1,t) = 0, \tag{9.41}$$

$$\hat{D}_i\partial_t\hat{w}_i(x,t) = \partial_x\hat{w}_i(x,t) - \hat{D}_i\bar{q}_i(x,t,\dot{\hat{D}})$$
$$-\hat{D}_ik_i\Phi(\hat{D}_ix,\hat{D}_l^v,0)LC\tilde{X}(t), \tag{9.42}$$

$$\hat{w}_i(1,t) = 0, \tag{9.43}$$

$$\hat{D}_i\partial_{xt}\hat{w}_i(x,t) = \partial_{xx}\hat{w}_i(x,t) - \hat{D}_i\partial_x\bar{q}_i(x,t,\dot{\hat{D}})$$
$$-\hat{D}_ik_i\partial_x\Phi(\hat{D}_ix,\hat{D}_l^v,0)LC\tilde{X}(t), \tag{9.44}$$

$$\partial_x\hat{w}_i(1,t) = \hat{D}_i\bar{q}_i(1,t,\dot{\hat{D}}) + \hat{D}_ik_i\Phi(\hat{D}_i,\hat{D}_l^v,0)LC\tilde{X}(t), \tag{9.45}$$

where $\bar{f}_i(x,t)$ and $\bar{q}_i(x,t,\dot{\hat{D}})$ are shown as (9.79) and (9.81) in section 9.4, respectively.

Then the adaptive controller in this section is shown in figure 9.1.

Theorem 9.2. *Consider the closed-loop system consisting of the ODE-PDE cascade (9.11)–(9.14), the PDE-state observer (9.16)–(9.17) and the ODE-state observer (9.20), the invertible backstepping transformation (9.28)–(9.29), the PDE boundary control (9.35), and the delay-adaptive update law (9.36)–(9.37). Then there exists $M^* > 0$ such that under all initial conditions satisfying*

$$|X(0)|^2 + |\hat{X}(0)|^2 + \sum_{i=1}^{m}\left(\|u_i(x,0)\|^2 + \|\hat{u}_i(x,0)\|^2\right.$$

$$+\|\partial_x\hat{u}_i(x,0)\|^2 + |\tilde{D}_i(0)|^2\Big) \leq M^*, \tag{9.46}$$

all the signals of the closed-loop system are bounded and the regulation is achieved, i.e., $\lim_{t\to\infty} X(t) = 0$.

Proof. We build a Lyapunov candidate encompassing all states of the closed-loop system (9.38)–(9.45):

$$V(t) = \hat{X}(t)^T P \hat{X}(t) + \tilde{X}(t)^T P_L \tilde{X}(t)$$

$$+ g_u \sum_{i=1}^{m} D_i \int_0^1 (1+x)\tilde{u}_i(x,t)^2 dx$$

$$+ g_w \sum_{i=1}^{m} \hat{D}_i(t) \left(\int_0^1 (1+x)\hat{w}_i(x,t)^2 dx \right.$$

$$+ \left. \int_0^1 (1+x)(\partial_x \hat{w}_i(x,t))^2 dx \right) + \sum_{i=1}^{m} \frac{1}{2}\tilde{D}_i(t)^2, \qquad (9.47)$$

where $g_u > 0$ and $g_w > 0$. If we take the time derivative of $V(t)$ along (9.38)–(9.45), following a similar procedure of the analysis in section 8.3, it is not hard to show the nonincreasing property of $V(t)$ by the initial condition (9.46), and consequently the local stability of the closed-loop system is obtained. $\qquad \square$

9.3 Local Stabilization under Uncertain Delays and Parameters

In this section, as shown in the last row of table 7.1, we further consider another case of uncertainty combination $(D_i, \theta, u_i(x,t))$ under Assumption 7.10 in section 7.4.

As shown in table 7.1, since the parameter vector θ is unknown, we do not employ the abbreviation (9.10) but focus on the system (9.6)–(9.9). By the certainty-equivalence principle, we use the parameter estimate $\hat{\theta}(t)$ to identify the unknown θ with the estimation error $\tilde{\theta}(t) = \theta - \hat{\theta}(t)$. Recalling (9.3), we have

$$A(\hat{\theta}) = A_0 + \sum_{h=1}^{p} \hat{\theta}_h A_h, \quad b_i(\hat{\theta}) = b_{i0} + \sum_{h=1}^{p} \hat{\theta}_h b_{ih}, \qquad (9.48)$$

$$A(\tilde{\theta}) = \sum_{h=1}^{p} \tilde{\theta}_h A_h, \quad b_i(\tilde{\theta}) = \sum_{h=1}^{p} \tilde{\theta}_h b_{ih}. \qquad (9.49)$$

If we replace θ and $u_i(x,t)$ in (7.29)–(7.30) with $\hat{\theta}(t)$ and $\hat{u}_i(x,t)$, the backstepping transformations have the forms

$$\hat{w}_i(x,t) = \hat{u}_i(x,t) - k_i(\hat{\theta})\Phi(\hat{D}_i x, \hat{D}_l^v, 0, \hat{\theta})X(t)$$

$$- \sum_{j=1}^{m} \hat{D}_j \int_0^{\phi_j(\hat{D}_i x, \hat{D}_j)} k_i(\hat{\theta})\Phi(\hat{D}_i x, \hat{D}_l^v, \hat{D}_j y, \hat{\theta})b_j(\hat{\theta})\hat{u}_j(y,t)dy, \qquad (9.50)$$

$$\hat{u}_i(x,t) = \hat{w}_i(x,t) + k_i(\hat{\theta})e^{A_m(\hat{\theta})\hat{D}_i x}X(t)$$

$$+ \sum_{j=1}^{m} \hat{D}_j \int_0^{\phi_j(\hat{D}_i x, \hat{D}_j)} k_i(\hat{\theta})e^{A_m(\hat{\theta})(\hat{D}_i x - \hat{D}_j y)}b_j(\hat{\theta})\hat{w}_j(y,t)dy, \qquad (9.51)$$

where for $j \in \{1,2,\cdots,m\}$, $\phi_j(\hat{D}_i x, \hat{D}_j)$ is exactly identical with (9.30) such that

$$\phi_j(\hat{D}_i x, \hat{D}_j) = \begin{cases} \frac{\hat{D}_i}{\hat{D}_j}x, & x \in \left[0, \frac{\hat{D}_j}{\hat{D}_i}\right] \\ 1, & x \in \left[\frac{\hat{D}_j}{\hat{D}_i}, \frac{\hat{D}_m}{\hat{D}_i}\right], \end{cases} \qquad (9.52)$$

while for $\frac{\hat{D}_{v-1}}{\hat{D}_j} \le y \le \frac{\hat{D}_i}{\hat{D}_j}x \le \frac{\hat{D}_v}{\hat{D}_j}$, $i,j \in \{1,2,\cdots,m\}$, $v \in \{1,2,\cdots,i\}$,

$$\Phi(\hat{D}_i x, \hat{D}_l^v, \hat{D}_j y, \hat{\theta}) = e^{A_{v-1}(\hat{\theta})(\hat{D}_i x - \hat{D}_j y)}, \qquad (9.53)$$

for $\frac{\hat{D}_{v-1}}{\hat{D}_i} \le x \le \frac{\hat{D}_v}{\hat{D}_i}$, $\frac{\hat{D}_{l-1}}{\hat{D}_j} \le y \le \frac{\hat{D}_l}{\hat{D}_j}$, $i,j \in \{1,2,\cdots,m\}$, $v,l \in \{1,2,\cdots,i\}$, $l < v$, $\hat{D}_l^v = (\hat{D}_{v-1}, \hat{D}_{v-2}, \cdots, \hat{D}_{l+1}, \hat{D}_l)$,

$$\Phi(\hat{D}_i x, \hat{D}_l^v, \hat{D}_j y, \hat{\theta}) = e^{A_{v-1}(\hat{\theta})(\hat{D}_i x - \hat{D}_{v-1})}e^{A_{v-2}(\hat{\theta})(\hat{D}_{v-1} - \hat{D}_{v-2})}$$

$$\cdots e^{A_l(\hat{\theta})(\hat{D}_{l+1} - \hat{D}_l)}e^{A_{l-1}(\hat{\theta})(\hat{D}_l - \hat{D}_j y)}, \qquad (9.54)$$

and specifically for (9.54), if $l = v - 1$,

$$\Phi(\hat{D}_i x, \hat{D}_l^v, \hat{D}_j y, \hat{\theta}) = e^{A_{v-1}(\hat{\theta})(\hat{D}_i x - \hat{D}_{v-1})}e^{A_{v-2}(\hat{\theta})(\hat{D}_{v-1} - \hat{D}_j y)}, \qquad (9.55)$$

with

$$\hat{D}_0 = 0, \quad \begin{cases} A_0(\hat{\theta}) = A(\hat{\theta}), \\ A_i(\hat{\theta}) = A_{i-1}(\hat{\theta}) + b_i(\hat{\theta})k_i(\hat{\theta}), i \in \{1,2,\cdots,m\} \\ A_m(\hat{\theta}) = (A + BK)(\hat{\theta}) = A(\hat{\theta}) + \sum_{i=1}^{m} b_i(\hat{\theta})k_i(\hat{\theta}). \end{cases} \qquad (9.56)$$

Subsequently, for $i = 1,2,\cdots,m$, if we substitute $x = 1$ into (9.50) to meet boundary condition $\hat{w}_i(1,t) = 0$, the prediction-based PDE boundary feedback law has the expression below:

$$U_i(t) = \hat{u}_i(1,t) = k_i(\hat{\theta})\Phi(\hat{D}_i, \hat{D}_l^v, 0, \hat{\theta})X(t)$$

$$+ \sum_{j=1}^{m} \hat{D}_j \int_0^{\phi_j(\hat{D}_i, \hat{D}_j)} k_i(\hat{\theta})\Phi(\hat{D}_i, \hat{D}_l^v, \hat{D}_j y, \hat{\theta})b_j(\hat{\theta})\hat{u}_j(y,t)dy. \qquad (9.57)$$

Then the ODE-PDE cascade (9.6)–(9.9) could be converted into the following closed-loop system:

$$\dot{X}(t) = (A + BK)(\hat{\theta})X(t) + \sum_{i=1}^{m} b_i(\hat{\theta})\hat{w}_i(0,t)$$

$$+ \sum_{h=1}^{p} \tilde{\theta}_h \left(A_h X(t) + \sum_{i=1}^{m} b_{ih}\hat{u}_i(0,t) \right)$$

$$+ \sum_{i=1}^{m} b_i(\theta)\tilde{u}_i(0,t), \tag{9.58}$$

$$D_i \partial_t \tilde{u}_i(x,t) = \partial_x \tilde{u}_i(x,t) - \tilde{D}_i \bar{\bar{f}}_i(x,t)$$

$$- \dot{\hat{D}}_i D_i(x-1)\bar{\bar{f}}_i(x,t), \tag{9.59}$$

$$\tilde{u}_i(1,t) = 0, \tag{9.60}$$

$$\hat{D}_i \partial_t w_i(x,t) = \partial_x w_i(x,t) - \hat{D}_i \bar{\bar{q}}_i(x,t,\dot{\hat{D}})$$

$$- \hat{D}_i k_i(\hat{\theta}) \Phi(\hat{D}_i x, \hat{D}_l^v, 0, \hat{\theta}) \sum_{j=1}^{m} b_j(\theta)\tilde{u}_j(0,t)$$

$$- \hat{D}_i \sum_{h=1}^{p} \tilde{\theta}_h \bar{\bar{r}}_{ih}(x,t) - \hat{D}_i \sum_{h=1}^{p} \dot{\hat{\theta}}_h \bar{\bar{s}}_{ih}(x,t), \tag{9.61}$$

$$w_i(1,t) = 0, \tag{9.62}$$

$$\hat{D}_i \partial_{xt} \hat{w}_i(x,t) = \partial_{xx} \hat{w}_i(x,t) - \hat{D}_i \partial_x \bar{\bar{q}}_i(x,t,\dot{\hat{D}})$$

$$- \hat{D}_i k_i(\hat{\theta}) \partial_x \Phi(\hat{D}_i x, \hat{D}_l^v, 0, \hat{\theta}) \sum_{j=1}^{m} b_j(\theta)\tilde{u}_j(0,t)$$

$$- \hat{D}_i \sum_{h=1}^{p} \tilde{\theta}_h \partial_x \bar{\bar{r}}_{ih}(x,t) - \hat{D}_i \sum_{h=1}^{p} \dot{\hat{\theta}}_h \partial_x \bar{\bar{s}}_{ih}(x,t), \tag{9.63}$$

$$\partial_x \hat{w}_i(1,t) = \hat{D}_i \bar{\bar{q}}_i(1,t,\dot{\hat{D}})$$

$$+ \hat{D}_i k_i(\hat{\theta}) \Phi(\hat{D}_i, \hat{D}_l^v, 0, \hat{\theta}) \sum_{j=1}^{m} b_j(\theta)\tilde{u}_j(0,t)$$

$$+ \hat{D}_i \sum_{h=1}^{p} \tilde{\theta}_h \bar{\bar{r}}_{ih}(1,t) + \hat{D}_i \sum_{h=1}^{p} \dot{\hat{\theta}}_h \bar{\bar{s}}_{ih}(1,t), \tag{9.64}$$

where $\bar{\bar{f}}_{ih}(x,t)$, $\bar{\bar{q}}_i(x,t,\dot{\hat{D}})$, $\bar{\bar{r}}_{ih}(x,t)$, $\bar{\bar{s}}_{ih}(x,t)$ are shown as (9.80), (9.82), (9.86), and (9.87) of section 9.4, respectively.

The delay-parameter-adaptive update laws are selected as follows:

$$i = 1, 2, \cdots, m,$$

$$\dot{\hat{D}}_i(t) = \gamma_D \mathrm{Proj}_{[\underline{D}_{i-1,i}, \overline{D}_{i,i+1}]} \tau_{D_i}(t), \quad \hat{D}_i(0) \in [\underline{D}_{i-1,i}, \overline{D}_{i,i+1}], \tag{9.65}$$

$$\tau_{D_i}(t) = -\sum_{h=1}^{m} \int_0^1 (1+x)\hat{w}_h(x,t)\bar{\bar{p}}_{hi}(x,t)dx, \tag{9.66}$$

$$i = 1, 2, \cdots, p,$$

$$\dot{\hat{\theta}}_i(t) = \gamma_\theta \mathrm{Proj}_\Theta \tau_{\theta_i}(t), \quad \hat{\theta}_i(0) \in \Theta, \quad \gamma_\theta > 0, \tag{9.67}$$

$$\tau_{\theta_i}(t) = 2X(t)^T P(\hat{\theta}) \Big(A_i X(t) + \sum_{h=1}^{m} b_{hi}\hat{u}_h(0,t)\Big)/g_w$$

$$- \sum_{h=1}^{m} \int_0^1 (1+x)\big(\hat{w}_h(x,t)\bar{\bar{r}}_{hi}(x,t) + \partial_x\hat{w}_h(x,t)\partial_x\bar{\bar{r}}_{hi}(x,t)\big)dx, \tag{9.68}$$

where $\bar{\bar{p}}_{ih}(x,t)$ is given in (9.72) of section 9.4.

Remark 9.3. Similar to Remark 9.1, as shown in (9.72), the delay update law (9.66) is a certainty-equivalence version of the numerator of (7.41) by replacing unknown θ and $u_i(x,t)$ with $\hat{\theta}$ and $\hat{u}_i(x,t)$. As the global stabilization cannot be derived, the normalization by the denominator of (7.41) is not employed in (9.66). The parameter update law (9.68) is chosen in order to cancel the parameter estimation error terms $\tilde{\theta}_i$ arising from (9.58), (9.61), and (9.63) in the Lyapunov-based analysis.

Then the adaptive controller in this section is shown in figure 9.2.

Theorem 9.4. *Consider the closed-loop system consisting of the ODE-PDE cascade (9.6)–(9.9), the PDE-state observer (9.16)–(9.17), the invertible backstepping transformation (9.50)–(9.51), the PDE boundary control (9.57), and the delay-parameter-adaptive update laws (9.65)–(9.68). Then there exists $M^* > 0$ such that under all initial conditions satisfying*

$$|X(0)|^2 + |\tilde{\theta}(0)|^2 + \sum_{i=1}^{m} \big(\|u_i(x,0)\|^2 + \|\hat{u}_i(x,0)\|^2$$

$$+ \|\partial_x\hat{u}_i(x,0)\|^2 + |\tilde{D}_i(0)|^2\big) \leq M^*, \tag{9.69}$$

all the signals of the closed-loop system are bounded and the regulation is achieved, i.e., $\lim_{t\to\infty} X(t) = 0$.

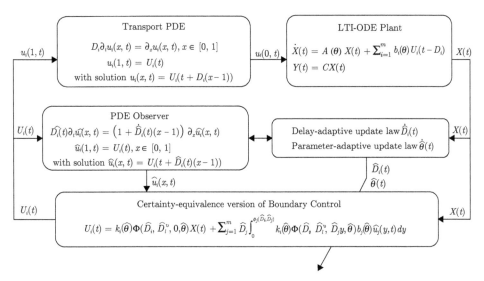

Figure 9.2. Adaptive control under uncertain delays, parameters, and PDE states

Proof. We build a Lyapunov candidate encompassing all states of the closed-loop system (9.58)–(9.64),

$$V(t) = X(t)^T P(\hat{\theta}) X(t) + g_u \sum_{i=1}^{m} D_i \int_0^1 (1+x)\tilde{u}_i(x,t)^2 dx$$

$$+ g_w \sum_{i=1}^{m} \hat{D}_i(t) \left(\int_0^1 (1+x)\hat{w}_i(x,t)^2 dx \right.$$

$$\left. + \int_0^1 (1+x)(\partial_x \hat{w}_i(x,t))^2 dx \right) + \sum_{i=1}^{m} \frac{1}{2}\tilde{D}_i(t)^2$$

$$+ \frac{g_w}{\gamma_\theta} \tilde{\theta}(t)^T \tilde{\theta}(t). \tag{9.70}$$

If we take the time derivative of $V(t)$ along (9.58)–(9.64), following a similar procedure of the proof in section 8.3, it is not hard to show the nonincreasing property of $V(t)$ by the initial condition (9.69), and consequently the proof of the theorem is completed. □

9.4   Auxiliary Calculations for Sections 9.2 and 9.3

The $\bar{p}_{ih}(x,t)$ in (9.37) and $\bar{\bar{p}}_{ih}(x,t)$ in (9.66) are listed as follows:

$$\bar{p}_{ih}(x,t) = k_i P_{\Phi(\hat{D}_i x, \hat{D}_l^v, 0)} \hat{X}(t) + k_i \Phi(\hat{D}_i x, \hat{D}_l^v, 0) b_h \hat{u}_h(0,t)$$

$$+ \sum_{j=1}^{m} \hat{D}_j \int_0^{\phi_j(\hat{D}_i x, \hat{D}_j)} k_i P_{\Phi(\hat{D}_i x, \hat{D}_l^v, \hat{D}_j y)} b_j \hat{u}_j(y, t) dy$$

$$+ \sum_{j=1}^{m} \rho_j k_i b_j \hat{u}_j(\phi_j(\hat{D}_i x, \hat{D}_j), t), \tag{9.71}$$

$$\bar{\bar{p}}_{ih}(x, t) = k_i(\hat{\theta}) P_{\Phi(\hat{D}_i x, \hat{D}_l^v, 0, \hat{\theta})} X(t) + k_i(\hat{\theta}) \Phi(\hat{D}_i x, \hat{D}_l^v, 0, \hat{\theta}) b_h(\hat{\theta}) \hat{u}_h(0, t)$$

$$+ \sum_{j=1}^{m} \hat{D}_j \int_0^{\phi_j(\hat{D}_i x, \hat{D}_j)} k_i(\hat{\theta}) P_{\Phi(\hat{D}_i x, \hat{D}_l^v, \hat{D}_j y, \hat{\theta})} b_j(\hat{\theta}) \hat{u}_j(y, t) dy$$

$$+ \sum_{j=1}^{m} \rho_j k_i(\hat{\theta}) b_j(\hat{\theta}) \hat{u}_j(\phi_j(\hat{D}_i x, \hat{D}_j), t), \tag{9.72}$$

where

$$\rho_j = \begin{cases} 1, & \text{if} \quad h = i, \phi_j(\hat{D}_i x, \hat{D}_j) = \frac{\hat{D}_i}{\hat{D}_j} x, i \neq j \\ 0, & \text{if} \quad h \neq i, j, \quad \text{or} \quad \phi_j(\hat{D}_i x, \hat{D}_j) = \frac{\hat{D}_i}{\hat{D}_j} x, i = j, \quad \text{or} \quad \phi_j(\hat{D}_i x, \hat{D}_j) = 1 \\ -1, & \text{if} \quad h = j, \phi_j(\hat{D}_i x, \hat{D}_j) = \frac{\hat{D}_i}{\hat{D}_j} x, i \neq j, \end{cases} \tag{9.73}$$

$$P_{\Phi(\hat{D}_i x, \hat{D}_l^v, 0)} = \begin{cases} \frac{\partial \Phi(\hat{D}_i x, \hat{D}_l^v, 0)}{\partial(\hat{D}_i x)}, & h = i \\ \frac{\partial \Phi(\hat{D}_i x, \hat{D}_l^v, 0)}{\partial \hat{D}_h}, & h \neq i \end{cases}, \quad P_{\Phi(\hat{D}_i x, \hat{D}_l^v, 0, \hat{\theta})} = \begin{cases} \frac{\partial \Phi(\hat{D}_i x, \hat{D}_l^v, 0, \hat{\theta})}{\partial(\hat{D}_i x)}, & h = i \\ \frac{\partial \Phi(\hat{D}_i x, \hat{D}_l^v, 0, \hat{\theta})}{\partial \hat{D}_h}, & h \neq i \end{cases}, \tag{9.74}$$

and $P_{\Phi(\hat{D}_i x, \hat{D}_l^v, \hat{D}_j y)}$ in (9.71) has exactly the same structure as $P_{\Phi(\hat{D}_i x, \hat{D}_l^v, \hat{D}_j y, \hat{\theta})}$ in (9.72) by substituting $\hat{\theta}$ with θ, while $P_{\Phi(\hat{D}_i x, \hat{D}_l^v, \hat{D}_j y, \hat{\theta})}$ has the form corresponding to $\Phi(\hat{D}_i x, \hat{D}_l^v, \hat{D}_j y, \hat{\theta})$ of (9.53)–(9.55) such that
For $\Phi(\hat{D}_i x, \hat{D}_l^v, \hat{D}_j y, \hat{\theta})$ of (9.53),

$$P_{\Phi(\hat{D}_i x, \hat{D}_l^v, \hat{D}_j y, \hat{\theta})} = \begin{cases} \frac{\partial \Phi(\hat{D}_i x, \hat{D}_l^v, \hat{D}_j y, \hat{\theta})}{\partial(\hat{D}_i x)}, & h = i, h \neq j \\ \frac{\partial \Phi(\hat{D}_i x, \hat{D}_l^v, \hat{D}_j y, \hat{\theta})}{\partial \hat{D}_h} = 0, & h \neq i, j \\ 0, & h = i = j \\ \frac{\partial \Phi(\hat{D}_i x, \hat{D}_l^v, \hat{D}_j y, \hat{\theta})}{\partial(\hat{D}_j y)}, & h = j, h \neq i. \end{cases} \tag{9.75}$$

For $\Phi(\hat{D}_i x, \hat{D}_l^v, \hat{D}_j y, \hat{\theta})$ of (9.54)–(9.55),

$$P_{\Phi(\hat{D}_i x, \hat{D}_l^v, \hat{D}_j y, \hat{\theta})}$$

$$= \begin{cases} A_{v-1}(\hat{\theta}) e^{A_{v-1}(\hat{\theta})\hat{D}_i x} \Gamma(\hat{D}_l^v, \hat{\theta}) e^{-A_{l-1}(\hat{\theta})\hat{D}_j y} \\ \quad + e^{A_{v-1}(\hat{\theta})\hat{D}_i x} \frac{\partial \Gamma(\hat{D}_l^v, \hat{\theta})}{\partial \hat{D}_h} e^{-A_{l-1}(\hat{\theta})\hat{D}_j y}, & h = i, h \neq j \\[2em] e^{A_{v-1}(\hat{\theta})\hat{D}_i x} \frac{\partial \Gamma(\hat{D}_l^v, \hat{\theta})}{\partial \hat{D}_h} e^{-A_{l-1}(\hat{\theta})\hat{D}_j y}, & h \neq i, j \\[2em] A_{v-1}(\hat{\theta}) e^{A_{v-1}(\hat{\theta})\hat{D}_i x} \Gamma(\hat{D}_l^v, \hat{\theta}) e^{-A_{l-1}(\hat{\theta})\hat{D}_j y} \\ \quad + e^{A_{v-1}(\hat{\theta})\hat{D}_i x} \frac{\partial \Gamma(\hat{D}_l^v, \hat{\theta})}{\partial \hat{D}_h} e^{-A_{l-1}(\hat{\theta})\hat{D}_j y} \\ \quad + e^{A_{v-1}(\hat{\theta})\hat{D}_i x} \Gamma(\hat{D}_l^v, \hat{\theta}) e^{-A_{l-1}(\hat{\theta})\hat{D}_j y} (-A_{l-1}(\hat{\theta})), & h = i = j \\[2em] e^{A_{v-1}(\hat{\theta})\hat{D}_i x} \frac{\partial \Gamma(\hat{D}_l^v, \hat{\theta})}{\partial \hat{D}_h} e^{-A_{l-1}(\hat{\theta})\hat{D}_j y} \\ \quad + e^{A_{v-1}(\hat{\theta})\hat{D}_i x} \Gamma(\hat{D}_l^v, \hat{\theta}) e^{-A_{l-1}(\hat{\theta})\hat{D}_j y} (-A_{l-1}(\hat{\theta})), & h = j, h \neq i, \end{cases} \tag{9.76}$$

with $\Gamma(\hat{D}_l^v, \hat{\theta})$ coming from (9.54)–(9.55) such that

$$\begin{aligned} \Phi(\hat{D}_i x, \hat{D}_l^v, \hat{D}_j y, \hat{\theta}) &= e^{A_{v-1}(\hat{\theta})(\hat{D}_i x - \hat{D}_{v-1})} e^{A_{v-2}(\hat{\theta})(\hat{D}_{v-1} - \hat{D}_{v-2})} \cdots \\ &\quad \times e^{A_l(\hat{\theta})(\hat{D}_{l+1} - \hat{D}_l)} e^{A_{l-1}(\hat{\theta})(\hat{D}_l - \hat{D}_j y)} \\ &= e^{A_{v-1}(\hat{\theta})\hat{D}_i x} \left(e^{-A_{v-1}(\hat{\theta})\hat{D}_{v-1}} e^{A_{v-2}(\hat{\theta})(\hat{D}_{v-1} - \hat{D}_{v-2})} \cdots \right. \\ &\quad \left. \times e^{A_l(\hat{\theta})(\hat{D}_{l+1} - \hat{D}_l)} e^{A_{l-1}(\hat{\theta})\hat{D}_l} \right) e^{-A_{l-1}(\hat{\theta})\hat{D}_j y} \\ &= e^{A_{v-1}(\hat{\theta})\hat{D}_i x} \Gamma(\hat{D}_l^v, \hat{\theta}) e^{-A_{l-1}(\hat{\theta})\hat{D}_j y}, \end{aligned} \tag{9.77}$$

and specifically, if $l = v - 1$,

$$\begin{aligned} \Phi(\hat{D}_i x, \hat{D}_l^v, \hat{D}_j y, \hat{\theta}) &= e^{A_{v-1}(\hat{\theta})(\hat{D}_i x - \hat{D}_{v-1})} e^{A_{v-2}(\hat{\theta})(\hat{D}_{v-1} - \hat{D}_j y)} \\ &= e^{A_{v-1}(\hat{\theta})\hat{D}_i x} \left(e^{-A_{v-1}(\hat{\theta})\hat{D}_{v-1}} e^{A_{v-2}(\hat{\theta})\hat{D}_{v-1}} \right) e^{-A_{v-2}(\hat{\theta})\hat{D}_j y} \\ &= e^{A_{v-1}(\hat{\theta})\hat{D}_i x} \Gamma(\hat{D}_l^v, \hat{\theta}) e^{-A_{v-2}(\hat{\theta})\hat{D}_j y}. \end{aligned} \tag{9.78}$$

The $\bar{f}_i(x, t)$ in (9.40), $\bar{\bar{f}}_i(x, t)$ in (9.59), $\bar{q}_i(x, t, \hat{D})$ in (9.42)–(9.45), and $\bar{\bar{q}}_i(x, t, \hat{D})$ in (9.61)–(9.64) are listed as follows:

$$\begin{aligned} \bar{f}_i(x, t) &= \frac{1}{\hat{D}_i} \partial_x \hat{u}_i(x, t) = \frac{1}{\hat{D}_i} \partial_x \hat{w}_i(x, t) + k_i A_m e^{A_m \hat{D}_i x} \hat{X}(t) \\ &\quad + \frac{1}{\hat{D}_i} \sum_{j=1}^m \hat{D}_j \partial_x \phi_j(\hat{D}_i x, \hat{D}_j) k_i b_j \hat{w}_j(\phi_j(\hat{D}_i x, \hat{D}_j), t) \end{aligned}$$

$$+\sum_{j=1}^{m}\hat{D}_j\int_0^{\phi_j(\hat{D}_ix,\hat{D}_j)}k_iA_me^{A_m(\hat{D}_ix-\hat{D}_jy)}b_j\hat{w}_j(y,t)dy,\qquad(9.79)$$

$$\bar{\bar{f}}_i(x,t)=\frac{1}{\hat{D}_i}\partial_x\hat{u}_i(x,t)=\frac{1}{\hat{D}_i}\partial_x\hat{w}_i(x,t)+k_i(\hat{\theta})A_m(\hat{\theta})e^{A_m(\hat{\theta})\hat{D}_ix}X(t)$$

$$+\frac{1}{\hat{D}_i}\sum_{j=1}^{m}\hat{D}_j\partial_x\phi_j(\hat{D}_ix,\hat{D}_j)k_i(\hat{\theta})b_j(\hat{\theta})\hat{w}_j(\phi_j(\hat{D}_ix,\hat{D}_j),t)$$

$$+\sum_{j=1}^{m}\hat{D}_j\int_0^{\phi_j(\hat{D}_ix,\hat{D}_j)}k_i(\hat{\theta})A_m(\hat{\theta})e^{A_m(\hat{\theta})(\hat{D}_ix-\hat{D}_jy)}b_j(\hat{\theta})\hat{w}_j(y,t)dy,$$

$$(9.80)$$

$$\dot{\bar{q}}_i(x,t,\dot{\hat{D}})=\dot{\hat{D}}_i\frac{(1-x)}{\hat{D}_i}\partial_x\hat{w}_i(x,t)+k_i\Phi(\hat{D}_ix,\hat{D}_l^v,0)\sum_{j=1}^{m}\dot{\hat{D}}_jb_j\hat{u}_j(0,t)$$

$$+k_iQ_{\Phi(\hat{D}_ix,\hat{D}_l^v,0)}(\dot{\hat{D}})\hat{X}(t)$$

$$+\sum_{j=1}^{m}\varepsilon_j(\dot{\hat{D}}_i-\dot{\hat{D}}_j)k_ib_j\hat{u}_j\left(\frac{\hat{D}_i}{\hat{D}_j}x,t\right)$$

$$+\sum_{j=1}^{m}\hat{D}_j\int_0^{\phi_j(\hat{D}_ix,\hat{D}_j)}k_iQ_{\Phi(\hat{D}_ix,\hat{D}_l^v,\hat{D}_jy)}(\dot{\hat{D}})b_j\hat{u}_j(y,t)dy,\qquad(9.81)$$

$$\dot{\bar{\bar{q}}}_i(x,t,\dot{\hat{D}})=\dot{\hat{D}}_i\frac{(1-x)}{\hat{D}_i}\partial_x\hat{w}_i(x,t)+k_i(\hat{\theta})\Phi(\hat{D}_ix,\hat{D}_l^v,0,\hat{\theta})\sum_{j=1}^{m}\dot{\hat{D}}_jb_j(\hat{\theta})\hat{u}_j(0,t)$$

$$+k_i(\hat{\theta})Q_{\Phi(\hat{D}_ix,\hat{D}_l^v,0,\hat{\theta})}(\dot{\hat{D}})X(t)$$

$$+\sum_{j=1}^{m}\varepsilon_j(\dot{\hat{D}}_i-\dot{\hat{D}}_j)k_i(\hat{\theta})b_j(\hat{\theta})\hat{u}_j\left(\frac{\hat{D}_i}{\hat{D}_j}x,t\right)$$

$$+\sum_{j=1}^{m}\hat{D}_j\int_0^{\phi_j(\hat{D}_ix,\hat{D}_j)}k_i(\hat{\theta})Q_{\Phi(\hat{D}_ix,\hat{D}_l^v,\hat{D}_jy,\hat{\theta})}(\dot{\hat{D}})b_j(\hat{\theta})\hat{u}_j(y,t)dy,$$

$$(9.82)$$

where $\epsilon_j=\begin{cases}1,&\phi_j(\hat{D}_ix,\hat{D}_j)=\frac{\hat{D}_i}{\hat{D}_j}x\\0,&\phi_j(\hat{D}_ix,\hat{D}_j)=1\end{cases}$.

In addition, $Q_{\Phi(\hat{D}_ix,\hat{D}_l^v,\hat{D}_jy)}(\dot{\hat{D}})$ and $Q_{\Phi(\hat{D}_ix,\hat{D}_l^v,\hat{D}_jy,\hat{\theta})}(\dot{\hat{D}})$ have the same structure with the only difference in θ and $\hat{\theta}$, and $Q_{\Phi(\hat{D}_ix,\hat{D}_l^v,\hat{D}_jy,\hat{\theta})}(\dot{\hat{D}})$ has the form corresponding to $\Phi(\hat{D}_ix,\hat{D}_l^v,\hat{D}_jy,\hat{\theta})$ of (9.53)–(9.55).

For $\Phi(\hat{D}_ix,\hat{D}_l^v,\hat{D}_jy,\hat{\theta})$ of (9.53),

$$Q_{\Phi(\hat{D}_i x, \hat{D}_l^v, \hat{D}_j y, \hat{\theta})}(\dot{D}) = A_{v-1}(\hat{\theta})(\dot{\hat{D}}_i - \dot{\hat{D}}_j)e^{A_{v-1}(\hat{\theta})(\hat{D}_i x - \hat{D}_j y)}. \qquad (9.83)$$

For $\Phi(\hat{D}_i x, \hat{D}_l^v, \hat{D}_j y, \hat{\theta})$ of (9.54),

$$
\begin{aligned}
Q_{\Phi(\hat{D}_i x, \hat{D}_l^v, \hat{D}_j y, \hat{\theta})}&(\dot{D}) \\
&= A_{v-1}(\hat{\theta})(\dot{\hat{D}}_i - \dot{\hat{D}}_{v-1})e^{A_{v-1}(\hat{\theta})(\hat{D}_i x - \hat{D}_{v-1})}e^{A_{v-2}(\hat{\theta})(\hat{D}_{v-1} - \hat{D}_{v-2})} \cdots \\
&\quad \times e^{A_l(\hat{\theta})(\hat{D}_{l+1} - \hat{D}_l)}e^{A_{l-1}(\hat{\theta})(\hat{D}_l - \hat{D}_j y)} \\
&\quad + e^{A_{v-1}(\hat{\theta})(\hat{D}_i x - \hat{D}_{v-1})}A_{v-2}(\hat{\theta})(\dot{\hat{D}}_{v-1} - \dot{\hat{D}}_{v-2})e^{A_{v-2}(\hat{\theta})(\hat{D}_{v-1} - \hat{D}_{v-2})} \cdots \\
&\quad \times e^{A_l(\hat{\theta})(\hat{D}_{l+1} - \hat{D}_l)}e^{A_{l-1}(\hat{\theta})(\hat{D}_l - \hat{D}_j y)} \\
&\quad + \quad \cdots \quad \cdots \\
&\quad + e^{A_{v-1}(\hat{\theta})(\hat{D}_i x - \hat{D}_{v-1})}e^{A_{v-2}(\hat{\theta})(\hat{D}_{v-1} - \hat{D}_{v-2})} \cdots \\
&\quad \times A_l(\hat{\theta})(\dot{\hat{D}}_{l+1} - \dot{\hat{D}}_l)e^{A_l(\hat{\theta})(\hat{D}_{l+1} - \hat{D}_l)}e^{A_{l-1}(\hat{\theta})(\hat{D}_l - \hat{D}_j y)} \\
&\quad + e^{A_{v-1}(\hat{\theta})(\hat{D}_i x - \hat{D}_{v-1})}e^{A_{v-2}(\hat{\theta})(\hat{D}_{v-1} - \hat{D}_{v-2})} \cdots \\
&\quad \times e^{A_l(\hat{\theta})(\hat{D}_{l+1} - \hat{D}_l)}e^{A_{l-1}(\hat{\theta})(\hat{D}_l - \hat{D}_j y)}A_{l-1}(\hat{\theta})(\dot{\hat{D}}_l - \dot{\hat{D}}_j). \qquad (9.84)
\end{aligned}
$$

For $\Phi(\hat{D}_i x, \hat{D}_l^v, \hat{D}_j y, \hat{\theta})$ of (9.55),

$$
\begin{aligned}
Q_{\Phi(\hat{D}_i x, \hat{D}_l^v, \hat{D}_j y, \hat{\theta})}(\dot{D}) = e^{A_{v-1}(\hat{\theta})(\hat{D}_i x - \hat{D}_{v-1})}&\left(A_{v-1}(\hat{\theta})(\dot{\hat{D}}_i - \dot{\hat{D}}_{v-1}) \right. \\
&\left. + A_{v-2}(\hat{\theta})(\dot{\hat{D}}_{v-1} - \dot{\hat{D}}_j) \right) e^{A_{v-2}(\hat{\theta})(\hat{D}_{v-1} - \hat{D}_j y)}. \quad (9.85)
\end{aligned}
$$

The $\bar{\bar{r}}_{ih}(x,t)$ and $\bar{\bar{s}}_{ih}(x,t)$ in (9.61)–(9.64) are listed as follows:

$$\bar{\bar{r}}_{ih}(x,t) = k_i(\hat{\theta})\Phi(\hat{D}_i x, \hat{D}_l^v, 0, \hat{\theta})\left(A_h X(t) + \sum_{j=1}^m b_{jh}\hat{u}_j(0,t) \right), \qquad (9.86)$$

$$
\begin{aligned}
\bar{\bar{s}}_{ih}(x,t) = \frac{\partial k_i(\hat{\theta})}{\partial \hat{\theta}_h}&\Phi(\hat{D}_i x, \hat{D}_l^v, 0, \hat{\theta})X(t) + k_i(\hat{\theta})\frac{\partial \Phi(\hat{D}_i x, \hat{D}_l^v, 0, \hat{\theta})}{\partial \hat{\theta}_h}X(t) \\
&+ \sum_{j=1}^m \hat{D}_j \int_0^{\phi_j(\hat{D}_i x, \hat{D}_j)} \left(\frac{\partial k_i(\hat{\theta})}{\partial \hat{\theta}_h}\Phi(\hat{D}_i x, \hat{D}_l^v, \hat{D}_j y, \hat{\theta})b_j(\hat{\theta}) \right. \\
&+ k_i(\hat{\theta})\frac{\partial \Phi(\hat{D}_i x, \hat{D}_l^v, \hat{D}_j y, \hat{\theta})}{\partial \hat{\theta}_h}b_j(\hat{\theta}) \\
&\left. + k_i(\hat{\theta})\Phi(\hat{D}_i x, \hat{D}_l^v, \hat{D}_j y, \hat{\theta})\frac{\partial b_j(\hat{\theta})}{\partial \hat{\theta}_h} \right) \hat{u}_j(y,t)dy, \qquad (9.87)
\end{aligned}
$$

and $\frac{\partial \Phi(\hat{D}_i x, \hat{D}_l^v, \hat{D}_j y, \hat{\theta})}{\partial \hat{\theta}_h}$ has the form corresponding to $\Phi(\hat{D}_i x, \hat{D}_l^v, \hat{D}_j y, \hat{\theta})$ of (9.53)–(9.55) such that:

For $\Phi(\hat{D}_i x, \hat{D}_l^v, \hat{D}_j y, \hat{\theta})$ of (9.53),

$$\frac{\partial \Phi(\hat{D}_i x, \hat{D}_l^v, \hat{D}_j y, \hat{\theta})}{\partial \hat{\theta}_h} = \frac{\partial A_{v-1}(\hat{\theta})}{\partial \hat{\theta}_h}(\hat{D}_i x - \hat{D}_j y) e^{A_{v-1}(\hat{\theta})(\hat{D}_i x - \hat{D}_j y)}, \quad (9.88)$$

$$\frac{\partial \Phi(\hat{D}_i x, \hat{D}_l^v, 0, \hat{\theta})}{\partial \hat{\theta}_h} = \frac{\partial A_{v-1}(\hat{\theta})}{\partial \hat{\theta}_h} \hat{D}_i x e^{A_{v-1}(\hat{\theta}) \hat{D}_i x}. \quad (9.89)$$

For $\Phi(\hat{D}_i x, \hat{D}_l^v, \hat{D}_j y, \hat{\theta})$ of (9.54),

$$\frac{\partial \Phi(\hat{D}_i x, \hat{D}_l^v, \hat{D}_j y, \hat{\theta})}{\partial \hat{\theta}_h}$$

$$= \frac{\partial A_{v-1}(\hat{\theta})}{\partial \hat{\theta}_h}(\hat{D}_i x - \hat{D}_{v-1}) e^{A_{v-1}(\hat{\theta})(\hat{D}_i x - \hat{D}_{v-1})} e^{A_{v-2}(\hat{\theta})(\hat{D}_{v-1} - \hat{D}_{v-2})} \cdots$$

$$\times e^{A_l(\hat{\theta})(\hat{D}_{l+1} - \hat{D}_l)} e^{A_{l-1}(\hat{\theta})(\hat{D}_l - \hat{D}_j y)}$$

$$+ e^{A_{v-1}(\hat{\theta})(\hat{D}_i x - \hat{D}_{v-1})} \frac{\partial A_{v-2}(\hat{\theta})}{\partial \hat{\theta}_h}(\hat{D}_{v-1} - \hat{D}_{v-2}) e^{A_{v-2}(\hat{\theta})(\hat{D}_{v-1} - \hat{D}_{v-2})} \cdots$$

$$\times e^{A_l(\hat{\theta})(\hat{D}_{l+1} - \hat{D}_l)} e^{A_{l-1}(\hat{\theta})(\hat{D}_l - \hat{D}_j y)}$$

$$+ \quad \cdots \quad \cdots$$

$$+ e^{A_{v-1}(\hat{\theta})(\hat{D}_i x - \hat{D}_{v-1})} e^{A_{v-2}(\hat{\theta})(\hat{D}_{v-1} - \hat{D}_{v-2})} \cdots$$

$$\times \frac{\partial A_l(\hat{\theta})}{\partial \hat{\theta}_h}(\hat{D}_{l+1} - \hat{D}_l) e^{A_l(\hat{\theta})(\hat{D}_{l+1} - \hat{D}_l)} e^{A_{l-1}(\hat{\theta})(\hat{D}_l - \hat{D}_j y)}$$

$$+ e^{A_{v-1}(\hat{D}_i x - \hat{D}_{v-1})} e^{A_{v-2}(\hat{D}_{v-1} - \hat{D}_{v-2})} \cdots$$

$$\times e^{A_l(\hat{\theta})(\hat{D}_{l+1} - \hat{D}_l)} e^{A_{l-1}(\hat{\theta})(\hat{D}_l - \hat{D}_j y)} \frac{\partial A_{l-1}(\hat{\theta})}{\partial \hat{\theta}_h}(\hat{D}_l - \hat{D}_j y), \quad (9.90)$$

$$\frac{\partial \Phi(\hat{D}_i x, \hat{D}_l^v, 0, \hat{\theta})}{\partial \hat{\theta}_h}$$

$$= \frac{\partial A_{v-1}(\hat{\theta})}{\partial \hat{\theta}_h}(\hat{D}_i x - \hat{D}_{v-1}) e^{A_{v-1}(\hat{\theta})(\hat{D}_i x - \hat{D}_{v-1})} e^{A_{v-2}(\hat{\theta})(\hat{D}_{v-1} - \hat{D}_{v-2})} \cdots$$

$$\times e^{A_l(\hat{\theta})(\hat{D}_{l+1} - \hat{D}_l)} e^{A_{l-1}(\hat{\theta}) \hat{D}_l}$$

$$+ e^{A_{v-1}(\hat{\theta})(\hat{D}_i x - \hat{D}_{v-1})} \frac{\partial A_{v-2}(\hat{\theta})}{\partial \hat{\theta}_h}(\hat{D}_{v-1} - \hat{D}_{v-2}) e^{A_{v-2}(\hat{\theta})(\hat{D}_{v-1} - \hat{D}_{v-2})} \cdots$$

$$\times e^{A_l(\hat{\theta})(\hat{D}_{l+1} - \hat{D}_l)} e^{A_{l-1}(\hat{\theta}) \hat{D}_l}$$

$$+ \quad \cdots \quad \cdots$$

$$+ e^{A_{v-1}(\hat{\theta})(\hat{D}_i x - \hat{D}_{v-1})} e^{A_{v-2}(\hat{\theta})(\hat{D}_{v-1} - \hat{D}_{v-2})} \cdots$$

$$\times \frac{\partial A_l(\hat{\theta})}{\partial \hat{\theta}_h} (\hat{D}_{l+1} - \hat{D}_l) e^{A_l(\hat{\theta})(\hat{D}_{l+1} - \hat{D}_l)} e^{A_{l-1}(\hat{\theta})\hat{D}_l}$$

$$+ e^{A_{v-1}(\hat{\theta})(\hat{D}_i x - \hat{D}_{v-1})} e^{A_{v-2}(\hat{\theta})(\hat{D}_{v-1} - \hat{D}_{v-2})} \cdots$$

$$\times e^{A_l(\hat{\theta})(\hat{D}_{l+1} - \hat{D}_l)} e^{A_{l-1}(\hat{\theta})\hat{D}_l} \frac{\partial A_{l-1}(\hat{\theta})}{\partial \hat{\theta}_h} \hat{D}_l. \tag{9.91}$$

For $\Phi(\hat{D}_i x, \hat{D}_l^v, \hat{D}_j y, \hat{\theta})$ of (9.55),

$$\frac{\partial \Phi(\hat{D}_i x, \hat{D}_l^v, \hat{D}_j y, \hat{\theta})}{\partial \hat{\theta}_h} = e^{A_{v-1}(\hat{\theta})(\hat{D}_i x - \hat{D}_{v-1})} \left(\frac{\partial A_{v-1}(\hat{\theta})}{\partial \hat{\theta}_h} (\hat{D}_i x - \hat{D}_{v-1}) \right.$$

$$\left. + \frac{\partial A_{v-2}(\hat{\theta})}{\partial \hat{\theta}_h} (\hat{D}_{v-1} - \hat{D}_j y) \right) e^{A_{v-2}(\hat{\theta})(\hat{D}_{v-1} - \hat{D}_j y)} \tag{9.92}$$

$$\frac{\partial \Phi(\hat{D}_i x, \hat{D}_l^v, 0, \hat{\theta})}{\partial \hat{\theta}_h} = e^{A_{v-1}(\hat{\theta})(\hat{D}_i x - \hat{D}_{v-1})} \left(\frac{\partial A_{v-1}(\hat{\theta})}{\partial \hat{\theta}_h} (\hat{D}_i x - \hat{D}_{v-1}) \right.$$

$$\left. + \frac{\partial A_{v-2}(\hat{\theta})}{\partial \hat{\theta}_h} \hat{D}_{v-1} \right) e^{A_{v-2}(\hat{\theta})\hat{D}_{v-1}}. \tag{9.93}$$

Part III

Distributed Input Delays

Chapter Ten

Predictor Feedback for Uncertainty-Free Systems

IN PARTS I AND II, we investigated predictor control problems of uncertain single-input or multi-input LTI systems with discrete input delays. In this part, we extend the predictor method to compensate for another big family of delays—distributed input delays.

The control designs for unknown discrete input delays in parts I and II are not applicable to the case of unknown distributed input delays, as the plant and actuator states are not in the strict feedback form. In this part, we presents a new systematic method to stabilize uncertain LTI systems with distributed input delays.

First of all, in order to lay a foundation for adaptive and robust control for uncertain systems in later chapters, chapter 10 contributes to a predictor framework in *rescaled unity-interval* notation for uncertainty-free systems.

10.1   Predictor Feedback for Uncertainty-Free Single-Input Systems

In this section, we propose a predictor feedback framework in rescaled unity-interval notation for uncertainty-free single-input systems.

Consider single-input linear systems with distributed input delay as follows:

$$\dot{X}(t) = AX(t) + \int_0^D B(D - \sigma)U(t - \sigma)d\sigma, \tag{10.1}$$

$$Y(t) = CX(t), \tag{10.2}$$

where $X(t) \in \mathbb{R}^n$ is the plant state, $Y(t) \in \mathbb{R}^q$ is the plant output, $U(\xi) \in \mathbb{R}$ for $\xi \in [t - D, t]$ is the actuator state, $U(t) \in \mathbb{R}$ is the control input, $D > 0$ is the constant delay, and A, B are the system matrix and input vector of appropriate dimensions, respectively. For notational simplicity, the system is assumed to be single input. The results of this section can be straightforwardly extended to the multi-input case, when the delays are the same in each individual input channel.

Through a multi-variable function in rescaled unity-interval notation

$$u(x,t) = U(t + D(x-1)) = U(t - \sigma), \quad x \in [0,1], \sigma \in [0,D], \tag{10.3}$$

the system (10.1)–(10.2) is converted into the ODE-PDE cascade

$$\dot{X}(t) = AX(t) + D \int_0^1 B(Dx)u(x,t)dx, \tag{10.4}$$

$$Y(t) = CX(t), \tag{10.5}$$

$$Du_t(x,t) = u_x(x,t), \quad x \in [0,1], \tag{10.6}$$

$$u(1,t) = U(t). \tag{10.7}$$

It is evident (10.3) is a solution of the transport PDE (10.6)–(10.7). The control objective is to stabilize (10.4)–(10.7).

Concentrating on (10.4)–(10.7), we have a single-input linear plant with distributed actuator delay that comes with the following five types of basic uncertainties:

- unknown delay D,
- unknown delay kernel $B(Dx)$,
- unknown parameters in the system matrix A,
- unmeasurable finite-dimensional plant state $X(t)$,
- unmeasurable infinite-dimensional actuator state $u(x,t)$.

In this section, we consider the simplest case that everything in the system is known. Namely, the finite-dimensional ODE state $X(t)$ and infinite-dimensional PDE state $u(x,t)$ for $x \in [0,1]$ are assumed to be measurable, and the delay D, the delay kernel $B(Dx)$, and the system matrix A are known.

The reduction-based change of variable is introduced as

$$Z(t) = X(t) + D^2 \int_0^1 \int_0^x e^{-AD(x-y)} B(Dy)dyu(x,t)dx. \tag{10.8}$$

Taking the time derivative of (10.8) along (10.4)–(10.7) and using the integration by parts, we get

$$\dot{Z}(t) = \dot{X}(t) + D^2 \int_0^1 \int_0^x e^{-AD(x-y)} B(Dy)dyu_t(x,t)dx$$

$$= AX(t) + D \int_0^1 B(Dx)u(x,t)dx$$

$$+ D \int_0^1 \int_0^x e^{-AD(x-y)} B(Dy)dyu_x(x,t)dx$$

$$= AX(t) + D \int_0^1 B(Dx)u(x,t)dx$$

$$+ D \int_0^1 e^{-AD(1-y)} B(Dy)dyu(1,t)$$

$$- D \int_0^1 B(Dx)u(x,t)dx$$

$$+ D^2 \int_0^1 \int_0^x A e^{-AD(x-y)} B(Dy)dyu(x,t)dx$$

$$= AZ(t) + D \int_0^1 e^{-AD(1-x)} B(Dx)dxU(t). \tag{10.9}$$

It is evident that through (10.8) the stabilization problem of linear systems with distributed delay (10.4) is reduced to the stabilization problem of a delay-free system (10.9).

Assumption 10.1. *For the system (10.4) and (10.9), the pair* $\left(A, D\int_0^1 e^{-AD(1-x)} \right.$ $\left. B(Dx)dx \right)$ *is stabilizable. There exist a vector* K *to make* $A + D \int_0^1 e^{-AD(1-x)}$ $B(Dx)dxK$ *Hurwitz. Namely, there exist matrices* $P = P^T > 0$ *and* $Q = Q^T > 0$ *such that*

$$\left(A + D \int_0^1 e^{-AD(1-x)} B(Dx)dxK \right)^T P$$

$$+ P \left(A + D \int_0^1 e^{-AD(1-x)} B(Dx)dxK \right) = -Q. \tag{10.10}$$

The controller is designed as

$$U(t) = u(1,t) = KZ(t). \tag{10.11}$$

Thus the system (10.9) becomes

$$\dot{Z}(t) = A_{cl}Z(t), \tag{10.12}$$

where

$$A_{cl} = A + D \int_0^1 e^{-AD(1-x)} B(Dx)dxK. \tag{10.13}$$

For stability analysis, the invertible backstepping-forwarding transformation is brought in as

$$w(x,t) = u(x,t) - Ke^{A_{cl}D(x-1)}Z(t). \tag{10.14}$$

Figure 10.1. Conversion between the original system (10.1)–(10.2) and target system (10.18)–(10.20)

Take the partial derivatives of (10.14) with respect to x and t, respectively:

$$w_x(x,t) = u_x(x,t) - Ke^{A_{cl}D(x-1)}A_{cl}DZ(t), \qquad (10.15)$$

$$w_t(x,t) = u_t(x,t) - Ke^{A_{cl}D(x-1)}A_{cl}Z(t). \qquad (10.16)$$

Multiplying (10.16) with D and subtracting (10.15), we have

$$Dw_t(x,t) = w_x(x,t). \qquad (10.17)$$

Combining (10.12) with (10.17), substituting $x=1$ into (10.14), and utilizing (10.11), we obtain the target system for analysis as follows:

$$\dot{Z}(t) = A_{cl}Z(t), \qquad (10.18)$$

$$Dw_t(x,t) = w_x(x,t), \quad x \in [0,1], \qquad (10.19)$$

$$w(1,t) = 0. \qquad (10.20)$$

Remark 10.2. As illustrated in figure 10.1, a few conversions are employed in the above control scheme. Firstly, through the multi-variable function (10.3), the original system with distributed input delay (10.1)–(10.2) is represented by the ODE-PDE cascade (10.4)–(10.7). Secondly, by the reduction-based change of variable (10.8), the stabilization problem of delayed plant (10.4) is reduced to the stabilization problem of the delay-free system (10.9). Finally, under the backstepping-forwarding transformation (10.14) and the control law (10.11), the ODE (10.9) and PDE (10.6)–(10.7) is transformed into the target system (10.18)–(10.20), which is convenient for stability analysis.

Remark 10.3. An alternative representation of the control scheme (10.3)–(10.20) in [6] is listed as follows:

$$u(x,t) = U(t+x-D) = U(t-\sigma), x \in [0,D], \sigma \in [0,D], \qquad (10.21)$$

$$\dot{X}(t) = AX(t) + \int_0^D B(x)u(x,t)dx, \qquad (10.22)$$

$$u_t(x,t) = u_x(x,t), \quad x \in [0,D], \qquad (10.23)$$

$$u(D,t) = U(t) = KZ(t), \qquad (10.24)$$

$$Z(t) = X(t) + \int_0^D \int_0^x e^{-A(x-y)}B(y)dyu(x,t)dx, \qquad (10.25)$$

$$w(x,t) = u(x,t) - Ke^{A_{cl}(x-D)}Z(t), \tag{10.26}$$

$$\dot{Z}(t) = A_{cl}Z(t), \quad A_{cl} = A + \int_0^D e^{-A(D-x)}B(x)dxK, \tag{10.27}$$

$$w_t(x,t) = w_x(x,t), \quad x \in [0,D], \tag{10.28}$$

$$w(D,t) = 0. \tag{10.29}$$

It is evident that (10.21)–(10.29) is equivalent to (10.3)–(10.20). The main difference is that (10.3)–(10.20) is parameterized in D, whereas (10.21)–(10.29) is nonparameterized in D. When the delay D is unknown and a time-varying signal $\hat{D}(t)$ is employed to estimate D, it is inconvenient for the adaptive control to be applied to (10.21)–(10.29). For example, a moving boundary $u(\hat{D}(t),t)$ appears in (10.26), which renders the boundary condition (10.29) nonhomogenous such that $w(\hat{D}(t),t) = u(\hat{D}(t),t) - u(D,t) \neq 0$. The estimate appears in the limit of integration of (10.25) such that $Z(t) = X(t) + \int_0^{\hat{D}(t)} \int_0^x e^{-A(x-y)}B(y)dyu(x,t)dx$, which makes it difficult to get the error $\tilde{D}(t) = D - \hat{D}(t)$ for the estimator design. That is the reason why the feedback scheme in rescaled unity-interval notation (10.3)–(10.20) is introduced to lay a foundation for delay-adaptive control, rather than applying adaptive control directly to (10.21)–(10.29).

Theorem 10.4. *The closed-loop system consisting of the plant (10.4)–(10.7) and the controller (10.11) is exponentially stable in the sense of the norm*

$$|X(t)|^2 + \int_0^1 u^2(x,t)dx. \tag{10.30}$$

Proof. Consider the Lyapunov candidate

$$V(t) = Z^T(t)PZ(t) + D\int_0^1 (1+x)w(x,t)^2 dx. \tag{10.31}$$

Bearing (10.10) and (10.20) in mind, taking the time derivative of (10.31) along the target system (10.18)–(10.20) and using the integration by parts, we get

$$\dot{V}(t) = Z^T(t)\left(A_{cl}^T P + PA_{cl}\right)Z(t)$$

$$+ D\int_0^1 2(1+x)w(x,t)w_t(x,t)dx$$

$$= -Z^T(t)QZ(t) + \int_0^1 2(1+x)w(x,t)w_x(x,t)dx$$

$$= -Z^T(t)QZ(t) + 4w(1,t)^2 - 2w(0,t)^2$$

$$- \int_0^1 2w(x,t)^2 dx - \int_0^1 2(1+x)w_x(x,t)w(x,t)dx$$

$$= -Z^T(t)QZ(t) - w(0,t)^2 - \int_0^1 w(x,t)^2 dx$$

$$\leq -\lambda_{\min}(Q)|Z(t)|^2 - \int_0^1 w(x,t)^2 dx$$

$$\leq -\frac{\lambda_{\min}(Q)}{\lambda_{\max}(P)}\lambda_{\max}(P)|Z(t)|^2 - \frac{1}{2D}D\int_0^1 (1+x)w(x,t)^2 dx$$

$$\leq -\min\left\{\frac{\lambda_{\min}(Q)}{\lambda_{\max}(P)}, \frac{1}{2D}\right\}V(t). \tag{10.32}$$

It is evident that (10.32) implies the exponential stability in the sense of the norm $|X(t)|^2 + \int_0^1 w(x,t)^2 dx$ by the comparison principle. Then making use of the inverse transformations of (10.8) and (10.14), the exponential stability in the sense of the norm $|X(t)|^2 + \int_0^1 u(x,t)^2 dx$ is derived. $\qquad\square$

10.2   Predictor Feedback for Uncertainty-Free Multi-Input Systems

In this section, we propose a predictor feedback framework in rescaled unity-interval notation for uncertainty-free multi-input systems.

Consider the multi-input linear systems with different distributed input delays

$$\dot{X}(t) = AX(t) + \sum_{i=1}^{2}\int_0^{D_i} B_i(\sigma)U_i(t-\sigma)d\sigma, \tag{10.33}$$

$$Y(t) = CX(t), \tag{10.34}$$

where $X(t) \in \mathbb{R}^n$ is the plant state, $Y(t) \in \mathbb{R}^q$ is the plant output, $U_1(t), U_2(t) \in \mathbb{R}$ are the control inputs, and $D_1, D_2 > 0$ are the distinct constant delays such that $D_1 \neq D_2$. For conceptual and notational clearness, we consider a two-input case. The same analysis and design could be straightforwardly carried out for an arbitrary number of inputs with different delays in each individual input channel.

To represent the actuator states $U_i(\xi) \in \mathbb{R}$ for $\xi \in [t - D_i, t]$, $i = 1, 2$, the multi-variable functions in rescaled unity-interval notation are introduced:

$$u_i(x,t) = U_i(t + D_i(x-1)) = U_i(t-\sigma), \quad x \in [0,1], \quad \sigma \in [0, D_i], \quad i = 1, 2. \tag{10.35}$$

Then the system (10.33)–(10.34) is transformed into the ODE-PDE cascade as follows:

$$\dot{X}(t) = AX(t) + \sum_{i=1}^{2} D_i \int_0^1 B_i(D_i(1-x))u_i(x,t)dx, \tag{10.36}$$

$$Y(t) = CX(t), \tag{10.37}$$

$$D_i \partial_t u_i(x,t) = \partial_x u_i(x,t), \quad x \in [0,1], \quad i = 1,2, \tag{10.38}$$

$$u_i(1,t) = U_i(t). \tag{10.39}$$

It is evident that (10.35) is a solution of the transport PDE (10.38)–(10.39).

If we compare (10.1)–(10.7) with (10.33)–(10.39), it is evident that (10.1)–(10.7) represent single-input systems, whereas (10.33)–(10.39) represent multi-input systems. Besides, (10.1) considers a specific form of delay kernel such that $\dot{X}(t) = AX(t) + \int_0^D B(D-\sigma)U(t-\sigma)d\sigma$, whereas (10.33) addresses a convolution form of delay kernel such that $\dot{X}(t) = AX(t) + \int_0^D B(\sigma)U(t-\sigma)d\sigma$ (when there is only one input) which is more popular in the literature [2], [11]. When (10.1) and (10.33) are represented in the PDE notation, one becomes $\dot{X}(t) = AX(t) + D\int_0^1 B(Dx)u(x,t)dx$ in (10.4), whereas the other is of the form $\dot{X}(t) = AX(t) + D\int_0^1 B(D(1-x))u(x,t)dx$ in (10.36). It is apparent that the system (10.4) is more elegant, and the subsequent control scheme is simpler.

If we concentrate on the system (10.36)–(10.39), a finite-dimensional multi-input linear plant with distributed actuator delays comes with five types of basic uncertainties:

- unknown and distinct delays D_i,
- unknown delay kernels $B_i(D_i(1-x))$,
- unknown system matrix A,
- unmeasurable finite-dimensional plant state $X(t)$,
- unmeasurable infinite-dimensional actuator state $u_i(x,t)$.

To lay a foundation for certainty-equivalence-based control schemes in later chapters, this section deals with the simplest case where all five variables above are known.

To stabilize (10.36), a reduction-based conversion is brought in as

$$Z(t) = X(t) + \sum_{i=1}^2 D_i^2 \int_0^1 \int_0^y e^{AD_i(\tau-y)} B_i(D_i(1-\tau))d\tau u_i(y,t)dy. \tag{10.40}$$

Differentiating (10.40) with respect to t along (10.36)–(10.39) and using the integration by parts in y, we get

$$\dot{Z}(t) = \dot{X}(t) + \sum_{i=1}^2 D_i \int_0^1 \int_0^y e^{AD_i(\tau-y)} B_i(D_i(1-\tau))d\tau \partial_y u_i(y,t)dy$$

$$= AX(t) + \sum_{i=1}^2 D_i \int_0^1 B_i(D_i(1-x))u_i(x,t)dx$$

$$+ \sum_{i=1}^2 D_i \int_0^1 e^{AD_i(\tau-1)} B_i(D_i(1-\tau))d\tau u_i(1,t)$$

$$-\sum_{i=1}^{2} D_i \int_0^1 B_i(D_i(1-y))u_i(y,t)dy$$

$$+\sum_{i=1}^{2} D_i^2 \int_0^1 \int_0^y Ae^{AD_i(\tau-y)}B_i(D_i(1-\tau))d\tau u_i(y,t)dy$$

$$= AZ(t) + \sum_{i=1}^{2} D_i \underbrace{\int_0^1 e^{AD_i(\tau-1)}B_i(D_i(1-\tau))d\tau}_{B_{D_i}} U_i(t). \tag{10.41}$$

It is apparent that, by the conversion (10.40), the stabilization problem of the delayed system (10.36) is reduced to the stabilization problem of a delay-free system (10.41).

Under the premise that the pair $(A,(B_{D_1},B_{D_2}))$ is stabilizable, the control laws are designed as

$$U_i(t) = u_i(1,t) = K_i e^{AD_i} Z(t), \quad i = 1,2, \tag{10.42}$$

where K_1 and K_2 are chosen to make

$$A_{cl} = A + \sum_{i=1}^{2} B_{D_i} K_i e^{AD_i} \tag{10.43}$$

Hurwitz, so that

$$A_{cl}^T P + P A_{cl} = -Q, \tag{10.44}$$

with $P = P^T > 0$ and $Q = Q^T > 0$. For the convenience of stability analysis, an invertible forwarding-backstepping transformation including both finite-dimensional plant state and infinite-dimensional actuator state is employed such that

$$w_i(x,t) = u_i(x,t) - K_i e^{AD_i x} Z(t) + D_i \int_x^1 K_i e^{AD_i(x-y)} B_{ie} u_i(y,t)dy$$

$$= u_i(x,t) - K_i e^{AD_i x} Z(t)$$

$$+ D_i^2 \int_x^1 \int_0^1 K_i e^{AD_i(x-y+\tau)} B_i(D_i(1-\tau))d\tau u_i(y,t)dy, \tag{10.45}$$

$$u_i(x,t) = w_i(x,t) + K_i e^{(A+B_{ie}K_i)D_i(x-1)} e^{AD_i} Z(t)$$

$$- D_i \int_x^1 K_i e^{(A+B_{ie}K_i)D_i(x-y)} B_{ie} w_i(y,t)dy, \tag{10.46}$$

where $B_{ie} = e^{AD_i} B_{D_i} = D_i \int_0^1 e^{AD_i\tau} B_i(D_i(1-\tau))d\tau$. The proof that (10.46) is an inverse transformation of (10.45) is given in section 10.2.1.

Differentiating (10.45) with respect to x and t along (10.38)–(10.39) and (10.41), respectively, and using the integration by parts in y, we have

$$\partial_x w_i(x,t) = \partial_x u_i(x,t) - K_i e^{A D_i x} A D_i Z(t)$$

$$- D_i^2 \int_0^1 K_i e^{A D_i \tau} B_i(D_i(1-\tau)) d\tau u_i(x,t)$$

$$+ D_i^3 \int_x^1 \int_0^1 K_i A e^{A D_i(x-y+\tau)} B_i(D_i(1-\tau)) d\tau u_i(y,t) dy, \quad (10.47)$$

$$\partial_t w_i(x,t) = \partial_t u_i(x,t) - K_i e^{A D_i x} \dot{Z}(t)$$

$$+ D_i \int_x^1 \int_0^1 K_i e^{A D_i(x-y+\tau)} B_i(D_i(1-\tau)) d\tau \partial_y u_i(y,t) dy$$

$$= \partial_t u_i(x,t) - K_i e^{A D_i x} \left(A Z(t) + \sum_{j=1}^2 B_{D_j} u_j(1,t) \right)$$

$$+ D_i \int_0^1 K_i e^{A D_i(x-1+\tau)} B_i(D_i(1-\tau)) d\tau u_i(1,t)$$

$$- D_i \int_0^1 K_i e^{A D_i \tau} B_i(D_i(1-\tau)) d\tau u_i(x,t)$$

$$+ D_i^2 \int_x^1 \int_0^1 K_i A e^{A D_i(x-y+\tau)} B_i(D_i(1-\tau)) d\tau u_i(y,t) dy. \quad (10.48)$$

Multiplying (10.48) with D_i and subtracting (10.47), substituting $x=1$ into (10.45) and substituting (10.42) into (10.41) and (10.45), we get the target closed-loop system as follows:

$$\dot{Z}(t) = \left(A + B_{D_1} K_1 e^{A D_1} + B_{D_2} K_2 e^{A D_2} \right) Z(t)$$
$$= A_{cl} Z(t), \quad (10.49)$$

$$D_1 \partial_t w_1(x,t) = \partial_x w_1(x,t) - D_1 K_1 e^{A D_1 x} B_{D_2} K_2 e^{A D_2} Z(t), \quad (10.50)$$

$$w_1(1,t) = 0, \quad x \in [0,1], \quad (10.51)$$

$$D_2 \partial_t w_2(x,t) = \partial_x w_2(x,t) - D_2 K_2 e^{A D_2 x} B_{D_1} K_1 e^{A D_1} Z(t), \quad (10.52)$$

$$w_2(1,t) = 0, \quad x \in [0,1]. \quad (10.53)$$

If the two-input system is reduced to the single-input system, i.e., $B_{D_2}=0$ in (10.50) or $B_{D_1}=0$ in (10.52), then (10.50)–(10.51) or (10.52)–(10.53) is a standard target PDE system of exponential stability.

As demonstrated in figure 10.2, a series of conversions are employed to convert the original system (10.33) into the target system (10.49)–(10.53).

Figure 10.2. A series of conversions between the original system (10.33)–(10.34) and the target system (10.49)–(10.53)

Remark 10.5. To compensate for distributed delays in uncertainty-free systems, there exist a couple of alternative representations of (10.40)–(10.53) that are in rescaled unity-interval notation: the control scheme in ODE notation that is proposed in [2], and the control design in PDE notation that is developed in [11]. For the reader's convenience, the predictor feedback framework in ODE notation is recapped as follows:

$$\dot{X}(t) = AX(t) + \sum_{i=1}^{2} \int_{0}^{D_i} B_i(\sigma)U_i(t - \sigma)d\sigma, \tag{10.54}$$

$$U_i(t) = K_i e^{AD_i} Z(t), \quad i = 1, 2, \tag{10.55}$$

$$Z(t) = X(t) + \sum_{i=1}^{2} \int_{t-D_i}^{t} \int_{t-s}^{D_i} e^{A(t-s-\sigma)} B_i(\sigma) d\sigma U_i(s) ds, \tag{10.56}$$

$$\dot{Z}(t) = AZ(t) + \sum_{i=1}^{2} \underbrace{\int_{0}^{D_i} e^{-A\sigma} B_i(\sigma) d\sigma}_{B_{D_i}} U_i(t)$$

$$= \left(A + B_{D_1} K_1 e^{AD_1} + B_{D_2} K_2 e^{AD_2} \right) Z(t), \tag{10.57}$$

whereas the predictor feedback framework in PDE notation is recapped as follows:

$$u_i(x, t) = U_i(t + x - D_i), \quad x \in [0, D_i], \quad i = 1, 2, \tag{10.58}$$

$$\dot{X}(t) = AX(t) + \sum_{i=1}^{2} \int_{0}^{D_i} B_i(D_i - x)u_i(x, t)dx, \tag{10.59}$$

$$\partial_t u_i(x, t) = \partial_x u_i(x, t), \quad x \in [0, D_i], \tag{10.60}$$

$$u_i(D_i, t) = U_i(t), \tag{10.61}$$

$$U_i(t) = K_i e^{AD_i} Z(t), \tag{10.62}$$

$$Z(t) = X(t) + \sum_{i=1}^{2} \int_{0}^{D_i} \int_{D_i-y}^{D_i} e^{A(D_i-y-\sigma)} B_i(\sigma) d\sigma u_i(y, t) dy, \tag{10.63}$$

$$w_i(x, t) = u_i(x, t) - K_i e^{Ax} Z(t)$$

$$+ \int_{x}^{D_i} \int_{0}^{D_i} K_i e^{A(D_i+x-y-\sigma)} B_i(\sigma) d\sigma u_i(y, t) dy, \tag{10.64}$$

$$u_i(x,t) = w_i(x,t) + K_i e^{(A+B_{ie}K_i)(x-D_i)} e^{AD_i} Z(t)$$

$$- \int_x^{D_i} K_i e^{(A+B_{ie}K_i)(x-y)} B_{ie} w_i(y,t)dy, \quad B_{ie} = e^{AD_i} B_{D_i}, \quad (10.65)$$

$$\dot{Z}(t) = \left(A + \sum_{i=1}^{2} \underbrace{\int_0^{D_i} e^{-A\sigma} B_i(\sigma)d\sigma\, K_i e^{AD_i}}_{B_{D_i}} \right) Z(t), \quad (10.66)$$

$$\partial_t w_1(x,t) = \partial_x w_1(x,t) - K_1 e^{Ax} B_{D_2} K_2 e^{AD_2} Z(t), \quad (10.67)$$
$$w_1(D_1,t) = 0, \quad (10.68)$$
$$\partial_t w_2(x,t) = \partial_x w_2(x,t) - K_2 e^{Ax} B_{D_1} K_1 e^{AD_1} Z(t), \quad (10.69)$$
$$w_2(D_2,t) = 0. \quad (10.70)$$

However, neither of the two methods can be directly extended to adaptive control for unknown delays, since they are not parameterized in delays D_i and a moving boundary condition will be produced when the unknown delay is replaced by its time-varying estimate. That is the reason why this section contributes to a predictor feedback scheme in rescaled unity-interval notation, which will be convenient for extending to delay-adaptive control later.

Theorem 10.6. *The closed-loop system consisting of the plant (10.36)–(10.39) and the controller (10.40)–(10.42) is exponentially stable in the sense of the norm $|X(t)|^2 + \sum_{i=1}^{2} \int_0^1 u_i^2(x,t)dx$.*

Proof. Build the Lyapunov candidate

$$V(t) = Z(t)^T P Z(t) + g \sum_{i=1}^{2} D_i \int_0^1 (1+x) w_i^2(x,t)dx, \quad (10.71)$$

where $g > 0$ is an analysis coefficient. Taking the time derivative of (10.71) along the target closed-loop system (10.49)–(10.53), and using Young's inequality and the integration by parts in x, we get

$$\dot{V}(t) = -Z(t)^T Q Z(t) + 2g \sum_{i=1}^{2} \int_0^1 (1+x) w_i(x,t) \partial_x w_i(x,t)dx$$

$$- 2g \int_0^1 (1+x) w_1(x,t) D_1 K_1 e^{AD_1 x} dx B_{D_2} K_2 e^{AD_2} Z(t)$$

$$- 2g \int_0^1 (1+x) w_2(x,t) D_2 K_2 e^{AD_2 x} dx B_{D_1} K_1 e^{AD_1} Z(t)$$

$$\leq -\lambda_{\min}(Q)|Z(t)|^2 - g \sum_{i=1}^{2} w_i(0,t)^2 - g \sum_{i=1}^{2} \|w_i(x,t)\|^2$$

$$+ g\gamma M_1 \|w_1(x,t)\|^2 + \frac{gM_1}{\gamma}|Z(t)|^2 + g\gamma M_2 \|w_2(x,t)\|^2 + \frac{gM_2}{\gamma}|Z(t)|^2$$

$$\leq -\left(\lambda_{\min}(Q) - \frac{gM_1}{\gamma} - \frac{gM_2}{\gamma}\right)|Z(t)|^2$$

$$- g\left(1 - \gamma M_1\right)\|w_1(x,t)\|^2 - g\left(1 - \gamma M_2\right)\|w_2(x,t)\|^2, \tag{10.72}$$

where $\gamma > 0$, $M_1 = 2D_1|K_1|e^{|A|D_1}|B_{D_2}K_2 e^{AD_2}|$, $M_2 = 2D_2|K_2|e^{|A|D_2}|B_{D_1}K_1$ $e^{AD_1}|$. If the coefficients are chosen to satisfy $\lambda_{\min}(Q) - \frac{gM_1}{\gamma} - \frac{gM_2}{\gamma} > 0$ and $\gamma < \min\{\frac{1}{M_1}, \frac{1}{M_2}\}$, it is easy to show that

$$\dot{V}(t) \leq -\mu V(t), \quad \mu > 0. \tag{10.73}$$

By the comparison principle, $Z(t)$, $w_1(x,t)$, $w_2(x,t)$ are exponentially stable. By inverse transformations (10.40) and (10.45)–(10.46), $X(t)$, $u_1(x,t)$, $u_2(x,t)$ are exponentially stable. Thus Theorem 10.6 is proved. $\qquad\square$

10.2.1   Auxiliary Calculations: Invertibility of Forwarding-Backstepping Transformation

In this section, we prove the invertibility of the forwarding-backstepping transformation (10.45)–(10.46) in section 10.2.

Firstly, we prove that the target system (10.50)–(10.53) can be inversely transformed into (10.38)–(10.39) by making use of the inverse transformation (10.46). Taking $i=2$, for instance, (10.46) is of the form

$$u_2(x,t) = w_2(x,t) + K_2 e^{(A+B_{2e}K_2)D_2(x-1)} e^{AD_2} Z(t)$$

$$- D_2 \int_x^1 K_2 e^{(A+B_{2e}K_2)D_2(x-y)} B_{2e} w_2(y,t)dy, \tag{10.74}$$

where $B_{2e} = e^{AD_2} B_{D_2}$.

Differentiating (10.74) with respect to x, we get

$$\partial_x u_2(x,t) = \partial_x w_2(x,t) + K_2 e^{(A+B_{2e}K_2)D_2(x-1)}(A+B_{2e}K_2)D_2 e^{AD_2} Z(t)$$

$$+ D_2 K_2 B_{2e} w_2(x,t)$$

$$- D_2 \int_x^1 K_2 e^{(A+B_{2e}K_2)D_2(x-y)}(A+B_{2e}K_2)D_2 B_{2e} w_2(y,t)dy. \tag{10.75}$$

Differentiating (10.74) with respect to t along (10.49), (10.52)–(10.53), and using the integration by parts in y, we get

$$\partial_t u_2(x,t) = \partial_t w_2(x,t) + K_2 e^{(A+B_{2e}K_2)D_2(x-1)} e^{AD_2}$$

$$\times \left(A + B_{D_1} K_1 e^{AD_1} + B_{D_2} K_2 e^{AD_2}\right) Z(t)$$

$$-\int_x^1 K_2 e^{(A+B_{2e}K_2)D_2(x-y)} B_{2e}$$

$$\times \left(\partial_y w_2(y,t) - D_2 K_2 e^{AD_2 y} B_{D_1} K_1 e^{AD_1} Z(t)\right) dy$$

$$= \partial_t w_2(x,t) + K_2 e^{(A+B_{2e}K_2)D_2(x-1)} e^{AD_2}$$

$$\times \left(A + B_{D_1} K_1 e^{AD_1} + B_{D_2} K_2 e^{AD_2}\right) Z(t) + K_2 B_{2e} w_2(x,t)$$

$$-\int_x^1 K_2 e^{(A+B_{2e}K_2)D_2(x-y)}(A+B_{2e}K_2)D_2 B_{2e} w_2(y,t) dy$$

$$+\int_x^1 K_2 e^{(A+B_{2e}K_2)D_2(x-y)} B_{2e} D_2 K_2 e^{AD_2 y} dy B_{D_1} K_1 e^{AD_1} Z(t). \tag{10.76}$$

Multiplying (10.76) with D_2 and subtracting (10.75), we get

$$D_2 \partial_t u_2(x,t) - \partial_x u_2(x,t) = -D_2 K_2 e^{AD_2 x} B_{D_1} K_1 e^{AD_1} Z(t)$$

$$+ D_2 K_2 e^{(A+B_{2e}K_2)D_2(x-1)} e^{AD_2} B_{D_1} K_1 e^{AD_1} Z(t)$$

$$+ D_2 \int_x^1 K_2 e^{(A+B_{2e}K_2)D_2(x-y)} B_{2e} D_2 K_2 e^{AD_2 y} dy B_{D_1} K_1 e^{AD_1} Z(t). \tag{10.77}$$

From (10.38), it is evident the right side of (10.77) should be equal to zero, which implies the following equality:

$$0 = -e^{AD_2 x} + e^{(A+B_{2e}K_2)D_2(x-1)} e^{AD_2}$$

$$+ \int_x^1 e^{(A+B_{2e}K_2)D_2(x-y)} B_{2e} D_2 K_2 e^{AD_2 y} dy, \tag{10.78}$$

where the third term on the right side of (10.78) is calculated as

$$\int_x^1 e^{(A+B_{2e}K_2)D_2(x-y)} B_{2e} D_2 K_2 e^{AD_2 y} dy$$

$$= D_2 \int_x^1 e^{(A+B_{2e}K_2)D_2(x-y)} (A+B_{2e}K_2 - A) e^{AD_2 y} dy$$

$$= D_2 \int_x^1 e^{(A+B_{2e}K_2)D_2(x-y)} (A+B_{2e}K_2) e^{AD_2 y} dy$$

$$- \int_x^1 e^{(A+B_{2e}K_2)D_2(x-y)} de^{AD_2 y}$$

$$= D_2 \int_x^1 e^{(A+B_{2e}K_2)D_2(x-y)} (A+B_{2e}K_2) e^{AD_2 y} dy$$

$$- e^{(A+B_{2e}K_2)D_2(x-1)} e^{AD_2} + e^{AD_2 x}$$

$$- \int_x^1 e^{(A+B_{2e}K_2)D_2(x-y)}(A+B_{2e}K_2)D_2 e^{AD_2 y} dy$$

$$= -e^{(A+B_{2e}K_2)D_2(x-1)}e^{AD_2} + e^{AD_2 x}. \tag{10.79}$$

If we substitute (10.79) into (10.78), it is evident the equality (10.78) is satisfied and the proof is completed.

Now we prove that (10.46) is an inverse of (10.45). Letting $i=1$ and substituting (10.46) into the first row of (10.45), we get

$$w_1(x,t) = w_1(x,t) + K_1 e^{(A+B_{1e}K_1)D_1(x-1)}e^{AD_1}Z(t)$$

$$- D_1 \int_x^1 K_1 e^{(A+B_{1e}K_1)D_1(x-y)}B_{1e}w_1(y,t)dy$$

$$- K_1 e^{AD_1 x}Z(t) + D_1 \int_x^1 K_1 e^{AD_1(x-y)}B_{1e}\Bigg[w_1(y,t)$$

$$+ K_1 e^{(A+B_{1e}K_1)D_1(y-1)}e^{AD_1}Z(t)$$

$$- D_1 \int_y^1 K_1 e^{(A+B_{1e}K_1)D_1(y-s)}B_{1e}w_1(s,t)ds\Bigg]dy. \tag{10.80}$$

In order to make (10.80) hold, the following two equalities should hold simultaneously:

$$0 = K_1 e^{(A+B_{1e}K_1)D_1(x-1)}e^{AD_1}Z(t) - K_1 e^{AD_1 x}Z(t)$$

$$+ D_1 \int_x^1 K_1 e^{AD_1(x-y)}B_{1e}K_1 e^{(A+B_{1e}K_1)D_1(y-1)}e^{AD_1}dy Z(t), \tag{10.81}$$

$$0 = -D_1 \int_x^1 K_1 e^{(A+B_{1e}K_1)D_1(x-y)}B_{1e}w_1(y,t)dy$$

$$+ D_1 \int_x^1 K_1 e^{AD_1(x-y)}B_{1e}w_1(y,t)dy$$

$$- D_1^2 \int_x^1 K_1 e^{AD_1(x-y)}B_{1e} \int_y^1 K_1 e^{(A+B_{1e}K_1)D_1(y-s)}B_{1e}w_1(s,t)dsdy. \tag{10.82}$$

Using the integration by parts in y, we calculate the third term on the right side of (10.81) as

$$D_1 \int_x^1 K_1 e^{AD_1(x-y)}B_{1e}K_1 e^{(A+B_{1e}K_1)D_1(y-1)}e^{AD_1}dy Z(t)$$

$$= -D_1 \int_x^1 K_1 e^{AD_1(x-y)}Ae^{(A+B_{1e}K_1)D_1(y-1)}e^{AD_1}dy Z(t)$$

$$+ D_1 \int_x^1 K_1 e^{AD_1(x-y)} (A + B_{1e}K_1) e^{(A+B_{1e}K_1)D_1(y-1)} e^{AD_1} dy\, Z(t)$$

$$= -D_1 \int_x^1 K_1 e^{AD_1(x-y)} A e^{(A+B_{1e}K_1)D_1(y-1)} e^{AD_1} dy\, Z(t)$$

$$+ K_1 e^{AD_1 x} \left(\int_x^1 e^{-AD_1 y} de^{(A+B_{1e}K_1)D_1(y-1)} \right) e^{AD_1} Z(t)$$

$$= -D_1 \int_x^1 K_1 e^{AD_1(x-y)} A e^{(A+B_{1e}K_1)D_1(y-1)} e^{AD_1} dy\, Z(t)$$

$$+ K_1 e^{AD_1 x} \left(e^{-AD_1} - e^{-AD_1 x} e^{(A+B_{1e}K_1)D_1(x-1)} \right.$$

$$\left. + \int_x^1 e^{-AD_1 y} AD_1 e^{(A+B_{1e}K_1)D_1(y-1)} dy \right) e^{AD_1} Z(t)$$

$$= K_1 e^{AD_1 x} Z(t) - K_1 e^{(A+B_{1e}K_1)D_1(x-1)} e^{AD_1} Z(t). \tag{10.83}$$

If we substitute (10.83) into (10.81), it is apparent that the equality (10.81) is satisfied.

Changing the sequence of the integration and using the integration by parts in y, we calculate the third term on the right side of (10.82) as

$$- D_1^2 \int_x^1 K_1 e^{AD_1(x-y)} B_{1e} \int_y^1 K_1 e^{(A+B_{1e}K_1)D_1(y-s)} B_{1e} w_1(s,t) ds\, dy$$

$$= -D_1^2 \int_x^1 K_1 e^{AD_1 x} \left(\int_x^s e^{-AD_1 y} B_{1e} K_1 e^{(A+B_{1e}K_1)D_1 y} dy \right)$$
$$\times e^{-(A+B_{1e}K_1)D_1 s} B_{1e} w_1(s,t) ds$$

$$= D_1^2 \int_x^1 K_1 e^{AD_1 x} \left(\int_x^s e^{-AD_1 y} A e^{(A+B_{1e}K_1)D_1 y} dy \right)$$
$$\times e^{-(A+B_{1e}K_1)D_1 s} B_{1e} w_1(s,t) ds$$

$$- D_1^2 \int_x^1 K_1 e^{AD_1 x} \left(\int_x^s e^{-AD_1 y} (A + B_{1e}K_1) e^{(A+B_{1e}K_1)D_1 y} dy \right)$$
$$\times e^{-(A+B_{1e}K_1)D_1 s} B_{1e} w_1(s,t) ds$$

$$= D_1^2 \int_x^1 K_1 e^{AD_1 x} \left(\int_x^s e^{-AD_1 y} A e^{(A+B_{1e}K_1)D_1 y} dy \right)$$
$$\times e^{-(A+B_{1e}K_1)D_1 s} B_{1e} w_1(s,t) ds$$

$$- D_1 \int_x^1 K_1 e^{AD_1 x} \left(\int_x^s e^{-AD_1 y} de^{(A+B_{1e}K_1)D_1 y} \right)$$
$$\times e^{-(A+B_{1e}K_1)D_1 s} B_{1e} w_1(s,t) ds$$

$$= D_1^2 \int_x^1 K_1 e^{AD_1 x} \left(\int_x^s e^{-AD_1 y} A e^{(A+B_{1e}K_1)D_1 y} dy \right)$$
$$\times e^{-(A+B_{1e}K_1)D_1 s} B_{1e} w_1(s,t) ds$$

$$- D_1 \int_x^1 K_1 e^{AD_1 x} \left(e^{-AD_1 s} e^{(A+B_{1e}K_1)D_1 s} \right.$$
$$\left. - e^{-AD_1 x} e^{(A+B_{1e}K_1)D_1 x} + \int_x^s e^{-AD_1 y} AD_1 e^{(A+B_{1e}K_1)D_1 y} dy \right)$$
$$\times e^{-(A+B_{1e}K_1)D_1 s} B_{1e} w_1(s,t) ds$$

$$= -D_1 \int_x^1 K_1 e^{AD_1(x-s)} B_{1e} w_1(s,t) ds$$
$$+ D_1 \int_x^1 K_1 e^{(A+B_{1e}K_1)D_1(x-s)} B_{1e} w_1(s,t) ds. \qquad (10.84)$$

If we substitute (10.84) into (10.82), it is obvious that the equality (10.82) is satisfied. Thus the proof is completed.

Chapter Eleven

Predictor Feedback of Uncertain Single-Input Systems

The single-input linear system with distributed input delay in section 10.1 is recapped as follows:

$$\dot{X}(t) = AX(t) + \int_0^D B(D-\sigma)U(t-\sigma)d\sigma, \tag{11.1}$$

$$Y(t) = CX(t), \tag{11.2}$$

which can be represented by the following ODE-PDE cascade:

$$\dot{X}(t) = AX(t) + D\int_0^1 B(Dx)u(x,t)dx, \tag{11.3}$$

$$Y(t) = CX(t), \tag{11.4}$$

$$Du_t(x,t) = u_x(x,t), \quad x \in [0,1], \tag{11.5}$$

$$u(1,t) = U(t). \tag{11.6}$$

As clarified in the last chapter, if we concentrate on (11.1)–(11.2) and its conversion (11.3)–(11.6), a single-input linear plant with distributed actuator delay comes with the following five types of basic uncertainties:

- unknown delay D,
- unknown delay kernel $B(Dx)$,
- unknown parameters in the system matrix A,
- unmeasurable finite-dimensional plant state $X(t)$,
- unmeasurable infinite-dimensional actuator state $u(x,t)$.

In this chapter, we present the predictor feedback for uncertain single-input systems, based on the predictor feedback framework for uncertainty-free single-input systems in section 10.1. We address five combinations of the five uncertainties above. To clearly describe the organization of chapter 11, we summarize different combinations of uncertainties considered in later sections in table 11.1.

Table 11.1. Uncertainty Collections of Single-Input Linear Systems with Distributed Input Delay

Literature	Delay D	Delay Kernel $B(Dx)$	Parameter A	Plant State $X(t)$	Actuator State $u(x,t)$
Section 10.1	known	known	known	known	known
Section 11.1	unknown	known	known	known	known
Section 11.2	unknown	unknown	known	known	known
Section 11.3	unknown	unknown	unknown	known	known
Section 11.4	unknown	unknown	known	known	unknown
Section 11.5	unknown	unknown	known	unknown	known

11.1 Adaptive State Feedback under Unknown Delay

In this section, we consider that the delay D is unknown. Since D is uncertain, the vector $B(Dx)$ for $x \in [0,1]$ in (11.3) may contain unknown variables so that it cannot be employed directly for control design. Two different cases are taken into account as follows:

- $B(Dx)$ is a constant vector independent of Dx,
- $B(Dx)$ is a continuous vector-valued function of Dx.

This section addresses the relatively simple case where the vector $B(Dx)$ for $x \in [0,1]$ in (11.3) $\big($i.e., $B(D-\sigma)$ for $\sigma \in [0,D]$ in (11.1)$\big)$ is a known constant vector independent of Dx such that

$$B(Dx) = B(D-\sigma) = B. \tag{11.7}$$

Then the system (11.1)–(11.2) is reduced to

$$\dot{X}(t) = AX(t) + \int_0^D BU(t-\sigma)d\sigma, \tag{11.8}$$

$$Y(t) = CX(t), \tag{11.9}$$

and the ODE-PDE cascade (11.3)–(11.6) is accordingly reduced to

$$\dot{X}(t) = AX(t) + D\int_0^1 Bu(x,t)dx, \tag{11.10}$$

$$Y(t) = CX(t), \tag{11.11}$$

$$Du_t(x,t) = u_x(x,t), \tag{11.12}$$

$$u(1,t) = U(t). \tag{11.13}$$

The control objective is to stabilize (11.10)–(11.13) when the delay D is unknown.

Remark 11.1. On the basis of the framework (10.8)–(10.20) in section 10.1, the key idea of certainty-equivalence-based adaptive control is to use an estimate to replace the unknown delay in (10.8), (10.11), and (10.14). And the Lyapunov-based update law is employed to cancel the estimation error term in the time derivative of the Lyapunov function. A nontrivial problem is to deal with the unknown D^2 in (10.8). One intuitive method is to estimate D^2 by $\hat{D}(t)^2$, where $\hat{D}(t)$ is an estimate of D. The other possible method is to treat D^2 as a whole such that $\eta = D^2$ and estimate η by $\hat{\eta}(t)$. However, both methods produce a mismatch term for update law design. The available method is to regard D^2 as a product of two parameters such that $D^2 = D \cdot D_1$, where $D_1 = D$, and estimate D and D_1 with different update laws, respectively.

To use projector operators in the update law later, we make the following assumption.

Assumption 11.2. *There exist known constants $\underline{D}$ and $\overline{D}$ such that*

$$0 < \underline{D} \leq D = D_1 \leq \overline{D}. \tag{11.14}$$

In order to stabilize (11.10)–(11.13), the following assumption is required.

Assumption 11.3. *For the system (11.10), the pair (A, β) is stabilizable where*

$$\beta = D \int_0^1 e^{-AD(1-x)} B\,dx = D_1 \int_0^1 e^{-AD(1-x)} B\,dx. \tag{11.15}$$

There exists a vector $K(\beta)$ to make $A + \beta K(\beta)$ Hurwitz. Namely, there exist matrices $P(\beta) = P(\beta)^T > 0$ and $Q(\beta) = Q(\beta)^T > 0$ such that

$$(A + \beta K(\beta))^T P(\beta) + P(\beta)(A + \beta K(\beta)) = -Q(\beta). \tag{11.16}$$

Denote $\hat{D}(t)$ and $\hat{D}_1(t)$ as the estimates of D and D_1 with estimation errors satisfying

$$\tilde{D}(t) = D - \hat{D}(t), \tag{11.17}$$

$$\tilde{D}_1(t) = D_1 - \hat{D}_1(t) = D - \hat{D}_1(t). \tag{11.18}$$

The delay-adaptive control scheme is listed as follows:
The control law is

$$U(t) = u(1, t) = K(\hat{\beta}(t))Z(t), \tag{11.19}$$

where $K(\hat{\beta}(t))$ is chosen to make

$$A_{cl}(\hat{\beta}(t)) = A + \underbrace{\hat{D}_1(t) \int_0^1 e^{-A\hat{D}(t)(1-x)} B dx\, K(\hat{\beta}(t))}_{\hat{\beta}(t)} \qquad (11.20)$$

Hurwitz and

$$Z(t) = X(t) + \hat{D}(t)\hat{D}_1(t) \int_0^1 \int_0^x e^{-A\hat{D}(t)(x-y)} B dy u(x,t) dx. \qquad (11.21)$$

The update laws are

$$\dot{\hat{D}}_1(t) = \gamma_{D_1} \mathrm{Proj}_{[\underline{D},\overline{D}]}\{\tau_{D_1}(t)\}, \quad \gamma_{D_1} > 0, \qquad (11.22)$$

$$\tau_{D_1}(t) = \frac{1/g Z^T(t) P(\hat{\beta}(t)) f_{D_1}(t) - \int_0^1 (1+x) w(x,t) h_{D_1}(x,t) dx}{1 + \Xi(t)}, \qquad (11.23)$$

$$\dot{\hat{D}}(t) = \gamma_D \mathrm{Proj}_{[\underline{D},\overline{D}]}\{\tau_D(t)\}, \quad \gamma_D > 0, \qquad (11.24)$$

$$\tau_D(t) = \frac{1/g Z^T(t) P(\hat{\beta}(t)) f_D(t) - \int_0^1 (1+x) w(x,t) h_D(x,t) dx}{1 + \Xi(t)}, \qquad (11.25)$$

where $P(\hat{\beta}(t))$ satisfies (11.16) and $g > 0$ is a designing coefficient:

$$w(x,t) = u(x,t) - K(\hat{\beta}(t)) e^{A_{cl}(\hat{\beta}(t))\hat{D}(t)(x-1)} Z(t), \qquad (11.26)$$

$$\Xi(t) = Z^T(t) P(\hat{\beta}(t)) Z(t) + g \int_0^1 (1+x) w(x,t)^2 dx, \qquad (11.27)$$

$$f_{D_1}(t) = \int_0^1 B u(x,t) dx, \qquad (11.28)$$

$$h_{D_1}(x,t) = K(\hat{\beta}(t)) e^{A_{cl}(\hat{\beta}(t))\hat{D}(t)(x-1)} f_{D_1}(t), \qquad (11.29)$$

$$f_D(t) = \hat{D}_1(t) \int_0^1 B u(x,t) dx$$

$$- \hat{D}_1(t) \int_0^1 e^{-A\hat{D}(t)(1-x)} B dx u(1,t)$$

$$- \hat{D}(t)\hat{D}_1(t) \int_0^1 \int_0^x A e^{-A\hat{D}(t)(x-y)} B dy u(x,t) dx, \qquad (11.30)$$

$$h_D(x,t) = K(\hat{\beta}(t)) e^{A_{cl}(\hat{\beta}(t))\hat{D}(t)(x-1)} \left(f_D(t) + A_{cl}(\hat{\beta}(t)) Z(t) \right), \qquad (11.31)$$

and the projector operator is defined as

$$
\text{Proj}_{[\underline{D},\overline{D}]}\{\tau\} =
\begin{cases}
0, & \hat{D}_1(t)=\underline{D} \quad \text{and} \quad \tau<0 \\
0, & \hat{D}_1(t)=\overline{D} \quad \text{and} \quad \tau>0 \\
\tau, & \text{else,}
\end{cases}
\tag{11.32}
$$

$$
\text{Proj}_{[\underline{D},\overline{D}]}\{\tau\} =
\begin{cases}
0, & \hat{D}(t)=\underline{D} \quad \text{and} \quad \tau<0 \\
0, & \hat{D}(t)=\overline{D} \quad \text{and} \quad \tau>0 \\
\tau, & \text{else.}
\end{cases}
\tag{11.33}
$$

Theorem 11.4. *Consider the closed-loop system consisting of the plant (11.10)–(11.13) and the adaptive controller (11.19)–(11.33). All the states $\left(X(t),\, u(x,t),\, \hat{D}(t),\, \hat{D}_1(t)\right)$ of the closed-loop system are globally bounded and the regulation of $X(t)$ and $U(t)$ such that $\lim_{t\to\infty} X(t)=\lim_{t\to\infty} U(t)=0$ is achieved.*

Proof. Taking the time derivative of (11.21) along (11.10)–(11.13), and using the integration by parts, we obtain

$$
\dot{Z}(t) = AX(t) + D \int_0^1 Bu(x,t)dx + \phi\left(\hat{D}(t), \hat{D}_1(t)\right)
$$

$$
+ \frac{\hat{D}(t)\hat{D}_1(t)}{D} \int_0^1 \int_0^x e^{-A\hat{D}(t)(x-y)} Bdyu_x(x,t)dx
$$

$$
= AX(t) + D \int_0^1 Bu(x,t)dx + \phi\left(\hat{D}(t), \hat{D}_1(t)\right)
$$

$$
+ \frac{\hat{D}(t)\hat{D}_1(t)}{D} \int_0^1 e^{-A\hat{D}(t)(1-y)} Bdyu(1,t)
$$

$$
- \frac{\hat{D}(t)\hat{D}_1(t)}{D} \int_0^1 Bu(x,t)dx
$$

$$
+ \frac{\hat{D}(t)\hat{D}_1(t)}{D} \int_0^1 \int_0^x A\hat{D}(t)e^{-A\hat{D}(t)(x-y)} Bdyu(x,t)dx
$$

$$
= AX(t) + D \int_0^1 Bu(x,t)dx + \phi\left(\hat{D}(t), \hat{D}_1(t)\right)
$$

$$
+ \left(1 - \frac{\tilde{D}(t)}{D}\right) \hat{D}_1(t) \int_0^1 e^{-A\hat{D}(t)(1-x)} Bdxu(1,t)
$$

$$
- \left(1 - \frac{\tilde{D}(t)}{D}\right) \hat{D}_1(t) \int_0^1 Bu(x,t)dx
$$

$$+ \left(1 - \frac{\tilde{D}(t)}{D}\right) \hat{D}_1(t)\hat{D}(t)$$

$$\times \int_0^1 \int_0^x Ae^{-A\hat{D}(t)(x-y)} Bdyu(x,t)dx, \tag{11.34}$$

where

$$\phi\left(\dot{\hat{D}}(t), \dot{\hat{D}}_1(t)\right)$$
$$= \int_0^1 \int_0^x \left[\left(\dot{\hat{D}}(t)\hat{D}_1(t) + \hat{D}(t)\dot{\hat{D}}_1(t)\right)I - \hat{D}(t)\hat{D}_1(t)A\dot{\hat{D}}(t)(x-y)\right]$$
$$\times e^{-A\hat{D}(t)(x-y)} Bdyu(x,t)dx. \tag{11.35}$$

Under the control law (11.19), the formula (11.34) becomes

$$\dot{Z}(t) = A_{cl}(\hat{\beta}(t))Z(t) + \tilde{D}_1(t)f_{D_1}(t) + \frac{\tilde{D}(t)}{D}f_D(t)$$
$$+ \phi\left(\dot{\hat{D}}(t), \dot{\hat{D}}_1(t)\right), \tag{11.36}$$

where $A_{cl}(\hat{\beta}(t))$ is given in (11.20), and $f_{D_1}(t)$ and $f_D(t)$ have been defined in (11.28) and (11.30).

Taking partial derivatives of (11.26) with respect to x and t, respectively, we get

$$w_x(x,t) = u_x(x,t) - K(\hat{\beta}(t))e^{A_{cl}(\hat{\beta}(t))\hat{D}(t)(x-1)}$$
$$\times A_{cl}(\hat{\beta}(t))\hat{D}(t)Z(t), \tag{11.37}$$

$$w_t(x,t) = u_t(x,t) - \psi\left(\dot{\hat{D}}(t), \dot{\hat{D}}_1(t)\right)$$
$$- K(\hat{\beta}(t))e^{A_{cl}(\hat{\beta}(t))\hat{D}(t)(x-1)}$$
$$\times \left(A_{cl}(\hat{\beta}(t))Z(t) + \tilde{D}_1(t)f_{D_1}(t) + \frac{\tilde{D}(t)}{D}f_D(t)\right), \tag{11.38}$$

where

$$\psi\left(\dot{\hat{D}}(t), \dot{\hat{D}}_1(t)\right)$$
$$= \left(\frac{\partial K(\hat{\beta}(t))}{\partial \hat{D}(t)}\dot{\hat{D}}(t) + \frac{\partial K(\hat{\beta}(t))}{\partial \hat{D}_1(t)}\dot{\hat{D}}_1(t)\right)e^{A_{cl}(\hat{\beta}(t))\hat{D}(t)(x-1)}Z(t)$$
$$+ K(\hat{\beta}(t))e^{A_{cl}(\hat{\beta}(t))\hat{D}(t)(x-1)}$$

$$\times \left[\left(\frac{\partial A_{cl}(\hat{\beta}(t))}{\partial \hat{D}(t)} \dot{D}(t) + \frac{\partial A_{cl}(\hat{\beta}(t))}{\partial \hat{D}_1(t)} \dot{D}_1(t) \right) \hat{D}(t) \right.$$

$$+ A_{cl}(\hat{\beta}(t))\dot{D}(t) \bigg] (x-1)Z(t)$$

$$+ K(\hat{\beta}(t))e^{A_{cl}(\hat{\beta}(t))\hat{D}(t)(x-1)}\phi\left(\dot{D}(t), \dot{D}_1(t)\right). \tag{11.39}$$

Multiplying (11.38) with D and subtracting (11.37), substituting $x=1$ into (11.26) and employing the control law (11.19), we get

$$Dw_t(x,t) = w_x(x,t) - D\psi\left(\dot{D}(t), \dot{D}_1(t)\right)$$

$$- D\tilde{D}_1(t)h_{D_1}(x,t) - \tilde{D}(t)h_D(x,t), \tag{11.40}$$

$$w(1,t) = 0, \quad x \in [0,1], \tag{11.41}$$

where $h_{D_1}(x,t)$ and $h_D(x,t)$ have been defined in (11.29) and (11.31). Build the Lyapunov candidate such that

$$V(t) = D\log\left(1 + \Xi(t)\right) + \frac{gD}{\gamma_{D_1}}\tilde{D}_1(t)^2 + \frac{g}{\gamma_D}\tilde{D}(t)^2, \tag{11.42}$$

where $\Xi(t)$ has been defined in (11.27).

Taking the time derivative of (11.42) along the target closed-loop system (11.36) and (11.40)–(11.41), we have

$$\dot{V}(t) = \frac{1}{1+\Xi(t)}\bigg[DZ^T(t)\big(P(\hat{\beta}(t))A_{cl}(\hat{\beta}(t))$$

$$+ A_{cl}^T(\hat{\beta}(t))P(\hat{\beta}(t))\big)Z(t)$$

$$+ DZ^T(t)\varphi\left(\dot{D}(t), \dot{D}_1(t)\right)Z(t)$$

$$+ 2Z^T(t)P(\hat{\beta}(t))\left(D\phi\left(\dot{D}(t), \dot{D}_1(t)\right)\right.$$

$$+ D\tilde{D}_1(t)f_{D_1}(t) + \tilde{D}(t)f_D(t)\bigg)$$

$$+ 2g\int_0^1 (1+x)w(x,t)w_x(x,t)dx$$

$$- 2Dg\int_0^1 (1+x)w(x,t)\psi\left(\dot{D}(t), \dot{D}_1(t)\right)dx$$

$$- 2g\int_0^1 (1+x)w(x,t)\left(D\tilde{D}_1(t)h_{D_1}(x,t)\right.$$

$$\left. + \tilde{D}(t) h_D(x,t) \right) dx \Bigg]$$

$$- \frac{2gD}{\gamma_{D_1}} \tilde{D}_1(t) \dot{\hat{D}}_1(t) - \frac{2g}{\gamma_D} \tilde{D}(t) \dot{\hat{D}}(t)$$

$$= \frac{1}{1+\Xi(t)} \Bigg[- DZ^T(t) Q(\hat{\beta}(t)) Z(t)$$

$$- gw(0,t)^2 - g\|w(x,t)\|^2$$

$$+ DZ^T(t) \varphi\left(\dot{\hat{D}}(t), \dot{\hat{D}}_1(t) \right) Z(t)$$

$$+ 2DZ^T(t) P(\hat{\beta}(t)) \phi\left(\dot{\hat{D}}(t), \dot{\hat{D}}_1(t) \right)$$

$$- 2Dg \int_0^1 (1+x) w(x,t) \psi\left(\dot{\hat{D}}(t), \dot{\hat{D}}_1(t) \right) dx \Bigg]$$

$$- \frac{2gD}{\gamma_{D_1}} \tilde{D}_1(t) \left(\dot{\hat{D}}_1(t) - \gamma_{D_1} \tau_{D_1}(t) \right)$$

$$- \frac{2g}{\gamma_D} \tilde{D}(t) \left(\dot{\hat{D}}(t) - \gamma_D \tau_D(t) \right), \tag{11.43}$$

where $\tau_{D_1}(t)$ and $\tau_D(t)$ have been defined in (11.23) and (11.25), and $\varphi\left(\dot{\hat{D}}(t), \dot{\hat{D}}_1(t) \right) = \frac{\partial P(\hat{\beta}(t))}{\partial \hat{D}(t)} \dot{\hat{D}}(t) + \frac{\partial P(\hat{\beta}(t))}{\partial \hat{D}_1(t)} \dot{\hat{D}}_1(t)$.

Please note that $\hat{D}_1(t)$ and $\hat{D}(t)$ are bounded as the projector operators ensure that they stay in the interval (11.14). If we make use of Young's and Cauchy-Schwarz inequalities, it is evident that the inverse transformation of (11.26) implies

$$\|u(x,t)\|^2 \le M_u \left(|Z(t)|^2 + \|w(x,t)\|^2 \right), \tag{11.44}$$

where $M_u > 0$ is a constant.

If we utilize (11.44) and inequalities $0 < \frac{|\epsilon|}{1+\epsilon^2} < 1$ and $0 < \frac{\epsilon^2}{1+\epsilon^2} < 1$, it is easy to show that

$$\left| \dot{\hat{D}}_1(t) \right| \le \gamma_{D_1} M_{D_1} \frac{|Z(t)|^2 + \|w(x,t)\|^2}{1+\Xi(t)} \le \gamma_{D_1} \bar{M}_{D_1}, \tag{11.45}$$

$$\left| \dot{\hat{D}}(t) \right| \le \gamma_D M_D \frac{|Z(t)|^2 + \|w(x,t)\|^2}{1+\Xi(t)} \le \gamma_D \bar{M}_D, \tag{11.46}$$

where M_{D_1}, M_D, $\bar{M}_{D_1}$, and $\bar{M}_D$ are positive constants.

Thus we have

$$\dot{V}(t) \le \frac{1}{1+\Xi(t)} \Big[-\underline{D}\lambda_{\min}(Q)|Z(t)|^2 - gw(0,t)^2 - g\|w(x,t)\|^2$$
$$+ (\gamma_{D_1} + \gamma_D)M \left(|Z(t)|^2 + \|w(x,t)\|^2 \right) \Big], \qquad (11.47)$$

where $M > 0$ is a constant. If we carefully select design coefficients $\lambda_{\min}(Q)$, g, γ_{D_1}, and γ_D, it is easy to get

$$\dot{V}(t) \le \frac{N}{1+\Xi(t)} \Big[-|Z(t)|^2 - w(0,t)^2 - \|w(x,t)\|^2 \Big], \qquad (11.48)$$

where $N > 0$ is a constant. Thus $Z(t)$ and $w(x,t)$ are bounded and converge to zero. By inverse conversions of (11.21) and (11.26), the original states $X(t)$ and $u(x,t)$ are bounded and converge to zero. Then the proof is completed. □

11.2 Adaptive State Feedback under Unknown Delay and Delay Kernel

This section is concerned with the more challenging case where the n-dimensional input vector $B(Dx)$ for $x \in [0,1]$ in (11.3) is a continuous function of Dx such that

$$B(Dx) = \begin{bmatrix} \rho_1(Dx) \\ \rho_2(Dx) \\ \vdots \\ \rho_n(Dx) \end{bmatrix}, \qquad (11.49)$$

where $\rho_i(Dx)$ for $i = 1, \cdots, n$ are unknown components of the vector-valued function $B(Dx)$.

On the basis of (11.49), we further denote

$$\mathscr{B}(x) = DB(Dx) = \sum_{i=1}^{n} D\rho_i(Dx)B_i = \sum_{i=1}^{n} b_i(x)B_i, \qquad (11.50)$$

where $b_i(x) = D\rho_i(Dx)$ for $i = 1, \cdots, n$ are unknown scalar continuous functions of x, and $B_i \in \mathbb{R}^n$ for $i = 1, \cdots, n$ are the unit vectors accordingly. A three-dimensional example of (11.50) is given below.

Example 11.5.

$$DB(Dx) = D \begin{bmatrix} \frac{1}{Dx+1} \\ \sin Dx \\ e^{Dx} \end{bmatrix} = \frac{D}{Dx+1} \begin{bmatrix} 1 \\ 0 \\ 0 \end{bmatrix} + D\sin Dx \begin{bmatrix} 0 \\ 1 \\ 0 \end{bmatrix}$$

$$+ De^{Dx} \begin{bmatrix} 0 \\ 0 \\ 1 \end{bmatrix}$$

$$= b_1(x)B_1 + b_2(x)B_2 + b_3(x)B_3. \tag{11.51}$$

As a result, with (11.50), the system (11.3)–(11.6) is rewritten as

$$\dot{X}(t) = AX(t) + \int_0^1 \mathscr{B}(x)u(x,t)dx, \tag{11.52}$$

$$Y(t) = CX(t), \tag{11.53}$$

$$Du_t(x,t) = u_x(x,t), \quad x \in [0,1], \tag{11.54}$$

$$u(1,t) = U(t). \tag{11.55}$$

To use projector operators later, the following assumption is assumed.

Assumption 11.6. *There exist known constants $\underline{D}$, $\overline{D}$, $\bar{b}_i$ and known continuous functions $b_i^*(x)$ such that*

$$0 < \underline{D} \le D \le \overline{D}, \quad 0 < \int_0^1 (b_i(x) - b_i^*(x))^2\, dx \le \bar{b}_i \tag{11.56}$$

for $i = 1, \cdots, n$.

To stabilize (11.52)–(11.55), the following assumption is required.

Assumption 11.7. *For the system (11.52), the pair (A, β) is stabilizable where*

$$\beta = \int_0^1 e^{-AD(1-x)} \mathscr{B}(x)dx. \tag{11.57}$$

There exists a vector $K(\beta)$ to make $A + \beta K(\beta)$ Hurwitz. Namely, there exist matrices $P(\beta) = P(\beta)^T > 0$ and $Q(\beta) = Q(\beta)^T > 0$ such that

$$(A + \beta K(\beta))^T P(\beta) + P(\beta)(A + \beta K(\beta)) = -Q(\beta). \tag{11.58}$$

Denote $\hat{D}(t)$ and $\hat{b}_i(x,t)$ as the estimates of D and $b_i(x)$ (for $i = 1, \cdots, n$), with estimation errors satisfying

$$\tilde{D}(t) = D - \hat{D}(t), \tag{11.59}$$

$$\tilde{b}_i(x,t) = b_i(x) - \hat{b}_i(x,t), \tag{11.60}$$

and

$$\hat{\mathscr{B}}(x,t) = \sum_{i=1}^{n} \hat{b}_i(x,t) B_i. \tag{11.61}$$

The delay-adaptive control scheme is designed as follows:
The control law is

$$U(t) = u(1,t) = K(\hat{\beta}(t))Z(t), \tag{11.62}$$

where $K(\hat{\beta}(t))$ is chosen to make

$$A_{cl}(\hat{\beta}(t)) = A + \int_0^1 e^{-A\hat{D}(t)(1-x)}\hat{\mathscr{B}}(x,t)dx K(\hat{\beta}(t))$$

$$= A + \underbrace{\int_0^1 e^{-A\hat{D}(t)(1-x)}\sum_{i=1}^{n}\hat{b}_i(x,t)B_i dx}_{\hat{\beta}(t)} K(\hat{\beta}(t)) \tag{11.63}$$

Hurwitz and

$$Z(t) = X(t) + \hat{D}(t)\int_0^1\int_0^x e^{-A\hat{D}(t)(x-y)}\hat{\mathscr{B}}(y,t)dy u(x,t)dx. \tag{11.64}$$

The update laws are

$$\dot{\hat{D}}(t) = \gamma_D \text{Proj}_{[\underline{D},\overline{D}]}\{\tau_D(t)\}, \quad \gamma_D > 0, \tag{11.65}$$

$$\tau_D(t) = \frac{1/gZ^T(t)P(\hat{\beta}(t))f_D(t) - \int_0^1(1+x)w(x,t)h_D(x,t)dx}{1+\Xi(t)}, \tag{11.66}$$

$$\dot{\hat{b}}_i(x,t) = \gamma_b \text{Proj}\{\tau_{b_i}(x,t)\}, \quad \gamma_b > 0, \tag{11.67}$$

$$\tau_{b_i}(x,t) = \frac{1/gZ^T(t)P(\hat{\beta}(t))f_{b_i}(x,t)}{1+\Xi(t)}$$

$$- \frac{\int_0^1(1+y)w(y,t)h_{b_i}(y,t)dy f_{b_i}(x,t)}{1+\Xi(t)}, \tag{11.68}$$

where $P(\hat{\beta}(t))$ satisfies (11.58) and $g > 0$ is a designing coefficient:

$$w(x,t) = u(x,t) - K(\hat{\beta}(t))e^{A_{cl}(\hat{\beta}(t))\hat{D}(t)(x-1)}Z(t), \tag{11.69}$$

$$\Xi(t) = Z^T(t)P(\hat{\beta}(t))Z(t) + g\int_0^1 (1+x)w(x,t)^2 dx, \tag{11.70}$$

$$f_D(t) = \int_0^1 \hat{\mathscr{B}}(x,t)u(x,t)dx$$

$$- \int_0^1 e^{-A\hat{D}(t)(1-x)}\hat{\mathscr{B}}(x,t)dxu(1,t)$$

$$- \hat{D}(t)\int_0^1 \int_0^x Ae^{-A\hat{D}(t)(x-y)}\hat{\mathscr{B}}(y,t)dyu(x,t)dx, \tag{11.71}$$

$$h_D(x,t) = K(\hat{\beta}(t))e^{A_{cl}(\hat{\beta}(t))\hat{D}(t)(x-1)}\left(f_D(t) + A_{cl}(\hat{\beta}(t))Z(t)\right), \tag{11.72}$$

$$f_{b_i}(x,t) = B_i u(x,t), \tag{11.73}$$

$$h_{b_i}(x,t) = K(\hat{\beta}(t))e^{A_{cl}(\hat{\beta}(t))\hat{D}(t)(x-1)}, \tag{11.74}$$

for $i = 1, \cdots, n$, and the projector operators are defined as

$$\text{Proj}_{[\underline{D},\overline{D}]}\{\tau\} = \begin{cases} 0, & \hat{D}(t) = \underline{D} \quad \text{and} \quad \tau < 0 \\ 0, & \hat{D}(t) = \overline{D} \quad \text{and} \quad \tau > 0 \\ \tau, & \text{else,} \end{cases} \tag{11.75}$$

$$\text{Proj}\{\tau(x)\} = \begin{cases} \tau(x) - \left(\hat{b}_i(x) - b_i^*(x)\right)\frac{\int_0^1 \left(\hat{b}_i(x) - b_i^*(x)\right)\tau(x)dx}{\int_0^1 \left(\hat{b}_i(x) - b_i^*(x)\right)^2 dx}, \\ \text{if} \quad \int_0^1 \left(\hat{b}_i(x) - b_i^*(x)\right)^2 dx = \bar{b}_i \\ \text{and} \quad \int_0^1 \left(\hat{b}_i(x) - b_i^*(x)\right)\tau(x)dx > 0, \\ \tau(x), \quad \text{else.} \end{cases} \tag{11.76}$$

Theorem 11.8. *Consider the closed-loop system consisting of the plant (11.52)–(11.55) and the adaptive controller (11.62)–(11.76). All the states $(X(t), u(x,t),$ $\hat{D}(t), \hat{b}_i(x,t))$ of the closed-loop system are globally bounded and the regulation of $X(t)$ and $U(t)$ such that $\lim_{t\to\infty} X(t) = \lim_{t\to\infty} U(t) = 0$ is achieved.*

Proof. Taking the time derivative of (11.64) along (11.52)–(11.55), and using the integration by parts, we obtain

$$\dot{Z}(t) = AX(t) + \int_0^1 \mathscr{B}(x)u(x,t)dx + \phi\left(\dot{\hat{D}}(t), \dot{\hat{\mathscr{B}}}(x,t)\right)$$

$$+ \frac{\hat{D}(t)}{D} \int_0^1 \int_0^x e^{-A\hat{D}(t)(x-y)} \dot{\hat{\mathscr{B}}}(y,t) dy u_x(x,t) dx$$

$$= AX(t) + \int_0^1 \mathscr{B}(x,t)u(x,t)dx + \phi\left(\dot{\hat{D}}(t), \dot{\hat{\mathscr{B}}}(x,t)\right)$$

$$+ \frac{\hat{D}(t)}{D} \int_0^1 e^{-A\hat{D}(t)(1-y)} \dot{\hat{\mathscr{B}}}(y,t) dy u(1,t)$$

$$- \frac{\hat{D}(t)}{D} \int_0^1 \dot{\hat{\mathscr{B}}}(x,t)u(x,t)dx$$

$$+ \frac{\hat{D}(t)}{D} \int_0^1 \int_0^x A\hat{D}(t)e^{-A\hat{D}(t)(x-y)} \dot{\hat{\mathscr{B}}}(y,t) dy u(x,t) dx$$

$$= AX(t) + \int_0^1 \mathscr{B}(x,t)u(x,t)dx + \phi\left(\dot{\hat{D}}(t), \dot{\hat{\mathscr{B}}}(x,t)\right)$$

$$+ \left(1 - \frac{\tilde{D}(t)}{D}\right) \int_0^1 e^{-A\hat{D}(t)(1-x)} \dot{\hat{\mathscr{B}}}(x,t) dx u(1,t)$$

$$- \left(1 - \frac{\tilde{D}(t)}{D}\right) \int_0^1 \dot{\hat{\mathscr{B}}}(x,t)u(x,t)dx$$

$$+ \left(1 - \frac{\tilde{D}(t)}{D}\right) \hat{D}(t) \int_0^1 \int_0^x Ae^{-A\hat{D}(t)(x-y)}$$

$$\times \dot{\hat{\mathscr{B}}}(y,t) dy u(x,t) dx, \tag{11.77}$$

where

$$\phi\left(\dot{\hat{D}}(t), \dot{\hat{\mathscr{B}}}(x,t)\right)$$

$$= \int_0^1 \int_0^x e^{-A\hat{D}(t)(x-y)} \left[\left(\dot{\hat{D}}(t)I - \hat{D}(t)A\dot{\hat{D}}(t)(x-y)\right) \right.$$

$$\left. \times \dot{\hat{\mathscr{B}}}(y,t) + \hat{D}(t) \sum_{i=1}^n \dot{\hat{b}}_i(y,t)B_i \right] dy u(x,t) dx. \tag{11.78}$$

Under the control law (11.62), the formula (11.77) becomes

$$\dot{Z}(t) = A_{cl}(\hat{\beta}(t))Z(t) + \phi\left(\dot{\hat{D}}(t), \dot{\hat{\mathscr{B}}}(x,t)\right) + \frac{\tilde{D}(t)}{D} f_D(t)$$

$$+ \int_0^1 \sum_{i=1}^n \tilde{b}_i(x,t) f_{b_i}(x,t) dx, \tag{11.79}$$

where $A_{cl}(\hat{\beta}(t))$ is given in (11.63), and $f_D(t)$ and $f_{b_i}(x,t)$ have been defined in (11.71) and (11.73).

Taking partial derivatives of (11.69) with respect to x and t, respectively, we get

$$
\begin{aligned}
w_x(x,t) = u_x(x,t) - K(\hat{\beta}(t))e^{A_{cl}(\hat{\beta}(t))\hat{D}(t)(x-1)} \\
\times A_{cl}(\hat{\beta}(t))\hat{D}(t)Z(t),
\end{aligned}
\tag{11.80}
$$

$$
\begin{aligned}
w_t(x,t) = u_t(x,t) - \psi\left(\dot{\hat{D}}(t), \dot{\hat{\mathscr{B}}}(x,t)\right) \\
- K(\hat{\beta}(t))e^{A_{cl}(\hat{\beta}(t))\hat{D}(t)(x-1)}\left(A_{cl}(\hat{\beta}(t))Z(t)\right. \\
\left. + \frac{\dot{\hat{D}}(t)}{D}f_D(t) + \int_0^1 \sum_{i=1}^n \tilde{b}_i(y,t)f_{b_i}(y,t)dy\right),
\end{aligned}
\tag{11.81}
$$

where

$$
\begin{aligned}
\psi\left(\dot{\hat{D}}(t), \dot{\hat{\mathscr{B}}}(x,t)\right) \\
= \left(\frac{\partial K(\hat{\beta}(t))}{\partial \hat{D}(t)}\dot{\hat{D}}(t) + \sum_{i=1}^n \frac{\partial K(\hat{\beta}(t))}{\partial \hat{b}_i(x,t)}\dot{\hat{b}}_i(x,t)\right)e^{A_{cl}(\hat{\beta}(t))\hat{D}(t)(x-1)} \\
\times Z(t) + K(\hat{\beta}(t))e^{A_{cl}(\hat{\beta}(t))\hat{D}(t)(x-1)}\left[\left(\frac{\partial A_{cl}(\hat{\beta}(t))}{\partial \hat{D}(t)}\dot{\hat{D}}(t)\right.\right. \\
\left.\left. + \sum_{i=1}^n \frac{\partial A_{cl}(\hat{\beta}(t))}{\partial \hat{b}_i(x,t)}\dot{\hat{b}}_i(x,t)\right)\hat{D}(t) + A_{cl}(\hat{\beta}(t))\dot{\hat{D}}(t)\right](x-1)Z(t) \\
+ K(\hat{\beta}(t))e^{A_{cl}(\hat{\beta}(t))\hat{D}(t)(x-1)}\phi\left(\dot{\hat{D}}(t), \dot{\hat{\mathscr{B}}}(x,t)\right).
\end{aligned}
\tag{11.82}
$$

Multiplying (11.81) with D and subtracting (11.80), and employing the control law (11.62), we get

$$
\begin{aligned}
Dw_t(x,t) = w_x(x,t) - D\psi\left(\dot{\hat{D}}(t), \dot{\hat{\mathscr{B}}}(x,t)\right) - \tilde{D}(t)h_D(x,t) \\
- Dh_{b_i}(x,t)\int_0^1 \sum_{i=1}^n \tilde{b}_i(y,t)f_{b_i}(y,t)dy,
\end{aligned}
\tag{11.83}
$$

$$
w(1,t) = 0, \quad x \in [0,1],
\tag{11.84}
$$

where $h_D(x,t)$ and $h_{b_i}(x,t)$ have been defined in (11.72) and (11.74).

Please note that $\hat{D}(t)$ and $\int_0^1 \hat{b}_i(x,t)dx$ are bounded, as the projector operators ensure that they stay in the interval (11.56). If we use Young's and Cauchy-Schwarz inequalities, it is evident that the inverse transformation of (11.69) implies

$$\|u(x,t)\|^2 \leq M_u \left(|Z(t)|^2 + \|w(x,t)\|^2 \right), \tag{11.85}$$

where $M_u > 0$ is a constant.

Utilizing (11.85) and inequalities $0 < \frac{|\epsilon|}{1+\epsilon^2} < 1$ and $0 < \frac{\epsilon^2}{1+\epsilon^2} < 1$, we can easily show that

$$\left| \dot{\tilde{D}}(t) \right| \leq \gamma_D M_D \frac{|Z(t)|^2 + \|w(x,t)\|^2}{1 + \Xi(t)} \leq \gamma_D \bar{M}_D \tag{11.86}$$

$$\left| \int_0^1 \dot{\tilde{b}}_i(x,t)dx \right| \leq \gamma_b M_{b_i} \frac{|Z(t)|^2 + \|w(x,t)\|^2}{1 + \Xi(t)} \leq \gamma_b \bar{M}_{b_i}, \tag{11.87}$$

where M_D, M_{b_i}, $\bar{M}_D$, $\bar{M}_{b_i}$ are positive constants.

Build the Lyapunov candidate such that

$$V(t) = D \log \left(1 + \Xi(t) \right) + + \frac{g}{\gamma_D} \tilde{D}(t)^2 + \sum_{i=1}^{n} \int_0^1 \frac{gD}{\gamma_b} \tilde{b}_i(x,t)^2 dx, \tag{11.88}$$

where $\Xi(t)$ has been defined in (11.70).

Taking the time derivative of (11.88) along the target closed-loop system (11.79), (11.83)–(11.84), employing (11.85)–(11.87), following a similar procedure of proof of Theorem 11.4, we get

$$\dot{V}(t) \leq \frac{N}{1 + \Xi(t)} \left[-|Z(t)|^2 - w(0,t)^2 - \|w(x,t)\|^2 \right]$$

$$- \frac{2g}{\gamma_D} \tilde{D}(t) \left(\dot{\tilde{D}}(t) - \gamma_D \tau_D(t) \right)$$

$$- \sum_{i=1}^{n} \int_0^1 \frac{2gD}{\gamma_b} \tilde{b}_i(x,t) \left(\dot{\tilde{b}}_i(x,t) - \gamma_b \tau_{b_i}(x,t) \right) dx$$

$$\leq \frac{N}{1 + \Xi(t)} \left[-|Z(t)|^2 - w(0,t)^2 - \|w(x,t)\|^2 \right], \tag{11.89}$$

where $N > 0$ is a constant, and $\tau_D(t)$ and $\tau_{b_i}(x,t)$ are given in (11.66) and (11.68). Thus $Z(t)$ and $w(x,t)$ are bounded and converge to zero. By inverse conversions of (11.64) and (11.69), the original states $X(t)$ and $u(x,t)$ are bounded and converge to zero. Thus the proof is proved. □

Remark 11.9. A special case of (11.49) is that the input vector $B(Dx)$ for $x \in [0,1]$ is a continuous function of Dx and is parameterizable in the delay D such that

$$B(Dx) = B_0 + \sum_{i=1}^{p} \rho_i(D)B_i(x), \tag{11.90}$$

where $B_0 \in \mathbb{R}^n$ is a known constant vector, $B_i(x): [0,1] \to \mathbb{R}^n$ for $i = 1, \cdots, p$ are known continuous vector functions of x, and $\rho_i(D)$ for $i = 1, \cdots, p$ are unknown constant parameters dependent upon D. A three-dimensional example of (11.90) is given below.

Example 11.10.

$$B(Dx) = \begin{bmatrix} Dx + \frac{1}{Dx+D} \\ \sqrt{Dx} + 1 \\ D^2 x^2 + 2 \end{bmatrix} = \begin{bmatrix} 0 \\ 1 \\ 2 \end{bmatrix} + D \begin{bmatrix} x \\ 0 \\ 0 \end{bmatrix} + \frac{1}{D} \begin{bmatrix} \frac{1}{x+1} \\ 0 \\ 0 \end{bmatrix}$$

$$+ \sqrt{D} \begin{bmatrix} 0 \\ \sqrt{x} \\ 0 \end{bmatrix} + D^2 \begin{bmatrix} 0 \\ 0 \\ x^2 \end{bmatrix}$$

$$= B_0 + \rho_1(D)B_1(x) + \rho_2(D)B_2(x)$$
$$+ \rho_3(D)B_3(x) + \rho_4(D)B_4(x). \tag{11.91}$$

On the basis of (11.90), we further denote

$$\mathcal{B}(x) = DB(Dx) = DB_0 + \sum_{i=1}^{p} D\rho_i(D)B_i(x) = \sum_{i=0}^{p} b_i B_i(x), \tag{11.92}$$

where $b_0 = D$ and $b_i = D\rho_i(D)$ for $i = 1, \cdots, p$ are unknown constant parameters with respect to D.

We assume there exist known constants $\underline{b}_i$, $\bar{b}_i$ such that $0 < \underline{b}_i \le b_i \le \bar{b}_i$. Instead of functional estimators in (11.60), we use scalar estimators $\hat{b}_i(t)$ for $i = 0, \cdots, p$ as the estimates of b_i, with estimation errors being

$$\tilde{b}_i(t) = b_i - \hat{b}_i(t) \tag{11.93}$$

and

$$\hat{\mathcal{B}}(x,t) = \sum_{i=0}^{p} \hat{b}_i(t)B_i(x). \tag{11.94}$$

The update law of functional adaptation (11.67)–(11.68) is reduced to the scalar adaptation below:

$$\dot{\hat{b}}_i(t) = \gamma_b \text{Proj}_{[\underline{b}_i, \bar{b}_i]} \{\tau_{b_i}(t)\}, \qquad \gamma_b > 0 \tag{11.95}$$

$$\tau_{b_i}(t) = \frac{1/g Z^T(t)P(\hat{\beta}(t))f_{b_i}(t) - \int_0^1 (1+x)w(x,t)h_{b_i}(x,t)dx}{1 + \Xi(t)}, \tag{11.96}$$

where everything else is defined the same as those in (11.68) except for

$$f_{b_i}(t) = \int_0^1 B_i(x)u(x,t)dx,$$
(11.97)

$$h_{b_i}(x,t) = K(\hat{\beta}(t))e^{A_{cl}(\hat{\beta}(t))\hat{D}(t)(x-1)}f_{b_i}(t),$$
(11.98)

and

$$\text{Proj}_{[\underline{b}_i, \bar{b}_i]}\{\tau\} = \begin{cases} 0, & \hat{b}_i(t) = \underline{b}_i \quad \text{and} \quad \tau < 0 \\ 0, & \hat{b}_i(t) = \bar{b}_i \quad \text{and} \quad \tau > 0 \\ \tau, & \text{else.} \end{cases}$$
(11.99)

11.3 Adaptive State Feedback under Unknown Delay, Delay Kernel, and Parameter

In this section, we consider the case of coexistent uncertainty in $(D, B(Dx), A)$. We employ the notation (11.49)–(11.55).

Since system matrix A contains unknown plant parameters, we denote that A is of the form

$$A = A_0 + \sum_{i=1}^{p} \theta_i A_i,$$
(11.100)

where $\theta_i \in \mathbb{R}$ for $i = 1, \cdots, p$ are unknown constant parameters and $A_i \in \mathbb{R}^{n \times n}$ for $i = 0, \cdots, p$ are known constant matrices. A three-dimensional example of (11.100) is given below.

Example 11.11.

$$A = \begin{bmatrix} a_{11} & a_{12} & a_{13} \\ a_{21} & a_{22} & a_{23} \\ a_{31} & a_{32} & a_{33} \end{bmatrix} = \begin{bmatrix} a_{11} & 0 & a_{13} \\ a_{21} & a_{22} & 0 \\ a_{31} & a_{32} & a_{33} \end{bmatrix} + a_{12} \begin{bmatrix} 0 & 1 & 0 \\ 0 & 0 & 0 \\ 0 & 0 & 0 \end{bmatrix}$$

$$+ a_{23} \begin{bmatrix} 0 & 0 & 0 \\ 0 & 0 & 1 \\ 0 & 0 & 0 \end{bmatrix},$$
(11.101)

where a_{ij} for $i, j \in \{1, 2, 3\}$ are known elements except for a_{12} and a_{23}.

To use projector operators for update laws later, the following assumption is assumed.

Assumption 11.12. *There exist known constants* $\underline{D}$, $\overline{D}$, $\underline{\theta}_i$, $\overline{\theta}_i$, $\overline{b}_i$ *and known continuous functions* $b_i^*(x)$ *such that*

$$\underline{D} \leq D \leq \overline{D}, \quad \underline{\theta}_i \leq \theta_i \leq \overline{\theta}_i, \tag{11.102}$$

$$\int_0^1 (b_i(x) - b_i^*(x))^2 \, dx \leq \overline{b}_i. \tag{11.103}$$

Denote $\hat{D}(t)$, $\hat{\theta}_i(t)$, and $\hat{b}_i(x,t)$ as the estimates of D, θ_i, and $b_i(x)$ with estimation errors satisfying

$$\tilde{D}(t) = D - \hat{D}(t), \quad \tilde{\theta}_i(t) = \theta_i - \hat{\theta}_i(t), \tag{11.104}$$

$$\tilde{b}_i(x,t) = b_i(x) - \hat{b}_i(x,t), \tag{11.105}$$

and accordingly

$$\hat{A}(t) = A_0 + \sum_{i=1}^p \hat{\theta}_i(t) A_i, \quad \hat{\mathscr{B}}(x,t) = \sum_{i=1}^n \hat{b}_i(x,t) B_i. \tag{11.106}$$

To stabilize (11.52)–(11.55), the certainty-equivalence-based adaptive control scheme is designed as follows:
The control law is

$$U(t) = u(1,t) = K(\hat{\theta}(t), \hat{\beta}(t)) Z(t), \tag{11.107}$$

where $K(\hat{\theta}(t), \hat{\beta}(t))$ is chosen to make

$$A_{cl}(\hat{\theta}(t), \hat{\beta}(t)) = \hat{A}(t)$$
$$+ \underbrace{\int_0^1 e^{-\hat{A}(t)\hat{D}(t)(1-x)} \hat{\mathscr{B}}(x,t) dx}_{\hat{\beta}(t)} K(\hat{\theta}(t), \hat{\beta}(t)) \tag{11.108}$$

Hurwitz and

$$Z(t) = X(t) + \hat{D}(t) \int_0^1 \int_0^x e^{-\hat{A}(t)\hat{D}(t)(x-y)} \hat{\mathscr{B}}(y,t) dy u(x,t) dx. \tag{11.109}$$

The update laws are

$$\dot{\hat{D}}(t) = \gamma_D \text{Proj}_{[\underline{D},\overline{D}]} \{\tau_D(t)\}, \quad \gamma_D > 0, \tag{11.110}$$

$$\tau_D(t) = \frac{1/g Z^T(t) P(\hat{\theta}(t), \hat{\beta}(t)) f_D(t)}{1 + \Xi(t)}$$
$$- \frac{\int_0^1 (1+x) w(x,t) h_D(x,t) dx}{1 + \Xi(t)}, \tag{11.111}$$

$$\dot{\hat{\theta}}_i(t) = \gamma_\theta \text{Proj}_{[\underline{\theta}_i, \bar{\theta}_i]}\{\tau_{\theta_i}(t)\}, \quad \gamma_\theta > 0, \tag{11.112}$$

$$\tau_{\theta_i}(t) = \frac{1/g Z^T(t) P(\hat{\theta}(t), \hat{\beta}(t)) f_{\theta_i}(t)}{1 + \Xi(t)}$$
$$- \frac{\int_0^1 (1+x) w(x,t) h_{\theta_i}(x,t) dx}{1 + \Xi(t)}, \tag{11.113}$$

$$\dot{\hat{b}}_i(x,t) = \gamma_b \text{Proj}\{\tau_{b_i}(x,t)\}, \quad \gamma_b > 0, \tag{11.114}$$

$$\tau_{b_i}(x,t) = \frac{1/g Z^T(t) P(\hat{\theta}(t), \hat{\beta}(t)) f_{b_i}(x,t)}{1 + \Xi(t)}$$
$$- \frac{\int_0^1 (1+y) w(y,t) h_{b_i}(y,t) dy f_{b_i}(x,t)}{1 + \Xi(t)}, \tag{11.115}$$

where $g > 0$ is a designing coefficient, and

$$w(x,t) = u(x,t) - K(\hat{\theta}(t), \hat{\beta}(t)) e^{A_{cl}(\hat{\theta}(t), \hat{\beta}(t)) \hat{D}(t)(x-1)} Z(t), \tag{11.116}$$

$$\Xi(t) = Z^T(t) P(\hat{\theta}(t), \hat{\beta}(t)) Z(t) + g \int_0^1 (1+x) w(x,t)^2 dx, \tag{11.117}$$

$$f_D(t) = \int_0^1 \hat{\mathscr{B}}(x,t) u(x,t) dx$$
$$- \int_0^1 e^{-\hat{A}(t) \hat{D}(t)(1-x)} \hat{\mathscr{B}}(x,t) dx u(1,t)$$
$$- \hat{D}(t) \int_0^1 \int_0^x \hat{A}(t) e^{-\hat{A}(t) \hat{D}(t)(x-y)} \hat{\mathscr{B}}(y,t) dy u(x,t) dx, \tag{11.118}$$

$$h_D(x,t) = K(\hat{\theta}(t), \hat{\beta}(t)) e^{A_{cl}(\hat{\theta}(t), \hat{\beta}(t)) \hat{D}(t)(x-1)}$$
$$\times \left(f_D(t) + A_{cl}(\hat{\theta}(t), \hat{\beta}(t)) Z(t) \right), \tag{11.119}$$

$$f_{\theta_i}(t) = A_i X(t), \tag{11.120}$$

$$h_{\theta_i}(x,t) = K(\hat{\theta}(t), \hat{\beta}(t)) e^{A_{cl}(\hat{\theta}(t), \hat{\beta}(t)) \hat{D}(t)(x-1)} f_{\theta_i}(t), \tag{11.121}$$

$$f_{b_i}(x,t) = B_i u(x,t), \tag{11.122}$$

$$h_{b_i}(x,t) = K(\hat{\theta}(t), \hat{\beta}(t)) e^{A_{cl}(\hat{\theta}(t), \hat{\beta}(t)) \hat{D}(t)(x-1)}, \tag{11.123}$$

and the ODE and PDE projector operators are of standard forms in [80, 122].

Theorem 11.13. *Consider the closed-loop system consisting of the plant (11.52)–(11.55) and the adaptive controller (11.107)–(11.123). All the states $(X(t), u(x,t), \hat{D}(t), \hat{\theta}_i(t), \hat{b}_i(x,t))$ of the closed-loop system are globally bounded and the regulation of $X(t)$ and $U(t)$ such that $\lim_{t\to\infty} X(t) = \lim_{t\to\infty} U(t) = 0$ is achieved.*

Proof. Differentiating (11.109) with respect to t along (11.52)–(11.55), and using the integration by parts in x, we obtain

$$\dot{Z}(t) = A_{cl}(\hat{\theta}(t), \hat{\beta}(t))Z(t) + \phi\left(\dot{\hat{D}}(t), \dot{\hat{\theta}}(t), \dot{\hat{\mathscr{B}}}(x,t)\right)$$

$$+ \frac{\tilde{D}(t)}{D} f_D(t) + \sum_{i=1}^{p} \tilde{\theta}_i(t) f_{\theta_i}(t) + \int_0^1 \sum_{i=1}^{n} \tilde{b}_i(x,t) f_{b_i}(x,t)dx, \quad (11.124)$$

where $A_{cl}(\hat{\theta}(t), \hat{\beta}(t))$ is given in (11.108), and $f_D(t)$, $f_{\theta_i}(t)$, and $f_{b_i}(x,t)$ have been defined in (11.118), (11.120), and (11.122), where

$$\phi\left(\dot{\hat{D}}(t), \dot{\hat{\theta}}(t), \dot{\hat{\mathscr{B}}}(x,t)\right)$$

$$= \int_0^1 \int_0^x e^{-\hat{A}(t)\hat{D}(t)(x-y)}\left[\left(\dot{\hat{D}}(t)I - \hat{D}(t)\left(\sum_{i=1}^{p} \dot{\hat{\theta}}_i(t)A_i\hat{D}(t)\right)\right.\right.$$

$$\left.\left. + \hat{A}(t)\dot{\hat{D}}(t)\right)(x-y)\right)\hat{\mathscr{B}}(y,t) + \hat{D}(t)\sum_{i=1}^{n} \dot{\hat{b}}_i(y,t)B_i\right]dy$$

$$\times u(x,t)dx. \quad (11.125)$$

Taking partial derivatives of (11.116) with respect to x and t, respectively, we get

$$Dw_t(x,t) = w_x(x,t) - D\psi\left(\dot{\hat{D}}(t), \dot{\hat{\theta}}(t), \dot{\hat{\mathscr{B}}}(x,t)\right)$$

$$- \tilde{D}(t)h_D(x,t) - D\sum_{i=1}^{p} \tilde{\theta}_i(t)h_{\theta_i}(x,t)$$

$$- Dh_{b_i}(x,t)\int_0^1 \sum_{i=1}^{n} \tilde{b}_i(y,t)f_{b_i}(y,t)dy, \quad (11.126)$$

$$w(1,t) = 0, \quad x \in [0,1], \quad (11.127)$$

where $h_D(x,t)$, $h_{\theta_i}(x,t)$, and $h_{b_i}(x,t)$ have been defined in (11.119), (11.121), and (11.123), and

$$\psi\left(\dot{\hat{D}}(t), \dot{\hat{\theta}}(t), \dot{\hat{\mathscr{B}}}(x,t)\right)$$

$$= \left(\frac{\partial K(\hat{\theta}(t), \hat{\beta}(t))}{\partial \hat{D}(t)} \dot{\hat{D}}(t) + \sum_{i=1}^{p} \frac{\partial K(\hat{\theta}(t), \hat{\beta}(t))}{\partial \hat{\theta}_i(t)} \dot{\hat{\theta}}_i(t) \right.$$

$$\left. + \sum_{i=1}^{n} \frac{\partial K(\hat{\theta}(t), \hat{\beta}(t))}{\partial \hat{b}_i(x,t)} \dot{\hat{b}}_i(x,t) \right) e^{A_{cl}(\hat{\theta}(t), \hat{\beta}(t))\hat{D}(t)(x-1)} Z(t)$$

$$+ K(\hat{\theta}(t), \hat{\beta}(t)) e^{A_{cl}(\hat{\theta}(t), \hat{\beta}(t))\hat{D}(t)(x-1)} \left[\left(\frac{\partial A_{cl}(\hat{\theta}(t), \hat{\beta}(t))}{\partial \hat{D}(t)} \dot{\hat{D}}(t) \right. \right.$$

$$\left. + \sum_{i=1}^{p} \frac{\partial A_{cl}(\hat{\theta}(t), \hat{\beta}(t))}{\partial \hat{\theta}_i(t)} \dot{\hat{\theta}}_i(t) + \sum_{i=1}^{n} \frac{\partial A_{cl}(\hat{\theta}(t), \hat{\beta}(t))}{\partial \hat{b}_i(x,t)} \dot{\hat{b}}_i(x,t) \right)$$

$$\left. \times \hat{D}(t) + A_{cl}(\hat{\theta}(t), \hat{\beta}(t))\dot{\hat{D}}(t) \right] (x-1)Z(t)$$

$$+ K(\hat{\theta}(t), \hat{\beta}(t)) e^{A_{cl}(\hat{\theta}(t), \hat{\beta}(t))\hat{D}(t)(x-1)} \phi \left(\dot{\hat{D}}(t), \dot{\hat{\theta}}(t), \dot{\hat{\mathcal{B}}}(x,t) \right). \qquad (11.128)$$

Please note that $\hat{D}(t)$, $\hat{\theta}_i(t)$, and $\int_0^1 \hat{b}_i(x,t)dx$ are bounded, as the projector operators ensure that they stay in the interval (11.102)–(11.103). Making use of Young's and Cauchy-Schwarz inequalities, it is evident that the inverse transformations of (11.109) and (11.116) imply

$$\|u(x,t)\|^2 \leq M_u \left(|Z(t)|^2 + \|w(x,t)\|^2 \right), \qquad (11.129)$$

$$|X(t)|^2 \leq M_X \left(|Z(t)|^2 + \|w(x,t)\|^2 \right), \qquad (11.130)$$

where $M_u > 0$ and $M_X > 0$ are constants, $\|u(x,t)\| = \left(\int_0^1 u(x,t)^2 dx \right)^{1/2}$, and $\|w(x,t)\| = \left(\int_0^1 w(x,t)^2 dx \right)^{1/2}$.

Utilizing (11.129)–(11.130) and inequalities $0 < \frac{|\epsilon|}{1+\epsilon^2} < 1$ and $0 < \frac{\epsilon^2}{1+\epsilon^2} < 1$, we can easily show that

$$\left| \dot{\hat{D}}(t) \right| \leq \gamma_D M_D \frac{|Z(t)|^2 + \|w(x,t)\|^2}{1+\Xi(t)} \leq \gamma_D \bar{M}_D, \qquad (11.131)$$

$$\left| \dot{\hat{\theta}}_i(t) \right| \leq \gamma_\theta M_{\theta_i} \frac{|Z(t)|^2 + \|w(x,t)\|^2}{1+\Xi(t)} \leq \gamma_\theta \bar{M}_{\theta_i}, \qquad (11.132)$$

$$\left| \int_0^1 \dot{\hat{b}}_i(x,t)dx \right| \leq \gamma_b M_{b_i} \frac{|Z(t)|^2 + \|w(x,t)\|^2}{1+\Xi(t)} \leq \gamma_b \bar{M}_{b_i}, \qquad (11.133)$$

where $M_D, M_\theta, M_{b_i}, \bar{M}_D, \bar{M}_\theta, \bar{M}_{b_i}$ are positive constants.

Build the Lyapunov candidate such that

$$V(t) = D \log \left(1 + \Xi(t)\right) + \frac{g}{\gamma_D} \tilde{D}(t)^2 + \frac{gD}{\gamma_\theta} \sum_{i=1}^{p} \tilde{\theta}_i(t)^2$$

$$+ \frac{gD}{\gamma_b} \sum_{i=1}^{n} \int_0^1 \tilde{b}_i(x,t)^2 dx, \tag{11.134}$$

where $\Xi(t)$ has been defined in (11.117).

Taking the time derivative of (11.134) along the target closed-loop system (11.124), (11.126)–(11.127), employing (11.129)–(11.133), we get

$$\dot{V}(t) \leq \frac{N}{1+\Xi(t)} \left[-|Z(t)|^2 - w(0,t)^2 - \|w(x,t)\|^2 \right]$$

$$- \frac{2g}{\gamma_D} \tilde{D}(t) \left(\dot{\tilde{D}}(t) - \gamma_D \tau_D(t) \right)$$

$$- \frac{2gD}{\gamma_\theta} \sum_{i=1}^{p} \tilde{\theta}_i(t) \left(\dot{\tilde{\theta}}_i(t) - \gamma_\theta \tau_{\theta_i}(t) \right)$$

$$- \frac{2gD}{\gamma_b} \sum_{i=1}^{n} \int_0^1 \tilde{b}_i(x,t) \left(\dot{\tilde{b}}_i(x,t) - \gamma_b \tau_{b_i}(x,t) \right) dx$$

$$\leq \frac{N}{1+\Xi(t)} \left[-|Z(t)|^2 - w(0,t)^2 - \|w(x,t)\|^2 \right], \tag{11.135}$$

where $N > 0$ is a constant, and $\tau_D(t)$, $\tau_{\theta_i}(t)$, and $\tau_{b_i}(x,t)$ are given in (11.111), (11.113), and (11.115). Thus $Z(t)$ and $w(x,t)$ are bounded and converge to zero. By inverse conversions of (11.109) and (11.116), the original states $X(t)$ and $u(x,t)$ are bounded and converge to zero. Thus the proof is completed. □

11.4   Robust Output Feedback under Unknown Delay, Delay Kernel, and PDE State

In this section, we consider the case of coexistent uncertainty in $(D, B(Dx), u(x,t))$.

As analyzed in chapter 8 of [80], when the actuator state $u(x,t)$ is unmeasurable and the delay value D is unknown at the same time, the adaptive control can only achieve local stabilization both in the initial state and in the initial parameter error. In other words, the initial delay estimate needs to be sufficiently close to the true delay. As a result, instead of adaptive control, in this section we use the constant estimate and rely on the robustness of the feedback law to small errors in the assumed delay value.

The notation (11.49)–(11.55) is employed. Denote $\hat{D}$ and $\hat{b}_i(x)$ as the constant estimates of D and $b_i(x)$ (for $i = 1, \cdots, n$) with estimation errors satisfying

$$\tilde{D} = D - \hat{D}, \quad \tilde{b}_i(x) = b_i(x) - \hat{b}_i(x) \tag{11.136}$$

and accordingly

$$\hat{\mathscr{B}}(x) = \sum_{i=1}^{n} \hat{b}_i(x) B_i, \quad \tilde{\mathscr{B}}(x) = \sum_{i=1}^{n} \tilde{b}_i(x) B_i. \tag{11.137}$$

Furthermore, we employ

$$\hat{u}(x,t) = U(t + \hat{D}(x-1)), \quad x \in [0,1] \tag{11.138}$$

to be an estimate of the actuator state $u(x,t)$, with the estimation error being

$$\tilde{u}(x,t) = u(x,t) - \hat{u}(x,t), \quad x \in [0,1]. \tag{11.139}$$

The PDE-observer-based control scheme is designed as follows:
The observer of actuator state is

$$\hat{D}\hat{u}_t(x,t) = \hat{u}_x(x,t), \quad x \in [0,1], \tag{11.140}$$

$$\hat{u}(1,t) = U(t). \tag{11.141}$$

The control law is

$$U(t) = \hat{u}(1,t) = K(\hat{\beta}) Z(t), \tag{11.142}$$

where $K(\hat{\beta})$ is chosen to make

$$A_{cl}(\hat{\beta}) = A + \int_0^1 e^{-A\hat{D}(1-x)} \hat{\mathscr{B}}(x) dx K(\hat{\beta})$$

$$= A + \underbrace{\int_0^1 e^{-A\hat{D}(1-x)} \sum_{i=1}^{n} \hat{b}_i(x) B_i dx}_{\hat{\beta}} K(\hat{\beta}) \tag{11.143}$$

Hurwitz and

$$Z(t) = X(t) + \hat{D} \int_0^1 \int_0^x e^{-A\hat{D}(x-y)} \hat{\mathscr{B}}(y) dy \hat{u}(x,t) dx. \tag{11.144}$$

The PDE (11.140)–(11.141) is an open-loop certainty equivalence observer of the PDE (11.54)–(11.55). The "certainty equivalence" refers to the fact that the unknown delay D is replaced by its estimate $\hat{D}$, whereas the "open-loop" refers to the fact that no output injection is used in the observer.

Theorem 11.14. *Consider the closed-loop system consisting of the plant (11.52)–(11.55) and the PDE-observer-based control scheme (11.140)–(11.144). There exists $\delta_1 > 0$ and $\delta_2 > 0$ such that for any $|\tilde{D}| < \delta_1$ and $\max_{x \in [0,1]} |\tilde{b}_i(x)| < \delta_2$ for $i = 1, \cdots, n$, the zero solution of the system $\big(X(t), u(x,t), \hat{u}(x,t)\big)$ is exponentially stable.*

Proof. The invertible backstepping-forwarding transformation is introduced as

$$\hat{w}(x,t) = \hat{u}(x,t) - K(\hat{\beta})e^{A_{cl}(\hat{\beta})\hat{D}(x-1)}Z(t). \qquad (11.145)$$

Differentiating (11.144) with respect to t along (11.52) and (11.140)–(11.141) and using integration by parts in x, we have

$$\dot{Z}(t) = \dot{X}(t) + \hat{D}\int_0^1 \int_0^x e^{-A\hat{D}(x-y)}\hat{\mathscr{B}}(y)dy\hat{u}_t(x,t)dx$$

$$= AX(t) + \int_0^1 \mathscr{B}(x)u(x,t)dx$$

$$\quad + \int_0^1 \int_0^x e^{-A\hat{D}(x-y)}\hat{\mathscr{B}}(y)dy\hat{u}_x(x,t)dx$$

$$= AX(t) + \int_0^1 \mathscr{B}(x)u(x,t)dx$$

$$\quad + \int_0^1 e^{-A\hat{D}(1-y)}\hat{\mathscr{B}}(y)dy\hat{u}(1,t) - \int_0^1 \hat{\mathscr{B}}(x)\hat{u}(x,t)dx$$

$$\quad + \int_0^1 \int_0^x A\hat{D}e^{-A\hat{D}(x-y)}\hat{\mathscr{B}}(y)dy\hat{u}(x,t)dx$$

$$= A_{cl}(\hat{\beta})Z(t) + \int_0^1 \tilde{\mathscr{B}}(x)\hat{u}(x,t)dx + \int_0^1 \mathscr{B}(x)\tilde{u}(x,t)dx$$

$$= A_{cl}(\hat{\beta})Z(t) + \int_0^1 \tilde{\mathscr{B}}(x)K(\hat{\beta})e^{A_{cl}(\hat{\beta})\hat{D}(x-1)}dxZ(t)$$

$$\quad + \int_0^1 \tilde{\mathscr{B}}(x)\hat{w}(x,t)dx + \int_0^1 \mathscr{B}(x)\tilde{u}(x,t)dx. \qquad (11.146)$$

The partial derivatives of (11.145) with respect to x and t are calculated below, respectively,

$$\hat{w}_x(x,t) = \hat{u}_x(x,t) - K(\hat{\beta})e^{A_{cl}(\hat{\beta})\hat{D}(x-1)}A_{cl}(\hat{\beta})\hat{D}Z(t), \qquad (11.147)$$

$$\hat{w}_t(x,t) = \hat{u}_t(x,t) - K(\hat{\beta})e^{A_{cl}(\hat{\beta})\hat{D}(x-1)}$$

$$\times \left(A_{cl}(\hat{\beta})Z(t) + \int_0^1 \tilde{\mathscr{B}}(y)\hat{u}(y,t)dy \right.$$

$$+ \left. \int_0^1 \mathscr{B}(y)\tilde{u}(y,t)dy \right)$$

$$= \hat{u}_t(x,t) - K(\hat{\beta})e^{A_{cl}(\hat{\beta})\hat{D}(x-1)}\left(A_{cl}(\hat{\beta})Z(t) \right.$$

$$+ \int_0^1 \tilde{\mathscr{B}}(y)K(\hat{\beta})e^{A_{cl}(\hat{\beta})\hat{D}(y-1)}dy\, Z(t)$$

$$+ \left. \int_0^1 \tilde{\mathscr{B}}(y)\hat{w}(y,t)dy + \int_0^1 \mathscr{B}(y)\tilde{u}(y,t)dy \right). \qquad (11.148)$$

Multiplying (11.148) with $\hat{D}$ and subtracting (11.147), substituting $x=1$ into (11.145), we get

$$\hat{D}\hat{w}_t(x,t) = \hat{w}_x(x,t) - \hat{D}K(\hat{\beta})e^{A_{cl}(\hat{\beta})\hat{D}(x-1)}$$

$$\times \left(\int_0^1 \tilde{\mathscr{B}}(y)\hat{u}(y,t)dy + \int_0^1 \mathscr{B}(y)\tilde{u}(y,t)dy \right)$$

$$= \hat{w}_x(x,t) - \hat{D}K(\hat{\beta})e^{A_{cl}(\hat{\beta})\hat{D}(x-1)}$$

$$\times \left(\int_0^1 \tilde{\mathscr{B}}(y)\hat{w}(y,t)dy + \int_0^1 \mathscr{B}(y)\tilde{u}(y,t)dy \right.$$

$$+ \left. \int_0^1 \tilde{\mathscr{B}}(y)K(\hat{\beta})e^{A_{cl}(\hat{\beta})\hat{D}(y-1)}dy\, Z(t) \right), \qquad (11.149)$$

$$\hat{w}(1,t) = 0. \qquad (11.150)$$

Further taking the partial derivatives of (11.149) with respect to x, we have

$$\hat{D}\hat{w}_{xt}(x,t) = \hat{w}_{xx}(x,t) - \hat{D}^2 K(\hat{\beta})e^{A_{cl}(\hat{\beta})\hat{D}(x-1)}A_{cl}(\hat{\beta})$$

$$\times \left(\int_0^1 \tilde{\mathscr{B}}(y)\hat{u}(y,t)dy + \int_0^1 \mathscr{B}(y)\tilde{u}(y,t)dy \right)$$

$$= \hat{w}_{xx}(x,t) - \hat{D}^2 K(\hat{\beta})e^{A_{cl}(\hat{\beta})\hat{D}(x-1)}A_{cl}(\hat{\beta})$$

$$\times \left(\int_0^1 \tilde{\mathscr{B}}(y)\hat{w}(y,t)dy + \int_0^1 \mathscr{B}(y)\tilde{u}(y,t)dy \right.$$

$$+ \left. \int_0^1 \tilde{\mathscr{B}}(y)K(\hat{\beta})e^{A_{cl}(\hat{\beta})\hat{D}(y-1)}dy\, Z(t) \right), \qquad (11.151)$$

$$\hat{w}_x(1,t) = \hat{D}K(\hat{\beta})\left(\int_0^1 \tilde{\mathscr{B}}(y)\hat{u}(y,t)dy + \int_0^1 \mathscr{B}(y)\tilde{u}(y,t)dy \right)$$

$$= \hat{D}K(\hat{\beta})\left(\int_0^1 \tilde{\mathcal{B}}(y)\hat{w}(y,t)dy + \int_0^1 \mathcal{B}(y)\tilde{u}(y,t)dy \right.$$

$$\left. + \int_0^1 \tilde{\mathcal{B}}(y)K(\hat{\beta})e^{A_{cl}(\hat{\beta})\hat{D}(y-1)}dyZ(t) \right). \tag{11.152}$$

Combing (11.54)–(11.55) with (11.140)–(11.141), and expressing $\hat{u}_x(x,t)$ in terms of $\hat{w}_x(x,t)$, we get the observer error system

$$D\tilde{u}_t(x,t) = \tilde{u}_x(x,t) - \frac{\tilde{D}}{\hat{D}}\hat{u}_x(x,t), \quad x \in [0,1],$$

$$= \tilde{u}_x(x,t) - \tilde{D}\left(\frac{1}{\hat{D}}\hat{w}_x(x,t) \right.$$

$$\left. + K(\hat{\beta})e^{A_{cl}(\hat{\beta})\hat{D}(x-1)}A_{cl}(\hat{\beta})Z(t) \right), \tag{11.153}$$

$$\tilde{u}(1,t) = 0. \tag{11.154}$$

Remark 11.15. The target ODE-PDE cascade given by the overall $\big(Z(t), \hat{w}(x,t), \tilde{u}(x,t)\big)$-system is obtained in (11.146), (11.149)–(11.150), (11.151)–(11.152), and (11.153)–(11.154). It is crucial to observe the interconnections among each subsystem, which are revealed in figure 11.1. The connections dependent upon $\tilde{D}$ and $\tilde{\mathcal{B}}(x)$ could be treated as "weak" when the estimation errors of delay and delay kernel are small. They disappear when $\hat{D} = D$ and $\hat{\mathcal{B}}(x) = \mathcal{B}(x)$. The connections $\tilde{u}(x,t) \to Z(t)$ and $\tilde{u}(x,t) \to \hat{w}(x,t)$ are "strong" and they are present even when $\hat{D} = D$ and $\hat{\mathcal{B}}(x) = \mathcal{B}(x)$, in which case we have two parallel cascades of exponentially stable subsystems $\tilde{u}$-$\hat{w}$ and $\tilde{u}$-Z. The analysis will capture the fact that the potentially destabilizing connections through $\tilde{D}$ and $\tilde{\mathcal{B}}(x)$ can be suppressed by making $\tilde{D}$ and $\tilde{\mathcal{B}}(x)$ small. Another potential problem is that a connection from $\hat{w}_x$ to $\tilde{u}$ exists. We will deal with it by including an H_1 norm in the stability analysis.

Consider the Lyapunov candidate:

$$V(t) = Z^T(t)P(\hat{\beta})Z(t) + g_1 D \int_0^1 (1+x)\tilde{u}(x,t)^2 dx$$

$$+ g_2\hat{D}\int_0^1 (1+x)\hat{w}(x,t)^2 dx + g_3\hat{D}\int_0^1 (1+x)\hat{w}_x(x,t)^2 dx, \tag{11.155}$$

where $g_1 > 0$, $g_2 > 0$, and $g_3 > 0$ are analysis coefficients. Differentiating $V(t)$ with respect to t along (11.146), (11.149)–(11.150), (11.151)–(11.152), and (11.153)–(11.154), and using integration by parts in x, we have

$$\dot{V}(t) = -Z^T(t)Q(\hat{\beta})Z(t)$$

$$+ 2Z^T(t)P(\hat{\beta})\int_0^1 \tilde{\mathcal{B}}(x)K(\hat{\beta})e^{A_{cl}(\hat{\beta})\hat{D}(x-1)}dxZ(t)$$

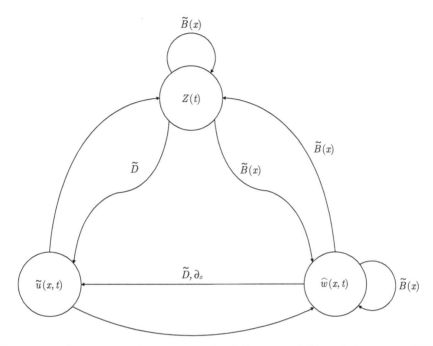

Figure 11.1. Interconnections among the different variables of the target ODE-PDE cascade $(Z, \hat{w}, \tilde{u})$-system

$$+ 2Z^T(t)P(\hat{\beta}) \int_0^1 \tilde{\mathscr{B}}(x)\hat{w}(x,t)dx$$

$$+ 2Z^T(t)P(\hat{\beta}) \int_0^1 \mathscr{B}(x)\tilde{u}(x,t)dx$$

$$- g_1\tilde{u}(0,t)^2 - g_1\|\tilde{u}(x,t)\|^2$$

$$- 2g_1\frac{\tilde{D}}{\hat{D}} \int_0^1 (1+x)\tilde{u}(x,t)\hat{w}_x(x,t)dx$$

$$- 2g_1\tilde{D} \int_0^1 (1+x)\tilde{u}(x,t)K(\hat{\beta})e^{A_{cl}(\hat{\beta})\hat{D}(x-1)}A_{cl}(\hat{\beta})dx\,Z(t)$$

$$- g_2\hat{w}(0,t)^2 - g_2\|\hat{w}(x,t)\|^2$$

$$- 2g_2 \int_0^1 (1+x)\hat{w}(x,t)\hat{D}K(\hat{\beta})e^{A_{cl}(\hat{\beta})\hat{D}(x-1)}dx$$

$$\times \left(\int_0^1 \tilde{\mathscr{B}}(y)\hat{w}(y,t)dy + \int_0^1 \mathscr{B}(y)\tilde{u}(y,t)dy \right.$$

$$\left. + \int_0^1 \tilde{\mathscr{B}}(y)K(\hat{\beta})e^{A_{cl}(\hat{\beta})\hat{D}(y-1)}dy\,Z(t) \right)$$

$$+ 2g_3\hat{w}_x(1,t)^2 - g_3\hat{w}_x(0,t)^2 - g_3\|\hat{w}_x(x,t)\|^2$$

$$- 2g_3 \int_0^1 (1+x)\hat{w}_x(x,t)\hat{D}^2 K(\hat{\beta})e^{A_{cl}(\hat{\beta})\hat{D}(x-1)}A_{cl}(\hat{\beta})dx$$

$$\times \left(\int_0^1 \tilde{\mathscr{B}}(y)\hat{w}(y,t)dy + \int_0^1 \mathscr{B}(y)\tilde{u}(y,t)dy \right.$$

$$\left. + \int_0^1 \tilde{\mathscr{B}}(y)K(\hat{\beta})e^{A_{cl}(\hat{\beta})\hat{D}(y-1)}dy Z(t) \right). \qquad (11.156)$$

Using Cauchy-Schwartz and Young's inequalities, we have

$$\dot{V}(t) \leq -\lambda_{\min}\left(Q(\hat{\beta})\right)|Z(t)|^2 + \delta_2 M_1 |Z(t)|^2$$

$$+ \delta_2 M_2 \left(\gamma |Z(t)|^2 + \frac{1}{\gamma}\|\hat{w}(x,t)\|^2 \right)$$

$$+ M_3 \left(\gamma |Z(t)|^2 + \frac{1}{\gamma}\|\tilde{u}(x,t)\|^2 \right)$$

$$- g_1 \tilde{u}(0,t)^2 - g_1 \|\tilde{u}(x,t)\|^2$$

$$+ g_1 \delta_1 M_4 \left(\gamma \|\tilde{u}(x,t)\|^2 + \frac{1}{\gamma}\|\hat{w}_x(x,t)\|^2 \right)$$

$$+ g_1 \delta_1 M_5 \left(\gamma \|\tilde{u}(x,t)\|^2 + \frac{1}{\gamma}|Z(t)|^2 \right)$$

$$- g_2 \hat{w}(0,t)^2 - g_2 \|\hat{w}(x,t)\|^2 + g_2 \delta_2 M_6 \|\hat{w}(x,t)\|^2$$

$$+ g_2 M_7 \left(\gamma \|\hat{w}(x,t)\|^2 + \frac{1}{\gamma}\|\tilde{u}(x,t)\|^2 \right)$$

$$+ g_2 \delta_2 M_8 \left(\gamma \|\hat{w}(x,t)\|^2 + \frac{1}{\gamma}|Z(t)|^2 \right)$$

$$- g_3 \hat{w}_x(0,t)^2 - g_3 \|\hat{w}_x(x,t)\|^2$$

$$+ g_3 \delta_2 M_9 \left(\gamma \|\hat{w}_x(x,t)\|^2 + \frac{1}{\gamma}\|\hat{w}(x,t)\|^2 \right)$$

$$+ g_3 M_{10} \left(\gamma \|\hat{w}_x(x,t)\|^2 + \frac{1}{\gamma}\|\tilde{u}(x,t)\|^2 \right)$$

$$+ g_3 \delta_2 M_{11} \left(\gamma \|\hat{w}_x(x,t)\|^2 + \frac{1}{\gamma}|Z(t)|^2 \right)$$

$$+ g_3 \delta_2^2 M_{12} \|\hat{w}(x,t)\|^2 + g_3 M_{13} \|\tilde{u}(x,t)\|^2 + g_3 \delta_2^2 M_{14}|Z(t)|^2, \qquad (11.157)$$

where

$$M_1 = 2\left|P(\hat{\beta})\right| \sum_{i=1}^n |B_i| \left|K(\hat{\beta})\right| e^{|A_{cl}(\hat{\beta})|\hat{D}},$$

$$M_2 = \left| P(\hat{\beta}) \right| \sum_{i=1}^{n} |B_i|, \quad M_3 = \left| P(\hat{\beta}) \right| \int_0^1 |\mathscr{B}(x)| \, dx,$$

$$M_4 = \frac{2}{\hat{D}}, \quad M_5 = 2 \left| K(\hat{\beta}) \right| e^{|A_{cl}(\hat{\beta})|\hat{D}} \left| A_{cl}(\hat{\beta}) \right|,$$

$$M_6 = 4\hat{D} \left| K(\hat{\beta}) \right| e^{|A_{cl}(\hat{\beta})|\hat{D}} \sum_{i=1}^{n} |B_i|,$$

$$M_7 = 2\hat{D} \left| K(\hat{\beta}) \right| e^{|A_{cl}(\hat{\beta})|\hat{D}} \int_0^1 |\mathscr{B}(x)| \, dx,$$

$$M_8 = 2\hat{D} \left| K(\hat{\beta}) \right| e^{|A_{cl}(\hat{\beta})|\hat{D}} \sum_{i=1}^{n} |B_i| \left| K(\hat{\beta}) \right| e^{|A_{cl}(\hat{\beta})|\hat{D}},$$

$$M_9 = 2\hat{D}^2 \left| K(\hat{\beta}) \right| e^{|A_{cl}(\hat{\beta})|\hat{D}} \left| A_{cl}(\hat{\beta}) \right| \sum_{i=1}^{n} |B_i|,$$

$$M_{10} = 2\hat{D}^2 \left| K(\hat{\beta}) \right| e^{|A_{cl}(\hat{\beta})|\hat{D}} \left| A_{cl}(\hat{\beta}) \right| \int_0^1 |\mathscr{B}(x)| \, dx,$$

$$M_{11} = 2\hat{D}^2 \left| K(\hat{\beta}) \right| e^{|A_{cl}(\hat{\beta})|\hat{D}} \left| A_{cl}(\hat{\beta}) \right| \sum_{i=1}^{n} |B_i| \left| K(\hat{\beta}) \right| e^{|A_{cl}(\hat{\beta})|\hat{D}},$$

$$M_{12} = 6\hat{D}^2 \left| K(\hat{\beta}) \right|^2 \sum_{i=1}^{n} |B_i|^2,$$

$$M_{13} = 6\hat{D}^2 \left| K(\hat{\beta}) \right|^2 \int_0^1 |\mathscr{B}(x)|^2 \, dx,$$

$$M_{14} = 6\hat{D}^2 \left| K(\hat{\beta}) \right|^2 \sum_{i=1}^{n} |B_i|^2 \left| K(\hat{\beta}) \right|^2 e^{2|A_{cl}(\hat{\beta})|\hat{D}}.$$

Grouping like terms, if the bounds of estimation error δ_1 and δ_2 satisfy

$$M_{15} = \lambda_{\min}\left(Q(\hat{\beta}) \right) - \delta_2 M_1 - \delta_2 M_2 \gamma - M_3 \gamma - \frac{g_1 \delta_1 M_5}{\gamma}$$
$$- \frac{g_2 \delta_2 M_8}{\gamma} - \frac{g_3 \delta_2 M_{11}}{\gamma} - g_3 \delta_2^2 M_{14} > 0, \tag{11.158}$$

$$M_{16} = g_1 - \frac{M_3}{\gamma} - g_1 \delta_1 M_4 \gamma - g_1 \delta_1 M_5 \gamma - \frac{g_2 M_7}{\gamma} - \frac{g_3 M_{10}}{\gamma}$$
$$- g_3 M_{13} > 0, \tag{11.159}$$

$$M_{17} = g_2 - \frac{\delta_2 M_2}{\gamma} - g_2 \delta_2 M_6 - g_2 M_7 \gamma - g_2 \delta_2 M_8 \gamma - \frac{g_3 \delta_2 M_9}{\gamma}$$
$$- g_3 \delta_2^2 M_{12} > 0, \tag{11.160}$$

$$M_{18} = g_3 - \frac{g_1 \delta_1 M_4}{\gamma} - g_3 \delta_2 M_9 \gamma - g_3 M_{10} \gamma - g_3 \delta_2 M_{11} \gamma > 0, \tag{11.161}$$

then we get

$$\dot{V}(t) \leq -M_{15}|Z(t)|^2 - M_{16}\|\tilde{u}(x,t)\|^2 - M_{17}\|\hat{w}(x,t)\|^2$$
$$- M_{18}\|\hat{w}_x(x,t)\|^2. \qquad (11.162)$$

Furthermore, if we make use of inverse transformations (11.144) and (11.145), the theorem is proved. □

11.5 Robust Output Feedback under Unknown Delay, Delay Kernel, and ODE State

In this section, we consider the case of coexistent uncertainty in $(D, B(Dx), X(t))$.

To stabilize (11.52)–(11.55), we have the following assumption.

Assumption 11.16. *The pair (A, C) is detectable. There exists a matrix L to make $A - LC$ Hurwitz such that*

$$(A - LC)^T P_L + P_L(A - LC) = Q_L, \qquad (11.163)$$

with $P_L^T = P_L > 0$ and $Q_L^T = Q_L > 0$.

When both input vector $B(Dx)$ and $X(t)$ are uncertain, the observer canonical form and "virtual" observer with Kreisselmeier-filters in chapter 10 of [89] are required for adaptive control, which highly complicates the design. What's more, the adaptive control problem may not be globally solvable in its most general form, as the relative degree plays an important role in the output feedback. Therefore, similar to section 11.4, we use the linear controller and make use of robustness of the control scheme to the delay and parameter mismatch.

We employ notation (11.49)–(11.55) and denote $\hat{X}(t)$ to be an estimate of the plant state $X(t)$ with estimation error

$$\tilde{X}(t) = X(t) - \hat{X}(t). \qquad (11.164)$$

The ODE-observer-based control scheme is designed as follows:
The observer of plant state is

$$\dot{\hat{X}}(t) = A\hat{X}(t) + \int_0^1 \hat{\mathscr{B}}(x)u(x,t)dx + L(Y(t) - C\hat{X}(t)). \qquad (11.165)$$

The control law is

$$U(t) = u(1,t) = K(\hat{\beta})\hat{Z}(t), \qquad (11.166)$$

where $K(\hat{\beta})$ is chosen to make

$$A_{cl}(\hat{\beta}) = A + \int_0^1 e^{-A\hat{D}(1-x)} \hat{\mathscr{B}}(x) dx K(\hat{\beta})$$

$$= A + \underbrace{\int_0^1 e^{-A\hat{D}(1-x)} \sum_{i=1}^n \hat{b}_i(x) B_i dx}_{\hat{\beta}} K(\hat{\beta}) \qquad (11.167)$$

Hurwitz and

$$\hat{Z}(t) = \hat{X}(t) + \hat{D} \int_0^1 \int_0^x e^{-A\hat{D}(x-y)} \hat{\mathscr{B}}(y) dy u(x, t) dx. \qquad (11.168)$$

Theorem 11.17. *Consider the closed-loop system consisting of the plant (11.52)–(11.55) and the ODE-observer-based control scheme (11.165)–(11.168). There exist $\delta_1 > 0$ and $\delta_2 > 0$ such that for any $|\tilde{D}| < \delta_1$ and $\max_{x \in [0,1]} |\tilde{b}_i(x)| < \delta_2$ for $i = 1, \cdots, n$, the zero solution of the system $(X(t), \hat{X}(t), u(x, t))$ is exponentially stable.*

Proof. If we subtract (11.52) by (11.165), the dynamics of estimation error satisfies

$$\dot{\tilde{X}}(t) = (A - LC)\tilde{X}(t) + \int_0^1 \tilde{\mathscr{B}}(x) u(x, t) dx. \qquad (11.169)$$

Differentiating (11.168) with respect to t along (11.165) and (11.54)–(11.55) and using integration by parts in x, we have

$$\dot{\hat{Z}}(t) = A_{cl}(\hat{\beta})\hat{Z}(t) + LC\tilde{X}(t) + \frac{\tilde{D}}{\hat{D}} \left(\int_0^1 \hat{\mathscr{B}}(x) u(x, t) dx \right.$$

$$- \int_0^1 e^{-A\hat{D}(1-x)} \hat{\mathscr{B}}(x) dx u(1, t)$$

$$\left. - \hat{D} \int_0^1 \int_0^x A e^{-A\hat{D}(x-y)} \hat{\mathscr{B}}(y) dy u(x, t) dx \right). \qquad (11.170)$$

The invertible backstepping-forwarding transformation is introduced as

$$w(x, t) = u(x, t) - K(\hat{\beta}) e^{A_{cl}(\hat{\beta})\hat{D}(x-1)} \hat{Z}(t). \qquad (11.171)$$

Taking partial derivatives of (11.171) with respect to t and x, respectively, we get

$$Dw_t(x,t) = w_x(x,t) - DK(\hat{\beta})e^{A_{cl}(\hat{\beta})\hat{D}(x-1)}LC\tilde{X}(t)$$

$$- \tilde{D}K(\hat{\beta})e^{A_{cl}(\hat{\beta})\hat{D}(x-1)}\left(A_{cl}(\hat{\beta})\hat{Z}(t) \right.$$

$$- \int_0^1 \hat{\mathscr{B}}(x)u(x,t)dx + \int_0^1 e^{-A\hat{D}(1-x)}\hat{\mathscr{B}}(x)dxu(1,t)$$

$$\left. + \hat{D}\int_0^1\int_0^x Ae^{-A\hat{D}(x-y)}\hat{\mathscr{B}}(y)dyu(x,t)dx \right), \qquad (11.172)$$

$$w(1,t) = 0. \qquad\qquad (11.173)$$

Consider the Lyapunov candidate

$$V(t) = D\hat{Z}^T(t)P(\hat{\beta})\hat{Z}(t) + \tilde{X}(t)P_L\tilde{X}(t)$$

$$+ gD\int_0^1 (1+x)w(x,t)^2 dx, \qquad (11.174)$$

where $g > 0$ is an analysis coefficient. Differentiating (11.174) along the target system (11.169), (11.170), and (11.172)–(11.173), utilizing the inverse transformations of (11.168) and (11.171), and following a procedure similar to the proof of Theorem 11.14, we can easily show that $\dot{V}(t)$ is nonpositive. Then Theorem 11.17 is proved. $\qquad\square$

Chapter Twelve

Predictor Feedback of Uncertain Multi-Input Systems

THE MULTI-INPUT LINEAR system with different distributed input delays in section 10.2 is recapped as follows:

$$\dot{X}(t) = AX(t) + \sum_{i=1}^{2} \int_0^{D_i} B_i(\sigma) U_i(t-\sigma) d\sigma, \qquad (12.1)$$

$$Y(t) = CX(t), \qquad (12.2)$$

which can be represented by the following ODE-PDE cascade:

$$\dot{X}(t) = AX(t) + \sum_{i=1}^{2} D_i \int_0^1 B_i(D_i(1-x)) u_i(x,t) dx, \qquad (12.3)$$

$$Y(t) = CX(t), \qquad (12.4)$$

$$D_i \partial_t u_i(x,t) = \partial_x u_i(x,t), \quad x \in [0,1], \quad i = 1,2, \qquad (12.5)$$

$$u_i(1,t) = U_i(t). \qquad (12.6)$$

If we concentrate on the system (12.3)–(12.6), a finite-dimensional multi-input linear system with distributed actuator delays comes with five types of basic uncertainties:

- unknown and distinct delays D_i,
- unknown delay kernels $B_i(D_i(1-x))$,
- unknown system matrix A,
- unmeasurable finite-dimensional plant state $X(t)$,
- unmeasurable infinite-dimensional actuator state $u_i(x,t)$.

Based on the predictor feedback framework for uncertainty-free multi-input systems in section 10.2, in this chapter, we present the predictor feedback for uncertain multi-input systems. We address four combinations of the five uncertainties above. To clearly describe the organization of chapter 12, different combinations of the above uncertainties considered in later sections are summarized in table 12.1.

Table 12.1. Uncertainty Collections of Multi-Input Linear Systems with Distributed Input Delays

Literature	Delays D_i	Delay Kernels $B_i(D(1-x))$	Parameter A	Plant State $X(t)$	Actuator State $u_i(x,t)$
Section 10.2	known	known	known	known	known
Section 12.1	unknown	known	known	known	known
Section 12.2	unknown	unknown	unknown	known	known
Section 12.3	unknown	unknown	known	unknown	known
Section 12.4	unknown	unknown	known	known	unknown

12.1 Adaptive State Feedback under Unknown Delays

As shown in table 12.1, this section solves the stabilization problem when delays D_1, D_2 are unknown. To focus on the delay uncertainty, the delay kernels $B_i(D_i(1-x))$ for $x \in [0,1]$ in (12.3) (i.e., $B_i(\sigma)$ for $\sigma \in [0, D_i]$ in (12.1)) are assumed to be constant vectors independent of $D_i(1-x)$ such that

$$B_i(D_i(1-x)) = B_i(\sigma) = B_i. \tag{12.7}$$

Thus (12.1)–(12.2) is simplified as

$$\dot{X}(t) = AX(t) + \sum_{i=1}^{2} B_i \int_0^{D_i} U_i(t-\sigma)d\sigma, \tag{12.8}$$

$$Y(t) = CX(t), \tag{12.9}$$

and (12.3)–(12.6) is simplified as

$$\dot{X}(t) = AX(t) + \sum_{i=1}^{2} D_i B_i \int_0^1 u_i(x,t)dx, \tag{12.10}$$

$$Y(t) = CX(t), \tag{12.11}$$

$$D_i \partial_t u_i(x,t) = \partial_x u_i(x,t), \quad x \in [0,1], \quad i=1,2, \tag{12.12}$$

$$u_i(1,t) = U_i(t). \tag{12.13}$$

Remark 12.1. On the basis of the framework in section 10.2, this section uses the certainty-equivalence-based adaptive control to stabilize the ODE-PDE cascade (12.10)–(12.13). The "certainty-equivalence" refers to the time-varying estimates used to replace the unknown delays in (10.40), (10.42), and (10.45), whereas the "adaptive" refers to the delay update laws that are Lyapunov-based to cancel the estimation error terms in the time derivative of the Lyapunov function. A nontrivial problem is to deal with the unknown D_i^2 in (10.40) and (10.45). One intuitive method is to estimate D_i^2 by $\hat{D}_i(t)^2$, where $\hat{D}_i(t)$ is an estimate of D_i. The other possible method is to treat D_i^2 as a whole such that $\eta_i = D_i^2$ and to estimate η_i by $\hat{\eta}_i(t)$. However, both of the above methods produce a mismatch term for the update law design. The available method is to regard D_i^2 as a product of two parameters such that $D_i^2 = D_i \cdot D_{ii}$, where $D_{ii} = D_i$, and estimate D_i and D_{ii} with different update laws, respectively.

To use projector operators for update laws later, the following assumption is made.

Assumption 12.2. *There exist known constants $\underline{D}$ and $\overline{D}$ such that*

$$0 < \underline{D} \le D_1 = D_{11} \le \overline{D}, \tag{12.14}$$

$$0 < \underline{D} \le D_2 = D_{22} \le \overline{D}. \tag{12.15}$$

In order to stabilize (12.10), the following assumption is needed.

Assumption 12.3. *For any $D_1 = D_{11}, D_2 = D_{22}$ satisfying (12.14)–(12.15), the pair $(A, (B_{D_1}, B_{D_2}))$ is stabilizable where*

$$B_{D_i} = D_i \int_0^1 e^{AD_i(\tau-1)} B_i d\tau = D_{ii} \int_0^1 e^{AD_i(\tau-1)} B_i d\tau. \tag{12.16}$$

For $i = 1, 2$, we denote $\hat{D}_i(t)$ and $\hat{D}_{ii}(t)$ as the estimates of D_i and D_{ii}, respectively, and the estimation errors accordingly satisfy

$$\tilde{D}_i(t) = D_i - \hat{D}_i(t), \tag{12.17}$$

$$\tilde{D}_{ii}(t) = D_{ii} - \hat{D}_{ii}(t) = D_i - \hat{D}_{ii}(t). \tag{12.18}$$

The delay-adaptive control scheme is listed as follows:
    The control laws are

$$U_i(t) = u_i(1, t) = \hat{K}_i(t) e^{A\hat{D}_i(t)} Z(t), \quad i = 1, 2, \tag{12.19}$$

where

$$Z(t) = X(t) + \sum_{i=1}^2 \hat{D}_i(t) \hat{D}_{ii}(t) \int_0^1 \int_0^y e^{A\hat{D}_i(t)(\tau-y)} B_i d\tau u_i(y, t) dy, \tag{12.20}$$

and $\hat{K}_1(t)$, $\hat{K}_2(t)$ are chosen to make

$$\hat{A}_{cl}(t) = A + \sum_{i=1}^{2} \underbrace{\hat{D}_{ii}(t) \int_0^1 e^{A\hat{D}_i(t)(\tau-1)} B_i d\tau \, \hat{K}_i(t) e^{A\hat{D}_i(t)}}_{\hat{B}_{D_i}(t)} \tag{12.21}$$

Hurwitz, so that

$$\hat{A}_{cl}^T(t)\hat{P}(t) + \hat{P}(t)\hat{A}_{cl}(t) = -\hat{Q}(t), \tag{12.22}$$

with $\hat{P}(t) = \hat{P}(t)^T > 0$ and $\hat{Q}(t) = \hat{Q}(t)^T > 0$. Please note that $\hat{B}_{D_1}(t)$, $\hat{K}_1(t)$ are functions of $(\hat{D}_1(t), \hat{D}_{11}(t))$, and $\hat{B}_{D_2}(t)$, $\hat{K}_2(t)$ are functions of $(\hat{D}_2(t), \hat{D}_{22}(t))$, while $\hat{A}_{cl}(t)$, $\hat{P}(t)$, $\hat{Q}(t)$ are functions of $(\hat{D}_1(t), \hat{D}_{11}(t), \hat{D}_2(t), \hat{D}_{22}(t))$.

The update laws are

$$\dot{\hat{D}}_1(t) = \gamma_D \text{Proj}_{[\underline{D},\overline{D}]}\{\tau_{D_1}(t)\}, \quad \gamma_D > 0, \tag{12.23}$$

$$\tau_{D_1}(t) = \frac{Z^T(t)\hat{P}(t)f_{D_1}(t)}{1 + \Xi(t)}$$

$$- \frac{g \int_0^1 (1+x)w_1(x,t)\hat{K}_1(t)e^{A\hat{D}_1(t)x}dx h_{D_1}(t)}{1 + \Xi(t)}$$

$$- \frac{g \int_0^1 (1+x)w_2(x,t)\hat{K}_2(t)e^{A\hat{D}_2(t)x}dx f_{D_1}(t)}{1 + \Xi(t)}, \tag{12.24}$$

$$\dot{\hat{D}}_2(t) = \gamma_D \text{Proj}_{[\underline{D},\overline{D}]}\{\tau_{D_2}(t)\}, \quad \gamma_D > 0, \tag{12.25}$$

$$\tau_{D_2}(t) = \frac{Z^T(t)\hat{P}(t)f_{D_2}(t)}{1 + \Xi(t)}$$

$$- \frac{g \int_0^1 (1+x)w_1(x,t)\hat{K}_1(t)e^{A\hat{D}_1(t)x}dx f_{D_2}(t)}{1 + \Xi(t)}$$

$$- \frac{g \int_0^1 (1+x)w_2(x,t)\hat{K}_2(t)e^{A\hat{D}_2(t)x}dx h_{D_2}(t)}{1 + \Xi(t)}, \tag{12.26}$$

$$\dot{\hat{D}}_{ii}(t) = \gamma_D \text{Proj}_{[\underline{D},\overline{D}]}\{\tau_{D_{ii}}(t)\}, \quad \gamma_D > 0, \quad i = 1, 2, \tag{12.27}$$

$$\tau_{D_{ii}}(t) = \frac{Z^T(t)\hat{P}(t)\int_0^1 B_i u_i(y,t)dy}{1 + \Xi(t)}$$

$$- \frac{g \sum_{j=1}^{2} \int_0^1 (1+x)w_j(x,t)\hat{K}_j(t)e^{A\hat{D}_j(t)x}dx \int_0^1 B_i u_i(y,t)dy}{1 + \Xi(t)}, \tag{12.28}$$

where

$$w_i(x,t) = u_i(x,t) - \hat{K}_i(t)e^{A\hat{D}_i(t)x}Z(t)$$

$$+ \hat{D}_i(t)\hat{D}_{ii}(t)\int_x^1\int_0^1 \hat{K}_i(t)e^{A\hat{D}_i(t)(x-y+\tau)}B_id\tau u_i(y,t)dy, \quad (12.29)$$

$$\Xi(t) = Z^T(t)\hat{P}(t)Z(t) + g\sum_{i=1}^2\int_0^1 (1+x)w_i(x,t)^2dx, \quad (12.30)$$

$$f_{D_i}(t) = \hat{D}_{ii}(t)\int_0^1 B_iu_i(y,t)dy - \hat{D}_{ii}(t)\int_0^1 e^{A\hat{D}_i(t)(\tau-1)}B_id\tau u_i(1,t)$$

$$- \hat{D}_i(t)\hat{D}_{ii}(t)\int_0^1\int_0^y Ae^{A\hat{D}_i(t)(\tau-y)}B_id\tau u_i(y,t)dy, \quad (12.31)$$

$$h_{D_i}(t) = f_{D_i}(t) + AZ(t) + \hat{B}_{D_i}(t)u_i(1,t), \quad (12.32)$$

and $\text{Proj}_{[\underline{D},\overline{D}]}(\cdot)$ is a standard projector operator like in that appendix E of [89].

Theorem 12.4. *Consider the closed-loop system consisting of the plant (12.10)–(12.13) and the adaptive controller (12.19)–(12.32). All the states $(X(t), u_1(x,t), u_2(x,t), \hat{D}_1(t), \hat{D}_2(t), \hat{D}_{11}(t), \hat{D}_{22}(t))$ of the closed-loop system are globally bounded and the regulation of $X(t)$ and $U_i(t)$ such that $\lim_{t\to\infty} X(t) = \lim_{t\to\infty} U_i(t) = 0$ is achieved.*

Proof. Calculating the time derivative of (12.20) along (12.10) and (12.12)–(12.13), using the integration by parts in y and the equality $\frac{\hat{D}_i(t)}{D_i} = 1 - \frac{\tilde{D}_i(t)}{D_i}$, we have

$$\dot{Z}(t) = \dot{X}(t) + \sum_{i=1}^2 \frac{\hat{D}_i(t)}{D_i}\hat{D}_{ii}(t)\int_0^1\int_0^y e^{A\hat{D}_i(t)(\tau-y)}B_id\tau\partial_y u_i(y,t)dy$$

$$+ \sum_{i=1}^2 \phi_i\big(\dot{\hat{D}}_i(t), \dot{\hat{D}}_{ii}(t)\big)$$

$$= AX(t) + \sum_{i=1}^2 D_i\int_0^1 B_iu_i(x,t)dx + \sum_{i=1}^2 \phi_i\big(\dot{\hat{D}}_i(t), \dot{\hat{D}}_{ii}(t)\big)$$

$$+ \sum_{i=1}^2 \left(1 - \frac{\tilde{D}_i(t)}{D_i}\right)\hat{D}_{ii}(t)\int_0^1 e^{A\hat{D}_i(t)(\tau-1)}B_id\tau u_i(1,t)$$

$$- \sum_{i=1}^2 \left(1 - \frac{\tilde{D}_i(t)}{D_i}\right)\hat{D}_{ii}(t)\int_0^1 B_iu_i(y,t)dy$$

$$+ \sum_{i=1}^2 \left(1 - \frac{\tilde{D}_i(t)}{D_i}\right)\hat{D}_{ii}(t)\int_0^1\int_0^y A\hat{D}_i(t)e^{A\hat{D}_i(t)(\tau-y)}B_id\tau u_i(y,t)dy$$

$$= AZ(t) + \sum_{i=1}^{2} \hat{D}_{ii}(t) \int_0^1 e^{A\hat{D}_i(t)(\tau-1)} B_i d\tau U_i(t) + \sum_{i=1}^{2} \phi_i\big(\dot{\hat{D}}_i(t), \dot{\hat{D}}_{ii}(t)\big)$$

$$+ \sum_{i=1}^{2} \tilde{D}_{ii}(t) \int_0^1 B_i u_i(y,t) dy + \sum_{i=1}^{2} \frac{\tilde{D}_i(t)}{D_i} \left[\hat{D}_{ii}(t) \int_0^1 B_i u_i(y,t) dy \right.$$

$$- \hat{D}_{ii}(t) \int_0^1 e^{A\hat{D}_i(t)(\tau-1)} B_i d\tau u_i(1,t)$$

$$\left. - \hat{D}_i(t)\hat{D}_{ii}(t) \int_0^1 \int_0^y A e^{A\hat{D}_i(t)(\tau-y)} B_i d\tau u_i(y,t) dy \right]. \tag{12.33}$$

Substituting the control law (12.19) into (12.33), we have

$$\dot{Z}(t) = \left(A + \hat{B}_{D_1}(t)\hat{K}_1(t)e^{A\hat{D}_1(t)} + \hat{B}_{D_2}(t)\hat{K}_2(t)e^{A\hat{D}_2(t)} \right) Z(t)$$

$$+ \tilde{D}_{11}(t) \int_0^1 B_1 u_1(y,t) dy + \tilde{D}_{22}(t) \int_0^1 B_2 u_2(y,t) dy$$

$$+ \frac{\tilde{D}_1(t)}{D_1} f_{D_1}(t) + \frac{\tilde{D}_2(t)}{D_2} f_{D_2}(t) + \phi_1\big(\dot{\hat{D}}_1(t), \dot{\hat{D}}_{11}(t)\big) + \phi_2\big(\dot{\hat{D}}_2(t), \dot{\hat{D}}_{22}(t)\big), \tag{12.34}$$

where $f_{D_1}(t), f_{D_2}(t)$ have been defined in (12.31) and

$$\phi_i\big(\dot{\hat{D}}_i(t), \dot{\hat{D}}_{ii}(t)\big) = \int_0^1 \int_0^y \left(\dot{\hat{D}}_i(t)\hat{D}_{ii}(t) + \hat{D}_i(t)\dot{\hat{D}}_{ii}(t) \right.$$

$$\left. + \hat{D}_i(t)\hat{D}_{ii}(t)A\dot{\hat{D}}_i(t)(\tau-y) \right) e^{A\hat{D}_i(t)(\tau-y)} B_i d\tau u_i(y,t) dy. \tag{12.35}$$

Taking the partial derivatives of (12.29) with respect to x and t along (12.12)–(12.13) and (12.34), respectively, and using the integration by parts in y, we get

$$\partial_x w_i(x,t) = \partial_x u_i(x,t) - \hat{K}_i(t)e^{A\hat{D}_i(t)x} A\hat{D}_i(t)Z(t)$$

$$- \hat{D}_i(t)\hat{D}_{ii}(t) \int_0^1 \hat{K}_i(t)e^{A\hat{D}_i(t)\tau} B_i d\tau u_i(x,t)$$

$$+ \hat{D}_i(t)\hat{D}_{ii}(t) \int_x^1 \int_0^1 \hat{K}_i(t)A\hat{D}_i(t)e^{A\hat{D}_i(t)(x-y+\tau)} B_i d\tau u_i(y,t) dy, \tag{12.36}$$

$$\partial_t w_i(x,t) = \partial_t u_i(x,t) + \psi_i\big(\dot{\hat{D}}_1(t), \dot{\hat{D}}_{11}(t), \dot{\hat{D}}_2(t), \dot{\hat{D}}_{22}(t)\big)$$

$$- \hat{K}_i(t)e^{A\hat{D}_i(t)x} \left(\left(A + \sum_{j=1}^{2} \hat{B}_{D_j}(t)\hat{K}_j(t)e^{A\hat{D}_j(t)} \right) Z(t) \right.$$

$$+ \sum_{j=1}^{2} \tilde{D}_{jj}(t) \int_0^1 B_j u_j(y,t) dy + \sum_{j=1}^{2} \frac{\tilde{D}_j(t)}{D_j} f_{D_j}(t) \Bigg)$$

$$+ \frac{\hat{D}_i(t)}{D_i} \hat{D}_{ii}(t) \int_0^1 \hat{K}_i(t) e^{A\hat{D}_i(t)(x-1+\tau)} B_i d\tau u_i(1,t)$$

$$- \frac{\hat{D}_i(t)}{D_i} \hat{D}_{ii}(t) \int_0^1 \hat{K}_i(t) e^{A\hat{D}_i(t)\tau} B_i d\tau u_i(x,t)$$

$$+ \frac{\hat{D}_i(t)}{D_i} \hat{D}_{ii}(t) \int_x^1 \int_0^1 \hat{K}_i(t) A\hat{D}_i(t) e^{A\hat{D}_i(t)(x-y+\tau)} B_i d\tau u_i(y,t) dy,$$

$$(12.37)$$

where

$$\psi_i\big(\dot{\hat{D}}_1(t), \dot{\hat{D}}_{11}(t), \dot{\hat{D}}_2(t), \dot{\hat{D}}_{22}(t)\big)$$

$$= -\left(\dot{\hat{K}}_i(t) + \hat{K}_i(t) A\dot{\hat{D}}_i(t)x\right) e^{A\hat{D}_i(t)x} Z(t)$$

$$+ \int_x^1 \int_0^1 \left(\left(\dot{\hat{D}}_i(t)\hat{D}_{ii}(t) + \hat{D}_i(t)\dot{\hat{D}}_{ii}(t)\right) \hat{K}_i(t) \right.$$

$$\left. + \hat{D}_i(t)\hat{D}_{ii}(t) \left(\dot{\hat{K}}_i(t) + K_i(t) A\dot{\hat{D}}_i(t)(x-y+\tau)\right) \right)$$

$$\times e^{A\hat{D}_i(t)(x-y+\tau)} B_i d\tau u_i(y,t) dy - \hat{K}_i(t) e^{A\hat{D}_i(t)x} \sum_{j=1}^{2} \phi_j\big(\dot{\hat{D}}_j(t), \dot{\hat{D}}_{jj}(t)\big),$$

$$(12.38)$$

with $\dot{\hat{K}}_i(t)$ being a function of $\big(\dot{\hat{D}}_i(t), \dot{\hat{D}}_{ii}(t)\big)$. Multiplying (12.37) with D_i and subtracting (12.36) and substituting $x=1$ into (12.29), we get

$$D_1 \partial_x w_1(x,t) = \partial_t w_1(x,t) + D_1 \psi_1\big(\dot{\hat{D}}_1(t), \dot{\hat{D}}_{11}(t), \dot{\hat{D}}_2(t), \dot{\hat{D}}_{22}(t)\big)$$

$$- D_1 \hat{K}_1(t) e^{A\hat{D}_1(t)x} \hat{B}_{D_2}(t) \hat{K}_2(t) e^{A\hat{D}_2(t)} Z(t)$$

$$- D_1 \hat{K}_1(t) e^{A\hat{D}_1(t)x} \left(\frac{\tilde{D}_1(t)}{D_1} h_{D_1}(t) + \frac{\tilde{D}_2(t)}{D_2} f_{D_2}(t) \right.$$

$$\left. + \tilde{D}_{11}(t) \int_0^1 B_1 u_1(y,t) dy + \tilde{D}_{22}(t) \int_0^1 B_2 u_2(y,t) dy \right), \quad (12.39)$$

$$w_1(1,t) = 0, \quad x \in [0,1], \qquad (12.40)$$

$$D_2 \partial_x w_2(x,t) = \partial_t w_2(x,t) + D_2 \psi_2\big(\dot{\hat{D}}_1(t), \dot{\hat{D}}_{11}(t), \dot{\hat{D}}_2(t), \dot{\hat{D}}_{22}(t)\big)$$

$$- D_2 \hat{K}_2(t) e^{A\hat{D}_2(t)x} \hat{B}_{D_1}(t) \hat{K}_1(t) e^{A\hat{D}_1(t)} Z(t)$$

$$- D_2 \hat{K}_2(t) e^{A\hat{D}_2(t)x} \left(\frac{\tilde{D}_2(t)}{D_2} h_{D_2}(t) + \frac{\tilde{D}_1(t)}{D_1} f_{D_1}(t) \right.$$

$$\left. + \tilde{D}_{11}(t) \int_0^1 B_1 u_1(y,t) dy + \tilde{D}_{22}(t) \int_0^1 B_2 u_2(y,t) dy \right), \quad (12.41)$$

$$w_2(1,t) = 0, \quad x \in [0,1], \quad (12.42)$$

where $h_{D_1}(t), h_{D_2}(t)$ have been defined in (12.32). Consider the Lyapunov candidate as follows:

$$V(t) = D_1 D_2 \log (1 + \Xi(t)) + \frac{D_2}{\gamma_D} \tilde{D}_1(t)^2 + \frac{D_1}{\gamma_D} \tilde{D}_2(t)^2$$

$$+ \frac{D_1 D_2}{\gamma_D} \tilde{D}_{11}(t)^2 + \frac{D_1 D_2}{\gamma_D} \tilde{D}_{22}(t)^2, \quad (12.43)$$

where $\Xi(t)$ has been defined in (12.30). Taking the time derivative of (12.43) along the target closed-loop system (12.34) and (12.39)–(12.42), we have

$$\dot{V}(t) = \frac{1}{1 + \Xi(t)} \left[- D_1 D_2 Z(t)^T \hat{Q}(t) Z(t) + D_1 D_2 Z(t)^T \dot{\hat{P}}(t) Z(t) \right.$$

$$+ 2D_1 D_2 Z(t)^T \hat{P}(t) \sum_{i=1}^{2} \phi_i \left(\dot{\hat{D}}_i(t), \dot{\hat{D}}_{ii}(t) \right)$$

$$+ 2D_1 D_2 \tilde{D}_{11}(t) Z(t)^T \hat{P}(t) \int_0^1 B_1 u_1(y,t) dy$$

$$+ 2D_1 D_2 \tilde{D}_{22}(t) Z(t)^T \hat{P}(t) \int_0^1 B_2 u_2(y,t) dy$$

$$+ 2D_2 \tilde{D}_1(t) Z(t)^T \hat{P}(t) f_{D_1}(t) + 2D_1 \tilde{D}_2(t) Z(t)^T \hat{P}(t) f_{D_2}(t)$$

$$- g D_2 w_1(0,t)^2 - g D_2 \|w_1(x,t)\|^2$$

$$+ 2g D_1 D_2 \int_0^1 (1+x) w_1(x,t) dx \psi_1 \left(\dot{\hat{D}}_1(t), \dot{\hat{D}}_{11}(t), \dot{\hat{D}}_2(t), \dot{\hat{D}}_{22}(t) \right)$$

$$- 2g D_1 D_2 \int_0^1 (1+x) w_1(x,t) \hat{K}_1(t) e^{A\hat{D}_1(t)x} dx \hat{B}_{D_2}(t) \hat{K}_2(t) e^{A\hat{D}_2(t)} Z(t)$$

$$- 2g D_2 \tilde{D}_1(t) \int_0^1 (1+x) w_1(x,t) \hat{K}_1(t) e^{A\hat{D}_1(t)x} dx h_{D_1}(t)$$

$$- 2g D_1 \tilde{D}_2(t) \int_0^1 (1+x) w_1(x,t) \hat{K}_1(t) e^{A\hat{D}_1(t)x} dx f_{D_2}(t)$$

$$- 2g D_1 D_2 \tilde{D}_{11}(t) \int_0^1 (1+x) w_1(x,t) \hat{K}_1(t) e^{A\hat{D}_1(t)x} dx \int_0^1 B_1 u_1(y,t) dy$$

$$
\begin{aligned}
&- 2gD_1D_2\tilde{D}_{22}(t)\int_0^1 (1+x)w_1(x,t)\hat{K}_1(t)e^{A\hat{D}_1(t)x}dx\int_0^1 B_2u_2(y,t)dy \\
&- gD_1w_2(0,t)^2 - gD_1\|w_2(x,t)\|^2 \\
&+ 2gD_1D_2\int_0^1 (1+x)w_2(x,t)dx\psi_2\big(\dot{\hat{D}}_1(t),\dot{\hat{D}}_{11}(t),\dot{\hat{D}}_2(t),\dot{\hat{D}}_{22}(t)\big) \\
&- 2gD_1D_2\int_0^1 (1+x)w_2(x,t)\hat{K}_2(t)e^{A\hat{D}_2(t)x}dx\hat{B}_{D_1}(t)\hat{K}_1(t)e^{A\hat{D}_1(t)}Z(t) \\
&- 2gD_1\tilde{D}_2(t)\int_0^1 (1+x)w_2(x,t)\hat{K}_2(t)e^{A\hat{D}_2(t)x}dxh_{D_2}(t) \\
&- 2gD_2\tilde{D}_1(t)\int_0^1 (1+x)w_2(x,t)\hat{K}_2(t)e^{A\hat{D}_2(t)x}dxf_{D_1}(t) \\
&- 2gD_1D_2\tilde{D}_{11}(t)\int_0^1 (1+x)w_2(x,t)\hat{K}_2(t)e^{A\hat{D}_2(t)x}dx\int_0^1 B_1u_1(y,t)dy \\
&\left.- 2gD_1D_2\tilde{D}_{22}(t)\int_0^1 (1+x)w_2(x,t)\hat{K}_2(t)e^{A\hat{D}_2(t)x}dx\int_0^1 B_2u_2(y,t)dy\right] \\
&- \frac{2D_2}{\gamma_D}\tilde{D}_1(t)\dot{\hat{D}}_1(t) - \frac{2D_1}{\gamma_D}\tilde{D}_2(t)\dot{\hat{D}}_2(t) \\
&- \frac{2D_1D_2}{\gamma_D}\tilde{D}_{11}(t)\dot{\hat{D}}_{11}(t) - \frac{2D_1D_2}{\gamma_D}\tilde{D}_{22}(t)\dot{\hat{D}}_{22}(t).
\end{aligned}
\tag{12.44}
$$

The inverse forwarding-backstepping transformation of (12.29) is introduced as

$$
\begin{aligned}
u_i(x,t) =\ & w_i(x,t) + \hat{K}_i(t)e^{\big(A+\hat{B}_{ie}(t)\hat{K}_i(t)\big)\hat{D}_i(t)(x-1)}e^{A\hat{D}_i(t)}Z(t) \\
&- \hat{D}_i(t)\int_x^1 \hat{K}_i(t)e^{\big(A+\hat{B}_{ie}(t)\hat{K}_i(t)\big)\hat{D}_i(t)(x-y)}\hat{B}_{ie}(t)w_i(y,t)dy,
\end{aligned}
\tag{12.45}
$$

where $\hat{B}_{ie}(t) = e^{A\hat{D}_i(t)}\hat{B}_{D_i}(t) = \hat{D}_{ii}(t)\int_0^1 e^{A\hat{D}_i(t)\tau}B_id\tau$.

Please note that $\hat{D}_i(t)$ and $\hat{D}_{ii}(t)$ stay in the interval (12.14)–(12.15) by the projector operators. If we make use of Young's and Cauchy-Schwarz inequalities, it is evident that the inverse transformation of (12.45) implies

$$
\|u_i(x,t)\|^2 \le M_u\left(|Z(t)|^2 + \|w_i(x,t)\|^2\right),
\tag{12.46}
$$

where $M_u > 0$ is a constant.

Utilizing (12.46) and inequalities $0 < \frac{|\epsilon|}{1+\epsilon^2} < 1$ and $0 < \frac{\epsilon^2}{1+\epsilon^2} < 1$, we can easily show that

$$
\left|\dot{\hat{D}}_i(t)\right| \le \gamma_D M_{D_i}\frac{|Z(t)|^2 + \sum_{j=1}^2 \|w_j(x,t)\|^2}{1+\Xi(t)} \le \gamma_D\bar{M}_{D_i},
\tag{12.47}
$$

$$\left|\dot{D}_{ii}(t)\right| \leq \gamma_D M_{D_{ii}} \frac{|Z(t)|^2 + \sum_{j=1}^{2} \|w_j(x,t)\|^2}{1+\Xi(t)} \leq \gamma_D \bar{M}_{D_{ii}}, \qquad (12.48)$$

where M_{D_i}, $M_{D_{ii}}$, $\bar{M}_{D_i}$, $\bar{M}_{D_{ii}}$ are positive constants.

If we make use of (12.46)–(12.48) and group like terms, (12.44) becomes

$$\dot{V}(t) \leq \frac{1}{1+\Xi(t)} \left[-\underline{D}^2 \inf_{t\geq0} \lambda_{\min}(Q(t))|Z(t)|^2 - g\underline{D}\sum_{i=1}^{2}\left(w_i(0,t)^2 + \|w_i(x,t)\|^2\right) \right.$$
$$\left. + \gamma_D \overline{D}^2 M\left(|Z(t)|^2 + \sum_{i=1}^{2}\|w_i(x,t)\|^2\right)\right]$$
$$- \frac{2D_2}{\gamma_D}\tilde{D}_1(t)\left(\dot{D}_1(t) - \gamma_D \tau_{D_1}(t)\right) - \frac{2D_1}{\gamma_D}\tilde{D}_2(t)\left(\dot{D}_2(t) - \gamma_D \tau_{D_2}(t)\right)$$
$$- \frac{2D_1 D_2}{\gamma_D}\tilde{D}_{11}(t)\left(\dot{D}_{11}(t) - \gamma_D \tau_{D_{11}}(t)\right)$$
$$- \frac{2D_1 D_2}{\gamma_D}\tilde{D}_{22}(t)\left(\dot{D}_{22}(t) - \gamma_D \tau_{D_{22}}(t)\right), \qquad (12.49)$$

where $M>0$ is a constant. By using the update laws in (12.23)–(12.28) and carefully selecting design coefficients $\lambda_{\min}(Q(t))$, g, and γ_D, we easily get

$$\dot{V}(t) \leq \frac{N}{1+\Xi(t)}\left[-|Z(t)|^2 - \sum_{i=1}^{2}w_i(0,t)^2 - \sum_{i=1}^{2}\|w_i(x,t)\|^2\right], \qquad (12.50)$$

where $N>0$ is a constant. Thus $Z(t)$, $w_1(x,t)$, and $w_2(x,t)$ are bounded and converge to zero (by Barbalats lemma in appendix A of [89] or an alternative version of Barbalat's lemma in Lemma 3.1 of [101]). By invertible conversions of (12.20), (12.29), and (12.45), the original states $X(t)$ and $u_1(x,t)$ and $u_2(x,t)$ are bounded and converge to zero. Then the proof of Theorem 12.4 is completed. $\square$

12.2 Adaptive State Feedback under Uncertain Delays, Delay Kernels, and Parameters

As shown in table 12.1, in this section we consider a much more challenging case where the delays D_i, the delay kernels $B_i(D_i(1-x))$, and the system matrix A are simultaneously uncertain.

The system matrix A contains unknown elements and is of the form

$$A = A_0 + \sum_{i=1}^{p}\theta_i A_i, \qquad (12.51)$$

where $\theta_i \in \mathbb{R}$ for $i = 1, \cdots, p$ are unknown constant plant parameters and $A_i \in \mathbb{R}^{n \times n}$ for $i = 0, \cdots, p$ are known constant matrices.

Example 12.5. A three-dimensional example of (12.51) is offered as follows:

$$A = \begin{bmatrix} a_{11} & a_{12} & a_{13} \\ a_{21} & a_{22} & a_{23} \\ a_{31} & a_{32} & a_{33} \end{bmatrix} = \begin{bmatrix} a_{11} & 0 & a_{13} \\ a_{21} & a_{22} & 0 \\ a_{31} & a_{32} & a_{33} \end{bmatrix} + a_{12} \begin{bmatrix} 0 & 1 & 0 \\ 0 & 0 & 0 \\ 0 & 0 & 0 \end{bmatrix} + a_{23} \begin{bmatrix} 0 & 0 & 0 \\ 0 & 0 & 1 \\ 0 & 0 & 0 \end{bmatrix},$$

$$\tag{12.52}$$

where $a_{ij} \in \mathbb{R}$ for $i, j \in \{1, 2, 3\}$ are the elements of the system matrix A, among which $\theta_1 = a_{12}$ and $\theta_2 = a_{23}$ are unknown.

The delay kernels $B_i(D_i(1-x))$ for $x \in [0, 1]$ and $i = 1, 2$ in (12.3) are n-dimensional continuous vector-valued functions of $D_i(1-x)$ such that

$$B_i(D_i(1-x)) = \begin{bmatrix} \rho_{1i}(D_i(1-x)) \\ \rho_{2i}(D_i(1-x)) \\ \vdots \\ \rho_{ni}(D_i(1-x)) \end{bmatrix} = \sum_{j=1}^{n} \rho_{ji}(D_i(1-x))B_{ji}, \tag{12.53}$$

where $\rho_{ji}(D_i(1-x)) : [0, 1] \to \mathbb{R}$ for $j = 1, \cdots, n$ are unknown components of $B_i(D_i(1-x))$, and $B_{ji} \in \mathbb{R}^n$ for $j = 1, \cdots, n$ are the unit vectors accordingly.

For notational brevity, we further denote

$$\mathscr{B}_i(x) = D_i B_i(D_i(1-x)) = \sum_{j=1}^{n} D_i \rho_{ji}(D_i(1-x))B_{ji} = \sum_{j=1}^{n} b_{ji}(x)B_{ji}, \quad i = 1, 2,$$

$$\tag{12.54}$$

where $b_{ji}(x) = D_i \rho_{ji}(D_i(1-x))$ for $j = 1, \cdots, n$ are unknown scalar continuous functions of x.

Thus the system (12.3)–(12.6) is rewritten as follows:

$$\dot{X}(t) = AX(t) + \sum_{i=1}^{2} \int_0^1 \mathscr{B}_i(x)u_i(x, t)dx, \tag{12.55}$$

$$Y(t) = CX(t), \tag{12.56}$$

$$D_i \partial_t u_i(x, t) = \partial_x u_i(x, t), \quad x \in [0, 1], \quad i = 1, 2, \tag{12.57}$$

$$u_i(1, t) = U_i(t). \tag{12.58}$$

To use projector operators for update laws later, the following assumption is made.

Assumption 12.6. *There exist known constants $\underline{D}$, $\overline{D}$, $\bar{\theta}$, $\bar{b}$, a known constant vector θ^*, and a known continuous function $b^*(x)$ such that*

$$0 < \underline{D} \leq D_1, D_2 \leq \overline{D}, \tag{12.59}$$

$$|\theta - \theta^*| \leq \bar{\theta}, \quad \theta = [\theta_1, \cdots, \theta_p]^T, \tag{12.60}$$

$$\int_0^1 (b_{ji}(x) - b^*(x))^2 \, dx \leq \bar{b}, \quad i = 1, 2, j = 1, \cdots, n. \tag{12.61}$$

Assumption 12.7. *For any D_i, θ_i, $b_{ji}(x)$ satisfying (12.59)–(12.61), the pair $\left(A, (B_{D_1}, B_{D_2})\right)$ is stabilizable where*

$$B_{D_i} = \int_0^1 e^{AD_i(\tau-1)} \mathscr{B}_i(\tau) d\tau. \tag{12.62}$$

Denote $\hat{D}_i(t)$, $\hat{\theta}_i(t)$, and $\hat{b}_{ji}(x,t)$ as the estimates of D_i, θ_i, and $b_{ji}(x)$ with estimation errors satisfying

$$\tilde{D}_i(t) = D_i - \hat{D}_i(t), \quad i = 1, 2, \tag{12.63}$$

$$\tilde{\theta}_i(t) = \theta_i - \hat{\theta}_i(t), \quad i = 1, \cdots, p, \tag{12.64}$$

$$\tilde{b}_{ji}(x,t) = b_{ji}(x) - \hat{b}_{ji}(x,t), \quad i = 1, 2, \quad j = 1, \cdots, n, \tag{12.65}$$

and accordingly

$$\hat{A}(t) = A_0 + \sum_{i=1}^p \hat{\theta}_i(t) A_i, \quad \tilde{A}(t) = \sum_{i=1}^p \tilde{\theta}_i(t) A_i, \tag{12.66}$$

$$\hat{\mathscr{B}}_i(x,t) = \sum_{j=1}^n \hat{b}_{ji}(x,t) B_{ji}, \quad \tilde{\mathscr{B}}_i(x,t) = \sum_{j=1}^n \tilde{b}_{ji}(x,t) B_{ji}, \tag{12.67}$$

$$\dot{\hat{A}}(t) = \sum_{i=1}^p \dot{\hat{\theta}}_i(t) A_i, \quad \dot{\hat{\mathscr{B}}}_i(x,t) = \sum_{j=1}^n \dot{\hat{b}}_{ji}(x,t) B_{ji}. \tag{12.68}$$

The adaptive control scheme is listed as follows:
The control laws are

$$U_i(t) = u_i(1,t) = \hat{K}_i(t) e^{\hat{A}(t)\hat{D}_i(t)} Z(t), \quad i = 1, 2, \tag{12.69}$$

where

$$Z(t) = X(t) + \sum_{i=1}^{2} \hat{D}_i(t) \int_0^1 \int_0^y e^{\hat{A}(t)\hat{D}_i(t)(\tau - y)} \hat{\mathscr{B}}_i(\tau, t) d\tau u_i(y, t) dy \qquad (12.70)$$

and $\hat{K}_1(t), \hat{K}_2(t)$ are chosen to make

$$\hat{A}_{cl}(t) = \hat{A}(t) + \sum_{i=1}^{2} \underbrace{\int_0^1 e^{\hat{A}(t)\hat{D}_i(t)(\tau - 1)} \hat{\mathscr{B}}_i(\tau, t) d\tau \, \hat{K}_i(t) e^{\hat{A}(t)\hat{D}_i(t)}}_{\hat{B}_{D_i}(t)} \qquad (12.71)$$

Hurwitz, so that

$$\hat{A}_{cl}^T(t)\hat{P}(t) + \hat{P}(t)\hat{A}_{cl}(t) = -\hat{Q}(t), \qquad (12.72)$$

with $\hat{P}(t) = \hat{P}(t)^T > 0$ and $\hat{Q}(t) = \hat{Q}(t)^T > 0$.

Please note that $\hat{B}_{D_1}(t)$, $\hat{K}_1(t)$ are functions of $(\hat{D}_1(t), \hat{A}(t), \hat{\mathscr{B}}_1(\tau, t))$, and $\hat{B}_{D_2}(t)$, $\hat{K}_2(t)$ are functions of $(\hat{D}_2(t), \hat{A}(t), \hat{\mathscr{B}}_2(\tau, t))$, while $\hat{A}_{cl}(t)$, $\hat{P}(t)$, $\hat{Q}(t)$ are functions of $(\hat{D}_1(t), \hat{D}_2(t), \hat{A}(t), \hat{\mathscr{B}}_1(\tau, t), \hat{\mathscr{B}}_2(\tau, t))$.

The update laws are

$$\dot{\hat{D}}_1(t) = \gamma_D \mathrm{Proj}\{\tau_{D_1}(t)\}, \quad \gamma_D > 0, \qquad (12.73)$$

$$\tau_{D_1}(t) = \frac{Z^T(t)\hat{P}(t)f_{D_1}(t)}{1 + \Xi(t)}$$

$$- \frac{g \int_0^1 (1+x)w_1(x, t)\hat{K}_1(t)e^{\hat{A}(t)\hat{D}_1(t)x} dx h_{D_1}(t)}{1 + \Xi(t)}$$

$$- \frac{g \int_0^1 (1+x)w_2(x, t)\hat{K}_2(t)e^{\hat{A}(t)\hat{D}_2(t)x} dx f_{D_1}(t)}{1 + \Xi(t)}, \qquad (12.74)$$

$$\dot{\hat{D}}_2(t) = \gamma_D \mathrm{Proj}\{\tau_{D_2}(t)\}, \quad \gamma_D > 0, \qquad (12.75)$$

$$\tau_{D_2}(t) = \frac{Z^T(t)\hat{P}(t)f_{D_2}(t)}{1 + \Xi(t)}$$

$$- \frac{g \int_0^1 (1+x)w_1(x, t)\hat{K}_1(t)e^{\hat{A}(t)\hat{D}_1(t)x} dx f_{D_2}(t)}{1 + \Xi(t)}$$

$$- \frac{g \int_0^1 (1+x)w_2(x, t)\hat{K}_2(t)e^{\hat{A}(t)\hat{D}_2(t)x} dx h_{D_2}(t)}{1 + \Xi(t)}, \qquad (12.76)$$

$$\dot{\hat{\theta}}_i(t) = \gamma_\theta \mathrm{Proj}\{\tau_{\theta_i}(t)\}, \quad \gamma_\theta > 0, \quad i = 1, \cdots, p \qquad (12.77)$$

$$\tau_{\theta_i}(t) = \frac{Z^T(t)\hat{P}(t)A_iX(t)}{1+\Xi(t)}$$

$$-\frac{g\sum_{j=1}^{2}\int_0^1(1+x)w_j(x,t)\hat{K}_j(t)e^{\hat{A}(t)\hat{D}_j(t)x}dxA_iX(t)}{1+\Xi(t)}, \qquad (12.78)$$

$$\dot{b}_{ji}(x,t) = \gamma_b\text{Proj}\{\tau_{b_{ji}}(x,t)\}, \quad \gamma_b>0, \quad i=1,2, \quad j=1,\cdots,n, \qquad (12.79)$$

$$\tau_{b_{ji}}(x,t) = \frac{Z^T(t)\hat{P}(t)B_{ji}u_i(x,t)}{1+\Xi(t)}$$

$$-\frac{g\sum_{j=1}^{2}\int_0^1(1+y)w_j(y,t)\hat{K}_j(t)e^{\hat{A}(t)\hat{D}_j(t)y}dyB_{ji}u_i(x,t)}{1+\Xi(t)}, \qquad (12.80)$$

where

$$w_i(x,t) = u_i(x,t) - \hat{K}_i(t)e^{\hat{A}(t)\hat{D}_i(t)x}Z(t)$$

$$+\hat{D}_i(t)\int_x^1\int_0^1\hat{K}_i(t)e^{\hat{A}(t)\hat{D}_i(t)(x-y+\tau)}\hat{\mathscr{B}}_i(\tau,t)d\tau u_i(y,t)dy, \qquad (12.81)$$

$$\Xi(t) = Z^T(t)\hat{P}(t)Z(t) + g\sum_{i=1}^{2}\int_0^1(1+x)w_i(x,t)^2dx, \qquad (12.82)$$

$$f_{D_i}(t) = \int_0^1\hat{\mathscr{B}}_i(y,t)u_i(y,t)dy - \int_0^1 e^{\hat{A}(t)\hat{D}_i(t)(\tau-1)}\hat{\mathscr{B}}_i(\tau,t)d\tau u_i(1,t)$$

$$-\hat{D}_i(t)\int_0^1\int_0^y\hat{A}(t)e^{\hat{A}(t)\hat{D}_i(t)(\tau-y)}\hat{\mathscr{B}}_i(\tau,t)d\tau u_i(y,t)dy, \qquad (12.83)$$

$$h_{D_i}(t) = f_{D_i}(t) + \hat{A}(t)Z(t) + \hat{B}_{D_i}(t)u_i(1,t), \qquad (12.84)$$

and $\text{Proj}\{\tau_{D_1}(t)\}$, $\text{Proj}\{\tau_{D_2}(t)\}$, $\text{Proj}\{\tau_{\theta_i}(t)\}$, $\text{Proj}\{\tau_{b_{ji}}(x,t)\}$ are standard ODE and PDE projector operators defined in (12.59)–(12.61) and are like those in [89] and [122].

Theorem 12.8. *Consider the closed-loop system consisting of the plant (12.55)–(12.58) and the adaptive controller (12.69)–(12.84). All the states $\left(X(t),\, u_i(x,t),\, \hat{D}_1(t),\, \hat{D}_2(t),\, \hat{\theta}_i(t),\, \hat{b}_{ji}(x,t)\right)$ of the closed-loop system are globally bounded and the regulation of $X(t)$ and $U_i(t)$ such that $\lim_{t\to\infty}X(t)=\lim_{t\to\infty}U_i(t)=0$ is achieved.*

Proof. Calculating the time derivative of (12.70) along (12.55) and (12.57)–(12.58), using the integration by parts in y and the equality $\frac{\hat{D}_i(t)}{D_i} = 1 - \frac{\tilde{D}_i(t)}{D_i}$, we have

$$\dot{Z}(t) = \left(\hat{A}(t) + \hat{B}_{D_1}(t)\hat{K}_1(t)e^{\hat{A}(t)\hat{D}_1(t)} + \hat{B}_{D_2}(t)\hat{K}_2(t)e^{\hat{A}(t)\hat{D}_2(t)}\right)Z(t)$$

$$+ \int_0^1 \tilde{\mathscr{B}}_1(y,t) u_1(y,t) dy + \int_0^1 \tilde{\mathscr{B}}_2(y,t) u_2(y,t) dy$$

$$+ \tilde{A}(t) X(t) + \frac{\tilde{D}_1(t)}{D_1} f_{D_1}(t) + \frac{\tilde{D}_2(t)}{D_2} f_{D_2}(t)$$

$$+ \phi_1\big(\dot{\hat{D}}_1(t), \dot{\hat{A}}(t), \dot{\hat{\mathscr{B}}}_1(y,t)\big) + \phi_2\big(\dot{\hat{D}}_2(t), \dot{\hat{A}}(t), \dot{\hat{\mathscr{B}}}_2(y,t)\big), \qquad (12.85)$$

where $f_{D_1}(t), f_{D_2}(t)$ have been defined in (12.83) and

$$\phi_i\big(\dot{\hat{D}}_i(t), \dot{\hat{A}}(t), \dot{\hat{\mathscr{B}}}_i(y,t)\big) = \int_0^1 \int_0^y \bigg(\dot{\hat{D}}_i(t) + \hat{D}_i(t)$$

$$\times \Big(\dot{\hat{A}}(t)\hat{D}_i(t) + \hat{A}(t)\dot{\hat{D}}_i(t) \Big) (\tau - y) \bigg) e^{\hat{A}(t)\hat{D}_i(t)(\tau - y)} \dot{\hat{\mathscr{B}}}_i(\tau,t) d\tau u_i(y,t) dy$$

$$+ \hat{D}_i(t) \int_0^1 \int_0^y e^{\hat{A}(t)\hat{D}_i(t)(\tau - y)} \dot{\hat{\mathscr{B}}}_i(\tau,t) d\tau u_i(y,t) dy. \qquad (12.86)$$

Taking the partial derivatives of (12.81) with respect to x and t along (12.57)–(12.58) and (12.85), using the integration by parts in y, we get

$$D_1 \partial_x w_1(x,t) = \partial_t w_1(x,t) + D_1 \psi_1\big(\dot{\hat{D}}(t), \dot{\hat{A}}(t), \dot{\hat{\mathscr{B}}}(\tau,t)\big)$$

$$- D_1 \hat{K}_1(t) e^{\hat{A}(t)\hat{D}_1(t)x} \hat{B}_{D_2}(t) \hat{K}_2(t) e^{\hat{A}(t)\hat{D}_2(t)} Z(t)$$

$$- D_1 \hat{K}_1(t) e^{\hat{A}(t)\hat{D}_1(t)x} \bigg(\frac{\tilde{D}_1(t)}{D_1} h_{D_1}(t) + \frac{\tilde{D}_2(t)}{D_2} f_{D_2}(t) + \tilde{A}(t) X(t)$$

$$+ \int_0^1 \tilde{\mathscr{B}}_1(y,t) u_1(y,t) dy + \int_0^1 \tilde{\mathscr{B}}_2(y,t) u_2(y,t) dy \bigg), \qquad (12.87)$$

$$w_1(1,t) = 0, \quad x \in [0,1], \qquad (12.88)$$

$$D_2 \partial_x w_2(x,t) = \partial_t w_2(x,t) + D_2 \psi_2\big(\dot{\hat{D}}(t), \dot{\hat{A}}(t), \dot{\hat{\mathscr{B}}}(\tau,t)\big)$$

$$- D_2 \hat{K}_2(t) e^{\hat{A}(t)\hat{D}_2(t)x} \hat{B}_{D_1}(t) \hat{K}_1(t) e^{\hat{A}(t)\hat{D}_1(t)} Z(t)$$

$$- D_2 \hat{K}_2(t) e^{\hat{A}(t)\hat{D}_2(t)x} \bigg(\frac{\tilde{D}_2(t)}{D_2} h_{D_2}(t) + \frac{\tilde{D}_1(t)}{D_1} f_{D_1}(t) + \tilde{A}(t) X(t)$$

$$+ \int_0^1 \tilde{\mathscr{B}}_1(y,t) u_1(y,t) dy + \int_0^1 \tilde{\mathscr{B}}_2(y,t) u_2(y,t) dy \bigg), \qquad (12.89)$$

$$w_2(1,t) = 0, \quad x \in [0,1], \qquad (12.90)$$

where $h_{D_1}(t), h_{D_2}(t)$ have been defined in (12.84), and

$$\psi_i\big(\dot{\hat{D}}(t), \dot{\hat{A}}(t), \dot{\hat{\mathscr{B}}}(\tau,t)\big)$$

$$= - \Big(\dot{\hat{K}}_i(t) + \hat{K}_i(t) \Big(\dot{\hat{A}}(t)\hat{D}_i(t) + \hat{A}(t)\dot{\hat{D}}_i(t) \Big) x \Big) e^{\hat{A}(t)\hat{D}_i(t)x} Z(t)$$

$$+ \int_x^1 \int_0^1 \left(\dot{\hat{D}}_i(t)\hat{K}_i(t) + \hat{D}_i(t)\left(\dot{\hat{K}}_i(t) \right. \right.$$

$$\left. + K_i(t)\left(\dot{\hat{A}}(t)\hat{D}_i(t) + \hat{A}(t)\dot{\hat{D}}_i(t) \right)(x - y + \tau) \right) \right)$$

$$\times e^{\hat{A}(t)\hat{D}_i(t)(x-y+\tau)}\dot{\hat{\mathscr{B}}}_i(\tau, t)d\tau u_i(y, t)dy$$

$$+ \hat{D}_i(t) \int_x^1 \int_0^1 \hat{K}_i(t)e^{\hat{A}(t)\hat{D}_i(t)(x-y+\tau)}\dot{\hat{\mathscr{B}}}_i(\tau, t)d\tau u_i(y, t)dy$$

$$- \hat{K}_i(t)e^{A\hat{D}_i(t)x}\sum_{j=1}^2 \phi_j \left(\dot{\hat{D}}_j(t), \dot{\hat{A}}(t), \dot{\hat{\mathscr{B}}}_j(y, t) \right). \tag{12.91}$$

The inverse forwarding-backstepping transformation of (12.81) is introduced as

$$u_i(x, t) = w_i(x, t) + \hat{K}_i(t)e^{\left(\hat{A}(t) + \hat{B}_{ie}(t)\hat{K}_i(t) \right)\hat{D}_i(t)(x-1)}e^{\hat{A}(t)\hat{D}_i(t)}Z(t)$$

$$- \hat{D}_i(t) \int_x^1 \hat{K}_i(t)e^{\left(\hat{A}(t) + \hat{B}_{ie}(t)\hat{K}_i(t) \right)\hat{D}_i(t)(x-y)}\hat{B}_{ie}(t)w_i(y, t)dy, \tag{12.92}$$

where $\hat{B}_{ie}(t) = e^{\hat{A}(t)\hat{D}_i(t)}\hat{B}_{D_i}(t) = \int_0^1 e^{\hat{A}(t)\hat{D}_i(t)\tau}\hat{\mathscr{B}}_i(\tau, t)d\tau$.

Consider the Lyapunov candidate as follows:

$$V(t) = D_1 D_2 \log\left(1 + \Xi(t)\right) + \frac{D_2}{\gamma_D}\tilde{D}_1(t)^2 + \frac{D_1}{\gamma_D}\tilde{D}_2(t)^2$$

$$+ \frac{D_1 D_2}{\gamma_\theta}\sum_{i=1}^p \tilde{\theta}_i(t)^2 + \frac{D_1 D_2}{\gamma_b}\sum_{i=1}^2\sum_{j=1}^n \int_0^1 \tilde{b}_{ji}(x, t)^2 dx, \tag{12.93}$$

where $\Xi(t)$ has been defined in (12.82).

Taking the time derivative of (12.93) along the target closed-loop system (12.85) and (12.87)–(12.90), following a similar procedure of the proof of Theorem 12.4, we can straightforwardly show that $\dot{V}(t) \le 0$ and Theorem 12.8 is proved. □

12.3 Robust Output Feedback under Unknown Delays, Delay Kernels, and ODE States

In this section, as shown in table 12.1, we consider the case of coexistent uncertainty in $\left(D_i, B_i(D_i(1 - x)), X(t) \right)$.

To stabilize (12.55)–(12.58), we have the following assumption.

Assumption 12.9. *The pair (A, C) is detectable. There exists a matrix L to make $A - LC$ Hurwitz such that*

$$(A - LC)^T P_L + P_L (A - LC) = Q_L, \tag{12.94}$$

with $P_L^T = P_L > 0$ and $Q_L^T = Q_L > 0$.

When both input vector $B_i(D_i(1 - x))$ and plant state $X(t)$ are uncertain at the same time, the observer canonical form and "virtual" observer with Kreisselmeier-filters in chapter 10 of [89] are required for adaptive control, which highly complicates the design. Besides, the adaptive control problem may not be globally solvable in its most general form, as the relative degree plays an important role in the output feedback. Therefore, we use the linear controller and rely on the robustness of the feedback law to small errors in the assumed values of delays and delay kernels.

We employ notation (12.53)–(12.58) and denote $\hat{D}_i$ and $\hat{b}_{ji}(x)$ as the constant estimates of D_i and $b_{ji}(x)$ with estimation errors satisfying

$$\tilde{D}_i = D_i - \hat{D}_i, \quad i = 1, 2, \tag{12.95}$$

$$\tilde{b}_{ji}(x) = b_{ji}(x) - \hat{b}_{ji}(x), \quad i = 1, 2, \quad j = 1, \cdots, n, \tag{12.96}$$

and accordingly

$$\hat{\mathscr{B}}_i(x) = \sum_{j=1}^{n} \hat{b}_{ji}(x) B_{ji}, \quad \tilde{\mathscr{B}}_i(x) = \sum_{j=1}^{n} \tilde{b}_{ji}(x) B_{ji}. \tag{12.97}$$

Denote $\hat{X}(t)$ to be an estimate of the plant state $X(t)$ with estimation error

$$\tilde{X}(t) = X(t) - \hat{X}(t). \tag{12.98}$$

The ODE-observer-based control scheme is designed as follows:
The observer of the plant state is

$$\dot{\hat{X}}(t) = A\hat{X}(t) + \sum_{i=1}^{2} \int_0^1 \hat{\mathscr{B}}_i(x) u_i(x, t) dx + L(Y(t) - C\hat{X}(t)). \tag{12.99}$$

The control law is

$$U(t) = u(1, t) = \hat{K}_i e^{A\hat{D}_i} \hat{Z}(t), \tag{12.100}$$

where

$$\hat{Z}(t) = \hat{X}(t) + \sum_{i=1}^{2} \hat{D}_i \int_0^1 \int_0^y e^{A\hat{D}_i(\tau - y)} \hat{\mathscr{B}}_i(\tau) d\tau u_i(y, t) dy, \tag{12.101}$$

and $\hat{K}_1$ and $\hat{K}_2$ are chosen to make

$$\hat{A}_{cl} = A + \sum_{i=1}^{2} \underbrace{\int_0^1 e^{A\hat{D}_i(\tau-1)}\mathscr{B}_i(\tau)d\tau}_{\hat{B}_{D_i}} \hat{K}_i e^{A\hat{D}_i} \tag{12.102}$$

be Hurwitz, so that

$$\hat{A}_{cl}^T \hat{P} + \hat{P}\hat{A}_{cl} = -\hat{Q}, \tag{12.103}$$

with $\hat{P} = \hat{P}^T > 0$ and $\hat{Q} = \hat{Q}^T > 0$.

Theorem 12.10. *Consider the closed-loop system consisting of the plant (12.55)–(12.58) and the ODE-observer-based control scheme (12.99)–(12.103). There exist $\delta_1 > 0$ and $\delta_2 > 0$ such that for any $|\tilde{D}_i| < \delta_1$ for $i = 1, 2$ and $\max_{x \in [0,1]} |\tilde{b}_{ji}(x)| < \delta_2$ for $i = 1, 2$ and $j = 1, \cdots, n$, the zero solution of the system $(X(t), \hat{X}(t), u_i(x,t))$ is exponentially stable.*

Proof. Consider the invertible forwarding-backstepping transformation as follows:

$$w_i(x,t) = u_i(x,t) - \hat{K}_i e^{A\hat{D}_i x}\hat{Z}(t) + \hat{D}_i \int_x^1 \hat{K}_i e^{A\hat{D}_i(x-y)}\hat{B}_{ie}u_i(y,t)dy$$

$$= u_i(x,t) - \hat{K}_i e^{A\hat{D}_i x}\hat{Z}(t) + \hat{D}_i \int_x^1 \int_0^1 \hat{K}_i e^{A\hat{D}_i(x-y+\tau)}\mathscr{B}_i(\tau)d\tau u_i(y,t)dy, \tag{12.104}$$

$$u_i(x,t) = w_i(x,t) + \hat{K}_i e^{(A+\hat{B}_{ie}\hat{K}_i)\hat{D}_i(x-1)}e^{A\hat{D}_i}\hat{Z}(t)$$

$$- \hat{D}_i \int_x^1 \hat{K}_i e^{(A+\hat{B}_{ie}\hat{K}_i)\hat{D}_i(x-y)}\hat{B}_{ie}w_i(y,t)dy, \tag{12.105}$$

where $\hat{B}_{ie} = e^{A\hat{D}_i}\hat{B}_{D_i} = \int_0^1 e^{A\hat{D}_i \tau}\mathscr{B}_i(\tau)d\tau$.

Calculating the time-derivative of the ODE-observer error (12.98) along (12.55) and (12.99), we have

$$\dot{\tilde{X}}(t) = (A - LC)\tilde{X}(t) + \sum_{i=1}^{2}\int_0^1 \tilde{\mathscr{B}}_i(x)u_i(x,t)dx$$

$$= (A - LC)\tilde{X}(t) + \sum_{i=1}^{2}\int_0^1 \tilde{\mathscr{B}}_i(x)w_i(x,t)dx$$

$$+ \sum_{i=1}^{2} \int_0^1 \tilde{\mathscr{B}}_i(x) \hat{K}_i e^{(A+\hat{B}_{ie}\hat{K}_i)\hat{D}_i(x-1)} e^{A\hat{D}_i} dx \hat{Z}(t)$$

$$- \sum_{i=1}^{2} \int_0^1 \tilde{\mathscr{B}}_i(x) \hat{D}_i \int_x^1 \hat{K}_i e^{(A+\hat{B}_{ie}\hat{K}_i)\hat{D}_i(x-y)} \hat{B}_{ie} w_i(y,t) dy dx. \quad (12.106)$$

Calculating the time derivative of (12.101) along (12.99) and (12.57)–(12.58), we have

$$\dot{\hat{Z}}(t) = \left(A + \hat{B}_{D_1} \hat{K}_1 e^{A\hat{D}_1} + \hat{B}_{D_2} \hat{K}_2 e^{A\hat{D}_2} \right) \hat{Z}(t) + LC\tilde{X}(t)$$

$$+ \sum_{i=1}^{2} \frac{\tilde{D}_i}{D_i} \left(\int_0^1 \hat{\mathscr{B}}_i(y) u_i(y,t) dy - \int_0^1 e^{A\hat{D}_i(\tau-1)} \hat{\mathscr{B}}_i(\tau) d\tau u_i(1,t) \right.$$

$$\left. - \hat{D}_i \int_0^1 \int_0^y A e^{A\hat{D}_i(\tau-y)} \hat{\mathscr{B}}_i(\tau) d\tau u_i(y,t) dy \right). \quad (12.107)$$

Substituting (12.105) into (12.107), we have

$$\dot{\hat{Z}}(t) = \hat{A}_{cl} \hat{Z}(t) + LC\tilde{X}(t) + \sum_{i=1}^{2} \frac{\tilde{D}_i}{D_i} f_i(\hat{Z}(t), w_i(y,t)), \quad (12.108)$$

where

$$f_i(\hat{Z}(t), w_i(y,t)) = \left(\int_0^1 \hat{\mathscr{B}}_i(y) \hat{K}_i e^{(A+\hat{B}_{ie}\hat{K}_i)\hat{D}_i(y-1)} e^{A\hat{D}_i} dy \hat{Z}(t) \right.$$

$$- \hat{B}_{D_i} \hat{K}_i e^{A\hat{D}_i} \hat{Z}(t) - \hat{D}_i \int_0^1 \int_0^y A e^{A\hat{D}_i(\tau-y)} \hat{\mathscr{B}}_i(\tau) d\tau$$

$$\times \hat{K}_i e^{(A+\hat{B}_{ie}\hat{K}_i)\hat{D}_i(y-1)} e^{A\hat{D}_i} dy \hat{Z}(t) + \int_0^1 \hat{\mathscr{B}}_i(y) w_i(y,t) dy$$

$$- \int_0^1 \hat{\mathscr{B}}_i(y) \hat{D}_i \int_y^1 \hat{K}_i e^{(A+\hat{B}_{ie}\hat{K}_i)\hat{D}_i(y-s)} \hat{B}_{ie} w_i(s,t) ds dy$$

$$- \hat{D}_i \int_0^1 \int_0^y A e^{A\hat{D}_i(\tau-y)} \hat{\mathscr{B}}_i(\tau) d\tau w_i(y,t) dy - \hat{D}_i \int_0^1 \int_0^y A e^{A\hat{D}_i(\tau-y)} \hat{\mathscr{B}}_i(\tau) d\tau$$

$$\times \hat{D}_i \int_y^1 \hat{K}_i e^{(A+\hat{B}_{ie}\hat{K}_i)\hat{D}_i(y-s)} \hat{B}_{ie} w_i(s,t) ds dy \right).$$

Calculating the partial derivatives of (12.104) with respect to x and t along (12.108) and (12.57)–(12.58), we get

$$D_1 \partial_x w_1(x,t) = \partial_t w_1(x,t) - D_1 \hat{K}_1 e^{A\hat{D}_1 x} \hat{B}_{D_2} \hat{K}_2 e^{A\hat{D}_2} \hat{Z}(t)$$

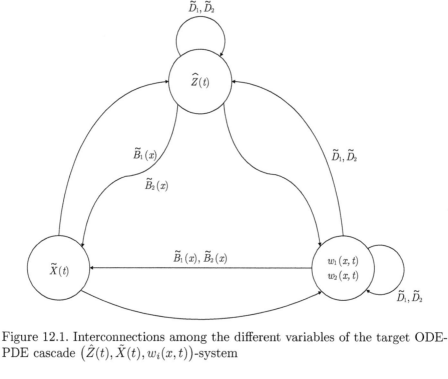

Figure 12.1. Interconnections among the different variables of the target ODE-PDE cascade $\left(\hat{Z}(t), \tilde{X}(t), w_i(x,t)\right)$-system

$$-D_1 \hat{K}_1 e^{A\hat{D}_1 x}\left(\frac{\tilde{D}_1}{D_1}\left(A\hat{Z}(t)+\hat{B}_{D_1}\hat{K}_1 e^{A\hat{D}_1 x}\hat{Z}(t)+f_1(\hat{Z}(t), w_1(y,t))\right)\right.$$

$$\left.+\frac{\tilde{D}_2}{D_2}f_2(\hat{Z}(t), w_2(y,t))\right)-D_1\hat{K}_1 e^{A\hat{D}_1 x}LC\tilde{X}(t), \qquad (12.109)$$

$$w_1(1,t)=0, \quad x\in[0,1], \qquad (12.110)$$

$$D_2\partial_x w_2(x,t)=\partial_t w_2(x,t)-D_2\hat{K}_2 e^{A\hat{D}_2 x}\hat{B}_{D_1}\hat{K}_1 e^{A\hat{D}_1}\hat{Z}(t)$$

$$-D_2\hat{K}_2 e^{A\hat{D}_2 x}\left(\frac{\tilde{D}_2}{D_2}\left(A\hat{Z}(t)+\hat{B}_{D_2}\hat{K}_2 e^{A\hat{D}_2 x}\hat{Z}(t)+f_2(\hat{Z}(t), w_2(y,t))\right)\right.$$

$$\left.+\frac{\tilde{D}_1}{D_1}f_1(\hat{Z}(t), w_1(y,t))\right)-D_2\hat{K}_2 e^{A\hat{D}_2 x}LC\tilde{X}(t), \qquad (12.111)$$

$$w_2(1,t)=0, \quad x\in[0,1]. \qquad (12.112)$$

Remark 12.11. The target ODE-PDE cascade given by the overall $\left(\hat{Z}(t), \tilde{X}(t), w_i(x,t)\right)$-system is obtained in (12.106), (12.108), and (12.109)–(12.112). It is crucial to observe the interconnections among the subsystems, which are revealed in figure 12.1. The connections dependent upon $\tilde{D}_i$ and $\tilde{\mathscr{B}}_i(x)$ could be treated as "weak" when the estimation errors of delays and delay kernels are small.

They disappear when $\hat{D}_i = D_i$ and $\hat{\mathscr{B}}_i(x) = \mathscr{B}_i(x)$. The connections $\tilde{X}(t) \to \hat{Z}(t)$, $\tilde{X}(t) \to w_i(x,t)$, and $\hat{Z}(t) \to w_i(x,t)$ are "strong" and they are present even when $\hat{D}_i = D_i$ and $\hat{\mathscr{B}}_i(x) = \mathscr{B}_i(x)$, in which case we have three parallel cascades of exponentially stable subsystems. The analysis should capture the fact that the potentially destabilizing connections through $\tilde{D}_i$ and $\tilde{\mathscr{B}}_i(x)$ can be suppressed by making $\tilde{D}_i$ and $\tilde{\mathscr{B}}_i(x)$ small.

Consider the Lyapunov candidate

$$V(t) = D_1 D_2 \hat{Z}^T(t)\hat{P}\hat{Z}(t) + D_1 D_2 \tilde{X}(t) P_L \tilde{X}(t)$$

$$+ g_1 D_1 D_2 \int_0^1 (1+x)w_1(x,t)^2 dx + g_2 D_1 D_2 \int_0^1 (1+x)w_2(x,t)^2 dx,$$

$$\tag{12.113}$$

where $g_1, g_2 > 0$ are analysis coefficients.

Differentiating (12.113) along the target system (12.106), (12.108), and (12.109)–(12.112) and utilizing Cauchy-Schwarz and Young's inequalities, we have

$$\dot{V}(t) \le -\overline{D}^2 \lambda_{\min}(\hat{Q})|\hat{Z}(t)|^2 - \overline{D}^2 \lambda_{\min}(Q_L)|\tilde{X}(t)|^2$$

$$- \sum_{i=1}^{2} \overline{D}\left(g_i \|w_i(x,t)\|^2 + g_i w_i(0,t)^2\right)$$

$$+ \sum_{i=1}^{2} |\tilde{D}_i| M_1\left(|\hat{Z}(t)|^2 + |\tilde{X}(t)|^2 + \|w_1(x,t)\|^2 + \|w_2(x,t)\|^2\right)$$

$$+ \sum_{i=1}^{2}\sum_{j=1}^{n} \max_{x\in[0,1]} |\tilde{b}_{ji}(x)| M_2\left(|\hat{Z}(t)|^2 + |\tilde{X}(t)|^2 + \|w_1(x,t)\|^2 + \|w_2(x,t)\|^2\right),$$

$$\tag{12.114}$$

where M_1, M_2 are positive constants.

From (12.114), as long as $|\tilde{D}_i| < \delta_1$ for $i = 1, 2$ and $\max_{x\in[0,1]} |\tilde{b}_{ji}(x)| < \delta_2$ for $i = 1, 2$ and $j = 1, \cdots, n$, where $\delta_1 > 0$ and $\delta_2 > 0$, it is not hard to show that $\dot{V}(t) \le -\mu V(t)$ with $\mu > 0$ being a constant. Then Theorem 12.10 is proved. $\square$

12.4 Robust Output Feedback under Unknown Delays, Delay Kernels, and PDE States

In this section, as shown in table 12.1, we consider the case of coexistent uncertainty in $\left(D_i, B_i(D_i(1-x)), u_i(x,t)\right)$.

As analyzed in chapter 8 of [80], when the actuator states $u_i(x,t)$ are unmeasurable and the delay values D_i are simultaneously unknown, the adaptive control can only achieve local stabilization both in the initial state and in the initial parameter error. In other words, the initial delay estimates need to be sufficiently close to the true delays. As a result, instead of adaptive control, in this section we employ the linear controller and rely on robustness of the feedback law to small errors in the assumed values of delays and delay kernels.

We employ the notation (12.53)–(12.58) and (12.95)–(12.97). Furthermore, we employ

$$\hat{u}_i(x,t) = U_i(t + \hat{D}_i(x-1)), \quad x \in [0,1], \quad i = 1, 2, \tag{12.115}$$

to be estimates of the actuator states $u_i(x,t)$, with the estimation error being

$$\tilde{u}_i(x,t) = u_i(x,t) - \hat{u}_i(x,t), \quad x \in [0,1]. \tag{12.116}$$

The PDE-observer-based control scheme is designed as follows: The observers for the actuator states in different control channels are

$$\hat{D}_i \partial_t \hat{u}_i(x,t) = \partial_x \hat{u}_i(x,t), \quad x \in [0,1], \tag{12.117}$$

$$\hat{u}_i(1,t) = U_i(t), \quad i = 1, 2. \tag{12.118}$$

The control law is

$$U_i(t) = \hat{u}_i(1,t) = \hat{K}_i e^{A\hat{D}_i} Z(t), \tag{12.119}$$

where

$$Z(t) = X(t) + \sum_{i=1}^{2} \hat{D}_i \int_0^1 \int_0^y e^{A\hat{D}_i(\tau-y)} \hat{\mathscr{B}}_i(\tau) d\tau \hat{u}_i(y,t) dy, \tag{12.120}$$

and $\hat{K}_1$ and $\hat{K}_2$ are chosen to make

$$\hat{A}_{cl} = A + \sum_{i=1}^{2} \underbrace{\int_0^1 e^{A\hat{D}_i(\tau-1)} \hat{\mathscr{B}}_i(\tau) d\tau \, \hat{K}_i e^{A\hat{D}_i}}_{\hat{B}_{D_i}} \tag{12.121}$$

Hurwitz, so that

$$\hat{A}_{cl}^T \hat{P} + \hat{P} \hat{A}_{cl} = -\hat{Q}, \tag{12.122}$$

with $\hat{P} = \hat{P}^T > 0$ and $\hat{Q} = \hat{Q}^T > 0$.

Theorem 12.12. *Consider the closed-loop system consisting of the plant (12.55)–(12.58) and the PDE-observer-based control scheme (12.117)–(12.122).*

There exist $\delta_1 > 0$ *and* $\delta_2 > 0$ *such that for any* $|\tilde{D}_i| < \delta_1$ *for* $i = 1, 2$ *and* $\max_{x \in [0,1]} |\tilde{b}_{ji}(x)| < \delta_2$ *for* $i = 1, 2$ *and* $j = 1, \cdots, n$, *the zero solution of the system* $(X(t), u_i(x,t), \hat{u}_i(x,t),)$ *is exponentially stable.*

Proof. Consider the invertible forwarding-backstepping transformation as follows:

$$\hat{w}_i(x,t) = \hat{u}_i(x,t) - \hat{K}_i e^{A\hat{D}_i x} Z(t) + \hat{D}_i \int_x^1 \hat{K}_i e^{A\hat{D}_i(x-y)} \hat{B}_{ie} \hat{u}_i(y,t) dy$$

$$= \hat{u}_i(x,t) - \hat{K}_i e^{A\hat{D}_i x} Z(t) + \hat{D}_i \int_x^1 \int_0^1 \hat{K}_i e^{A\hat{D}_i(x-y+\tau)} \tilde{\mathscr{B}}_i(\tau) d\tau \hat{u}_i(y,t) dy, \tag{12.123}$$

$$\hat{u}_i(x,t) = \hat{w}_i(x,t) + \hat{K}_i e^{(A+\hat{B}_{ie}\hat{K}_i)\hat{D}_i(x-1)} e^{A\hat{D}_i} Z(t)$$

$$- \hat{D}_i \int_x^1 \hat{K}_i e^{(A+\hat{B}_{ie}\hat{K}_i)\hat{D}_i(x-y)} \hat{B}_{ie} \hat{w}_i(y,t) dy, \tag{12.124}$$

where $\hat{B}_{ie} = e^{A\hat{D}_i} \hat{B}_{D_i} = \int_0^1 e^{A\hat{D}_i\tau} \tilde{\mathscr{B}}_i(\tau) d\tau$.

Subtracting (12.57)–(12.58) from (12.117)–(12.118), we get the dynamics of the PDE-observer error as follows:

$$D_i \partial_t \tilde{u}_i(x,t) = \partial_x \tilde{u}_i(x,t) - \frac{\tilde{D}_i}{\hat{D}_i} \partial_x \hat{u}_i(x,t), \quad x \in [0,1],$$

$$= \partial_x \tilde{u}_i(x,t) - \tilde{D}_i \left(\frac{1}{\hat{D}_i} \partial_x \hat{w}_i(x,t) + \hat{K}_i \hat{B}_{ie} \hat{w}_i(x,t) \right.$$

$$+ \hat{K}_i (A + \hat{B}_{ie} \hat{K}_i) e^{(A+\hat{B}_{ie}\hat{K}_i)\hat{D}_i(x-1)} e^{A\hat{D}_i} Z(t)$$

$$\left. - \hat{D}_i \int_x^1 \hat{K}_i (A + \hat{B}_{ie}\hat{K}_i) e^{(A+\hat{B}_{ie}\hat{K}_i)\hat{D}_i(x-y)} \hat{B}_{ie} \hat{w}_i(y,t) dy \right), \tag{12.125}$$

$$\tilde{u}_i(1,t) = 0. \tag{12.126}$$

Taking the time derivative of (12.120) along (12.55) and (12.117)–(12.118), and expressing $\hat{u}_i(x,t)$ by $\hat{w}_i(x,t)$ through (12.124), we get

$$\dot{Z}(t) = \left(A + \hat{B}_{D_1} \hat{K}_1 e^{A\hat{D}_1} + \hat{B}_{D_2} \hat{K}_2 e^{A\hat{D}_2} \right) Z(t)$$

$$+ \sum_{i=1}^2 \int_0^1 \mathscr{B}_i(x) \tilde{u}_i(x,t) dx + \sum_{i=1}^2 \int_0^1 \tilde{\mathscr{B}}_i(x) \hat{u}_i(x,t) dx$$

$$= \hat{A}_{cl} Z(t) + \sum_{i=1}^2 \int_0^1 \mathscr{B}_i(x) \tilde{u}_i(x,t) dx + \sum_{i=1}^2 \int_0^1 \tilde{\mathscr{B}}_i(x) q_i(Z(t), \hat{w}_i(x,t), x) dx, \tag{12.127}$$

where

$$\sum_{i=1}^{2} \int_{0}^{1} \tilde{\mathscr{B}}_{i}(x) q_{i}(Z(t), \hat{w}_{i}(x,t), x) dx$$

$$= \sum_{i=1}^{2} \int_{0}^{1} \tilde{\mathscr{B}}_{i}(x) \hat{K}_{i} e^{(A+\hat{B}_{ie}\hat{K}_{i})\hat{D}_{i}(x-1)} e^{A\hat{D}_{i}} dx Z(t) + \sum_{i=1}^{2} \int_{0}^{1} \tilde{\mathscr{B}}_{i}(x) \hat{w}_{i}(x,t) dx$$

$$- \sum_{i=1}^{2} \int_{0}^{1} \tilde{\mathscr{B}}_{i}(x) \hat{D}_{i} \int_{x}^{1} \hat{K}_{i} e^{(A+\hat{B}_{ie}\hat{K}_{i})\hat{D}_{i}(x-y)} \hat{B}_{ie} \hat{w}_{i}(y,t) dy dx.$$

Calculating the partial derivatives of (12.123) with respect to x and t along (12.127) and (12.117)–(12.118), we get

$$\hat{D}_{1} \partial_{t} \hat{w}_{1}(x,t) = \partial_{x} \hat{w}_{1}(x,t) - \hat{D}_{1} \hat{K}_{1} e^{A\hat{D}_{1}x} \hat{B}_{D_{2}} \hat{K}_{2} e^{A\hat{D}_{2}} Z(t)$$

$$- \hat{D}_{1} \hat{K}_{1} e^{A\hat{D}_{1}x} \left(\sum_{i=1}^{2} \int_{0}^{1} \mathscr{B}_{i}(y) \tilde{u}_{i}(y,t) dy \right.$$

$$+ \sum_{i=1}^{2} \int_{0}^{1} \tilde{\mathscr{B}}_{i}(y) q_{i}(Z(t), \hat{w}_{i}(y,t), y) dy \right), \qquad (12.128)$$

$$\hat{w}_{1}(1,t) = 0, \quad x \in [0,1], \qquad (12.129)$$

$$\hat{D}_{2} \partial_{t} \hat{w}_{2}(x,t) = \partial_{x} \hat{w}_{2}(x,t) - \hat{D}_{2} \hat{K}_{2} e^{A\hat{D}_{2}x} \hat{B}_{D_{1}} \hat{K}_{1} e^{A\hat{D}_{1}} Z(t)$$

$$- \hat{D}_{2} \hat{K}_{2} e^{A\hat{D}_{2}x} \left(\sum_{i=1}^{2} \int_{0}^{1} \mathscr{B}_{i}(y) \tilde{u}_{i}(y,t) dy \right.$$

$$+ \sum_{i=1}^{2} \int_{0}^{1} \tilde{\mathscr{B}}_{i}(y) q_{i}(Z(t), \hat{w}_{i}(y,t), y) dy \right), \qquad (12.130)$$

$$\hat{w}_{2}(1,t) = 0, \quad x \in [0,1]. \qquad (12.131)$$

Remark 12.13. The target ODE-PDE cascade given by the overall $(Z(t), \tilde{u}_{i}(x,t), \hat{w}_{i}(x,t))$-system is obtained in (12.125)–(12.126), (12.127), and (12.128)–(12.131). It is crucial to observe the interconnections in the subsystem, which is revealed in figure 12.2. The connections dependent upon $\tilde{D}_{i}$ and $\tilde{\mathscr{B}}_{i}(x)$ could be treated as "weak" when the estimation errors of delays and delay kernels are small. They disappear when $\hat{D}_{i} = D_{i}$ and $\hat{\mathscr{B}}_{i}(x) = \mathscr{B}_{i}(x)$. The connections $\tilde{u}_{i}(x,t) \to Z(t)$, $\tilde{u}_{i}(x,t) \to \hat{w}_{i}(x,t)$, and $Z(t) \to \hat{w}_{i}(x,t)$ are "strong" and they are present even when $\hat{D}_{i} = D_{i}$ and $\hat{\mathscr{B}}_{i}(x) = \mathscr{B}_{i}(x)$, in which case we have three parallel cascades of exponentially stable subsystems. The analysis should capture the fact that the potentially destabilizing connections through $\tilde{D}_{i}$ and $\tilde{\mathscr{B}}_{i}(x)$ can be suppressed by making $\tilde{D}_{i}$ and $\tilde{\mathscr{B}}_{i}(x)$ small.

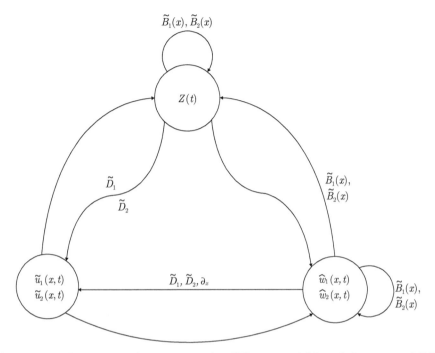

Figure 12.2. Interconnections among the different variables of the target ODE-PDE cascade $\big(Z(t), \tilde{u}_i(x,t), \hat{w}_i(x,t)\big)$-system

Consider the Lyapunov candidate

$$V(t) = Z^T(t)\hat{P}Z(t) + g_{u_1}D_1 \int_0^1 (1+x)\tilde{u}_1(x,t)^2 dx + g_{u_2}D_2 \int_0^1 (1+x)\tilde{u}_2(x,t)^2 dx$$

$$+ g_{w_1}\hat{D}_1 \int_0^1 (1+x)\left(\hat{w}_1(x,t)^2 + (\partial_x\hat{w}_1(x,t))^2\right) dx$$

$$+ g_{w_2}\hat{D}_2 \int_0^1 (1+x)\left(\hat{w}_2(x,t)^2 + (\partial_x\hat{w}_2(x,t))^2\right) dx, \tag{12.132}$$

where $g_{u_1} > 0$, $g_{u_2} > 0$, and $g_{w_1} > 0$, $g_{w_2} > 0$ are analysis coefficients. We differentiate $V(t)$ in (12.132) with respect to t along the target system (12.125)–(12.126), (12.127), and (12.128)–(12.131), and utilize Cauchy-Schwarz and Young's inequalities. If we follow a procedure similar to the proof of Theorem 12.10, as long as $|\tilde{D}_i| < \delta_1$ for $i = 1, 2$ and $\max_{x \in [0,1]} |\tilde{b}_{ji}(x)| < \delta_2$ for $i = 1, 2$ and $j = 1, \cdots, n$, where $\delta_1 > 0$ and $\delta_2 > 0$, it is not hard to show that $\dot{V}(t) \leq -\mu V(t)$ with $\mu > 0$ being a constant. Then Theorem 12.12 is proved. $\qquad\square$

Appendix A

A.1 Basic Inequalities

In this appendix we review a few inequalities for basic Sobolev spaces of functions of one variable. Let us first recall two elementary well-known inequalities:

Young's inequality (elementary version)

$$\boxed{ab \le \frac{\gamma}{2}a^2 + \frac{1}{2\gamma}b^2, \quad \forall \gamma > 0.}$$

(A.1)

Cauchy-Schwartz inequality

$$\boxed{\int_0^1 u(x)w(x)dx \le \sqrt{\int_0^1 u(x)^2 dx}\sqrt{\int_0^1 w(x)^2 dx}.}$$

(A.2)

The following lemma establishes the relationship between the L_2 norms of w and w_x.

Lemma A.1 (Poincaré Inequality). *For any w, continuously differentiable on $[0, 1]$,*

$$\int_0^1 w(x)^2 dx \le 2w^2(1) + 4 \int_0^1 w_x(x)^2 dx, \qquad (\text{A.3})$$

$$\int_0^1 w(x)^2 dx \le 2w^2(0) + 4 \int_0^1 w_x(x)^2 dx. \qquad (\text{A.4})$$

Proof. We start with the L_2 norm,

$$\int_0^1 w^2 dx = xw^2|_0^1 - 2\int_0^1 xww_x dx \quad (\text{integration by parts})$$

$$= w^2(1) - 2 \int_0^1 x w w_x dx$$

$$\leq w^2(1) + \frac{1}{2} \int_0^1 w^2 dx + 2 \int_0^1 x^2 w_x^2 dx.$$

Subtracting the second term from both sides of the inequality, we get the first inequality (A.3):

$$\frac{1}{2} \int_0^1 w^2 dx \leq w^2(1) + 2 \int_0^1 x^2 w_x^2 dx$$

$$\leq w^2(1) + 2 \int_0^1 w_x^2 dx. \tag{A.5}$$

The second inequality (A.4) is obtained in a similar fashion.   □

The inequalities (A.3)–(A.4) are conservative. The tight version of them is given next, which is sometimes called "a variation of Wirtingers inequality" [56].

Lemma A.2.
$$\int_0^1 (w(x) - w(0))^2 dx \leq \frac{4}{\pi^2} \int_0^1 w_x^2(x) dx. \tag{A.6}$$

Equality holds only for $w(x) = A + B \sin \frac{\pi x}{2}$.

The proof of (A.6) is far more complicated than the proof of (A.3)–(A.4).

Now we turn to reviewing the basic relationships between the L_2 and H_1 Sobolev norms and the maximum norm. The H_1 norm can be defined in more than one way. We define it as

$$\|w\|_{H_1}^2 := \int_0^1 w^2 dx + \int_0^1 w_x^2 dx. \tag{A.7}$$

Note also that by using Poincaré inequality, it is possible to drop the first integral in (A.7) whenever the function is zero at at least one of the boundaries.

Lemma A.3 (Agmon's Inequality). *For a function* $w \in H_1$, *the following inequalities hold:*

$$\max_{x \in [0,1]} |w(x)|^2 \leq w^2(0) + 2\|w\|\|w_x\|, \tag{A.8}$$

$$\max_{x \in [0,1]} |w(x)|^2 \leq w^2(1) + 2\|w\|\|w_x\|. \tag{A.9}$$

Proof. We begin with

$$\int_0^x w w_x dx = \int_0^x d \frac{1}{2} w^2$$

$$= \frac{1}{2} w^2 |_0^x$$

$$= \frac{1}{2} w(x)^2 - \frac{1}{2} w(0)^2, \qquad (A.10)$$

which gives

$$\frac{1}{2}|w(x)^2| \leq \int_0^x |w||w_x| dx + \frac{1}{2} w(0)^2. \qquad (A.11)$$

Using the fact that an integral of a positive function is an increasing function of its upper limit, we can rewrite the last inequality as

$$|w(x)^2| \leq 2 \int_0^1 |w(x)||w_x(x)| dx + w(0)^2. \qquad (A.12)$$

The right-hand side of this inequality does not depend on x; therefore,

$$\max_{x \in [0,1]} |w(x)^2| \leq 2 \int_0^1 |w(x)||w_x(x)| dx + w(0)^2. \qquad (A.13)$$

Using the Cauchy-Schwartz inequality, we get the first inequality of (A.8). The second inequality is obtained in a similar fashion. □

Appendix B

B.1   Input-to-Output Stability

In addition to a review of basic input-output stability results, we give several technical lemmas used in the book.

For a function $x : \mathbb{R}_+ \to \mathbb{R}^n$, we define the L_p norm, $p \in [1, \infty]$, as

$$
\|x\|_p = \begin{cases} \left(\displaystyle\int_0^\infty |x(t)|^p dt \right)^{1/p}, & p \in [1, \infty) \\[2mm] \displaystyle\sup_{t \geq 0} |x(t)|, & p = \infty, \end{cases}
\tag{B.1}
$$

and the $L_{p,e}$ norm (truncated L_p norm) as

$$
\|x_t\|_p = \begin{cases} \left(\displaystyle\int_0^t |x(\tau)|^p d\tau \right)^{1/p}, & p \in [1, \infty) \\[2mm] \displaystyle\sup_{\tau \in [0,t]} |x(\tau)|, & p = \infty. \end{cases}
\tag{B.2}
$$

Lemma B.1 (Hölder's Inequality). *If $p, q \in [1, \infty]$ and $\frac{1}{p} + \frac{1}{q} = 1$, then*

$$
\|(fg)_t\|_1 \leq \|f_t\|_p \|g_t\|_q, \qquad \forall t \geq 0.
\tag{B.3}
$$

We consider an LTI causal system described by the convolution

$$
y(t) = h \star u = \int_0^t h(t - \tau) u(\tau) d\tau,
\tag{B.4}
$$

where $u : \mathbb{R}_+ \to \mathbb{R}$ is the input, $y : \mathbb{R}_+ \to \mathbb{R}$ is the output, and $h : \mathbb{R} \to \mathbb{R}$ is the system's impulse response, which is defined to be zero for negative values of its argument.

Theorem B.2 (Young's Convolution Theorem). *If $h \in L_{1,e}$, then*

$$
\|(h \star u)_t\|_p \leq \|h_t\|_1 \|u_t\|_p, \qquad p \in [1, \infty].
\tag{B.5}
$$

Proof. Let $y = h \star u$. Then, for $p \in [1, \infty)$, we have

$$|y(t)| \leq \int_0^t |h(t-\tau)| \, |u(\tau)| d\tau$$

$$= \int_0^t |h(t-\tau)|^{\frac{p-1}{p}} \, |h(t-\tau)|^{\frac{1}{p}} \, |u(\tau)| d\tau$$

$$\leq \left(\int_0^t |h(t-\tau)| d\tau \right)^{\frac{p-1}{p}} \left(\int_0^t |h(t-\tau)| \, |u(\tau)|^p d\tau \right)^{\frac{1}{p}}$$

$$= \|h_t\|_1^{\frac{p-1}{p}} \left(\int_0^t |h(t-\tau)| \, |u(\tau)|^p d\tau \right)^{\frac{1}{p}}, \tag{B.6}$$

where the second inequality is obtained by applying Hölder's inequality. Raising (B.6) to power p and integrating from 0 to t, we get

$$\|y_t\|_p^p \leq \int_0^t \|h_t\|_1^{p-1} \left(\int_0^\tau |h(\tau-s)| \, |u(s)|^p ds \right) d\tau$$

$$= \|h_t\|_1^{p-1} \int_0^t \left(\int_s^t |h(\tau-s)| \, |u(s)|^p d\tau \right) ds$$

$$= \|h_t\|_1^{p-1} \int_0^t \left(\int_0^t |h(\tau-s)| \, |u(s)|^p d\tau \right) ds$$

$$= \|h_t\|_1^{p-1} \int_0^t |u(s)|^p \left(\int_0^t |h(\tau-s)| d\tau \right) ds$$

$$\leq \|h_t\|_1^{p-1} \int_0^t |u(s)|^p \left(\int_0^t |h(\tau)| d\tau \right) ds$$

$$\leq \|h_t\|_1^{p-1} \|h\|_1 \|u_t\|_p^p$$

$$\leq \|h_t\|_1^p \|u_t\|_p^p, \tag{B.7}$$

where the second line is obtained by changing the sequence of integration, and the third line by using the causality of h. The proof for the case $p = \infty$ is immediate by taking a supremum of u over $[0, t]$ in the convolution. $\qquad\square$

Lemma B.3. *Let v and ρ be real-valued functions defined on $\mathbb{R}_+$, and let b and c be positive constants. If they satisfy the differential inequality*

$$\dot{v} \leq -cv + b\rho(t)^2, \qquad v(0) \geq 0, \tag{B.8}$$

(i) then the following integral inequality holds:

$$v(t) \leq v(0)e^{-ct} + b \int_0^t e^{-c(t-\tau)} \rho(\tau)^2 d\tau. \tag{B.9}$$

(ii) If, in addition, $\rho \in L_2$, then $v \in L_1$ and

$$\|v\|_1 \le \frac{1}{c}\left(v(0) + b\|\rho\|_2^2\right). \tag{B.10}$$

Proof. (i) Upon multiplication of (B.8) by e^{ct}, it becomes

$$\frac{d}{dt}\left(v(t)e^{ct}\right) \le b\rho(t)^2 e^{ct}. \tag{B.11}$$

Integrating (B.11) over $[0, t]$, we arrive at (B.9).
    (ii) By integrating (B.9) over $[0, t]$, we get

$$\int_0^t v(\tau)d\tau \le \int_0^t v(0)e^{-c\tau}d\tau + b\int_0^t \left[\int_0^\tau e^{-c(\tau-s)}\rho(s)^2 ds\right]d\tau$$

$$\le \frac{1}{c}v(0) + b\int_0^t \left[\int_0^\tau e^{-c(\tau-s)}\rho(s)^2 ds\right]d\tau. \tag{B.12}$$

Noting that the second term is $b\|(h \star \rho^2)_t\|_1$, where

$$h(t) = e^{-ct}, \qquad t \ge 0, \tag{B.13}$$

we apply Theorem B.2. Since

$$\|h\|_1 = \frac{1}{c}, \tag{B.14}$$

we obtain (B.10). □

Lemma B.4. *Let v, l_1, and l_2 be real-valued functions defined on $\mathbb{R}_+$ and let c be a positive constant. If l_1 and l_2 are nonnegative and in L_1 and satisfy the differential inequality*

$$\dot{v} \le -cv + l_1(t)v + l_2(t), \quad v(0) \ge 0, \tag{B.15}$$

then $v \in L_\infty \cap L_1$ and

$$v(t) \le \left(v(0)e^{-ct} + \|l_2\|_1\right)e^{\|l_1\|_1}, \tag{B.16}$$

$$\|v\|_1 \le \frac{1}{c}\left(v(0) + \|l_2\|_1\right)e^{\|l_1\|_1}. \tag{B.17}$$

Proof. Using the facts that

$$v(t) \le w(t), \tag{B.18}$$

$$\dot{w} = -cw + l_1(t)w + l_2(t), \tag{B.19}$$

$$w(0) = v(0) \tag{B.20}$$

(the comparison principle), and applying the variation-of-constants formula, the differential inequality (B.15) is rewritten as

$$v(t) \le v(0)e^{\int_0^t [-c+l_1(s)]ds} + \int_0^t e^{\int_\tau^t [-c+l_1(s)]ds} l_2(\tau)d\tau$$
$$\le v(0)e^{-ct}e^{\int_0^\infty l_1(s)ds} + \int_0^t e^{-c(t-\tau)}l_2(\tau)d\tau e^{\int_0^\infty l_1(s)ds}$$
$$\le \left[v(0)e^{-ct} + \int_0^t e^{-c(t-\tau)}l_2(\tau)d\tau \right] e^{\|l_1\|_1}. \tag{B.21}$$

By taking a supremum of $e^{-c(t-\tau)}$ over $[0,\infty]$, we obtain (B.16). Integrating (B.21) over $[0,\infty]$, we get

$$\int_0^t v(\tau)d\tau \le \left(\frac{1}{c}v(0) + \int_0^t \left[\int_0^\tau e^{-c(\tau-s)}l_2(s)ds \right] d\tau \right) e^{\|l_1\|_1}. \tag{B.22}$$

Applying Theorem B.2 to the double integral, we arrive at (B.17). □

Remark B.5. An alternative proof that $v \in L_\infty \cap L_1$ in Lemma B.4 is using Gronwall's lemma (Lemma B.8). However, with Gronwall's lemma, the estimates of the bounds (B.16) and (B.17) are more conservative:

$$v(t) \le \left(v(0)e^{-ct} + \|l_2\|_1 \right) \left(1 + \|l_1\|_1 e^{\|l_1\|_1} \right), \tag{B.23}$$
$$\|v\|_1 \le \frac{1}{c} \left(v(0) + \|l_2\|_1 \right) \left(1 + \|l_1\|_1 e^{\|l_1\|_1} \right), \tag{B.24}$$

because

$$e^x < (1 + xe^x), \qquad \forall x > 0. \tag{B.25}$$

Note that the ratio between the bounds (B.23) and (B.16) and that between the bounds (B.24) and (B.17) are of the order $\|l_1\|_1$ when $\|l_1\|_1 \to \infty$. ◇

For cases where l_1 and l_2 are functions of time that converge to zero but are not in L_p for any $p \in [1, \infty)$, we have the following lemma.

Lemma B.6. *Consider the differential inequality*

$$\dot{v} \le -(c - \beta_1(r_0, t))v + \beta_2(r_0, t) + \rho, \quad v(0) = v_0 \ge 0, \tag{B.26}$$

where $c > 0$ and $r_0 \ge 0$ are constants, and β_1 and β_2 are class-$\mathcal{KL}$ functions. Then there exist a class-$\mathcal{KL}$ function β_v and a class-$\mathcal{K}$ function γ_v such that

$$v(t) \le \beta_v(v_0 + r_0, t) + \gamma_v(\rho), \quad t \ge 0. \tag{B.27}$$

Moreover, if

$$\beta_i(r,t)=\alpha_i(r)e^{-\sigma_i t}, \quad i=1,2, \tag{B.28}$$

where $\alpha_i \in \mathscr{K}$ *and* $\sigma_i > 0$, *then there exist* $\alpha_\nu \in \mathscr{K}$ *and* σ_ν *such that*

$$\beta_\nu(r,t)=\alpha_\nu(r)e^{-\sigma_\nu t}. \tag{B.29}$$

Proof. We start by introducing

$$\tilde{v}=v-\frac{\rho}{c} \tag{B.30}$$

and rewriting (B.26) as

$$\dot{\tilde{v}} \leq -[c-\beta_1(r_0,t)]\tilde{v}+\frac{\rho}{c}\beta_1(r_0,t)+\beta_2(r_0,t). \tag{B.31}$$

It then follows that

$$v(t) \leq v_0 e^{\int_0^t [\beta_1(r_0,s)-c]ds} + \int_0^t \left[\frac{\rho}{c}\beta_1(r_0,\tau)+\beta_2(r_0,\tau)\right] e^{\int_\tau^t [\beta_1(r_0,s)-c]ds} d\tau + \frac{\rho}{c}. \tag{B.32}$$

We note that

$$e^{\int_\tau^t [\beta_1(r_0,s)-c]ds} \leq k(r_0)e^{-\frac{c}{2}(t-\tau)}, \qquad \forall \tau \in [0,t], \tag{B.33}$$

where k is a positive, continuous, increasing function. To get an estimate of the overshoot coefficient $k(r_0)$, we provide a proof of (B.33). For each c, there exists a class-$\mathscr{K}$ function $T_c : \mathbb{R}_+ \to \mathbb{R}_+$ such that

$$\beta_1(r_0,s) \leq \frac{c}{2}, \qquad \forall s \geq T_c(r_0). \tag{B.34}$$

Therefore, for $0 \leq \tau \leq T_c(r_0) \leq t$, we have

$$\int_\tau^t [\beta_1(r_0,s)-c]\,ds \leq \int_\tau^{T_c(r_0)} [\beta_1(r_0,s)-c]\,ds + \int_{T_c(r_0)}^t \left(-\frac{c}{2}\right) ds$$

$$\leq (\beta_1(r_0,0)-c)(T_c(r_0)-\tau)-\frac{c}{2}(t-T_c(r_0))$$

$$\leq T_c(r_0)\beta_1(r_0,0)-\frac{c}{2}(t-\tau), \tag{B.35}$$

so the overshoot coefficient in (B.33) is given by

$$k(r_0) \stackrel{\triangle}{=} e^{T_c(r_0)\beta_1(r_0,0)}. \tag{B.36}$$

For the other two cases, $t \leq T_c(r_0)$ and $T_c(r_0) \leq \tau$, getting (B.33) with $k(r_0)$ as in (B.36) is immediate. Now substituting (B.33) into (B.32), we get

$$v(t) \leq v_0 k(r_0) e^{-\frac{c}{2}t} + k(r_0) \int_0^t \left[\frac{\rho}{c} \beta_1(r_0, \tau) + \beta_2(r_0, \tau) \right] e^{-\frac{c}{2}(t-\tau)} d\tau + \frac{\rho}{c} .$$
$$\text{(B.37)}$$

To complete the proof, we show that a class-$\mathscr{KL}$ function β convolved with an exponentially decaying kernel is bounded by another class-$\mathscr{KL}$ function:

$$\int_0^t e^{-\frac{c}{2}(t-\tau)} \beta(r_0, \tau) d\tau = \int_0^{t/2} e^{-\frac{c}{2}(t-\tau)} \beta(r_0, \tau) d\tau + \int_{t/2}^t e^{-\frac{c}{2}(t-\tau)} \beta(r_0, \tau) d\tau$$

$$\leq \beta(r_0, 0) \int_0^{t/2} e^{-\frac{c}{2}(t-\tau)} d\tau + \beta(r_0, t/2) \int_{t/2}^t e^{-\frac{c}{2}(t-\tau)} d\tau$$

$$\leq \frac{2}{c} \left[\beta(r_0, 0) e^{-\frac{c}{4}t} + \beta(r_0, t/2) \right] . \qquad \text{(B.38)}$$

Thus, (B.37) becomes

$$v(t) \leq k(r_0) \left\{ \left[v_0 + \frac{2\rho}{c^2} \beta_1(r_0, 0) + \frac{2}{c} \beta_2(r_0, 0) \right] e^{-\frac{c}{4}t} \right.$$

$$\left. + \frac{2\rho}{c^2} \beta_1(r_0, t/2) + \frac{2}{c} \beta_2(r_0, t/2) \right\} + \frac{\rho}{c} . \qquad \text{(B.39)}$$

By applying Young's inequality to the terms

$$k(r_0) \frac{2\rho}{c^2} \beta_1(r_0, 0) e^{-\frac{c}{4}t} \qquad \text{(B.40)}$$

and

$$k(r_0) \frac{2\rho}{c^2} \beta_1(r_0, t/2) , \qquad \text{(B.41)}$$

we obtain (B.27) with

$$\beta_v(r, t) = k(r) \left\{ \left[r + \frac{k(r)}{c^2} \beta_1(r, 0)^2 + \frac{2}{c} \beta_2(r, 0) \right] e^{-\frac{c}{4}t} \right.$$

$$\left. + \frac{k(r)}{c^2} \beta_1(r, t/2)^2 + \frac{2}{c} \beta_2(r, t/2) \right\} , \qquad \text{(B.42)}$$

$$\gamma_v(r) = \frac{r}{c} + \frac{r^2}{c^2} . \qquad \text{(B.43)}$$

The last statement of the lemma is immediate by substitution into (B.42). □

The proof of the following Lemma can be found in [55] (Lemma 3.4).

Lemma B.7 (Comparison Principle). *Consider the scalar differential equation*

$$\dot{u} = f(t, u), \quad u(t_0) = u_0, \tag{B.44}$$

where $f(t, u)$ is continuous in t and locally Lipschitz in u, for all $t \geq 0$ and all $u \in \mathbb{J} \subset \mathbb{R}$. Let $[t_0, T)$ (T could be infinity) be the maximal interval of existence of the solution $u(t)$, and suppose $u(t) \in \mathbb{J}$ for all $t \in [t_0, T)$. Let $v(t)$ be a continuous function whose upper right-hand derivative $D^+ v(t)$ satisfies the differential inequality

$$D^+ v(t) \leq f(t, v(t)), \quad v(t_0) \leq u(t_0), \tag{B.45}$$

with $v(t) \in \mathbb{J}$ for all $t \in [t_0, T)$. Then $v(t) \leq u(t)$ for all $t \in [t_0, T)$.

Now we give a version of Gronwall's lemma.

Lemma B.8 (Gronwall). *Consider the continuous functions $\lambda : \mathbb{R}_+ \to \mathbb{R}$, $\mu : \mathbb{R}_+ \to \mathbb{R}_+$, and $\nu : \mathbb{R}_+ \to \mathbb{R}_+$, where μ and ν are also nonnegative. If a continuous function $y : \mathbb{R}_+ \to \mathbb{R}$ satisfies the inequality*

$$y(t) \leq \lambda(t) + \mu(t) \int_{t_0}^{t} \nu(s) y(s) ds, \quad \forall t \geq t_0 \geq 0, \tag{B.46}$$

then

$$y(t) \leq \lambda(t) + \mu(t) \int_{t_0}^{t} \lambda(s) \nu(s) e^{\int_s^t \mu(\tau) \nu(\tau) d\tau} ds, \quad \forall t \geq t_0 \geq 0. \tag{B.47}$$

In particular, if $\lambda(t) \equiv \lambda$ is a constant and $\mu(t) \equiv 1$, then

$$y(t) \leq \lambda e^{\int_{t_0}^{t} \nu(\tau) d\tau}, \quad \forall t \geq t_0 \geq 0. \tag{B.48}$$

Appendix C

C.1 Lyapunov Stability and $\mathcal{K}_\infty$ Functions

Consider the nonautonomous ODE system

$$\dot{x} = f(x,t), \tag{C.1}$$

where $f : \mathbb{R}^n \times \mathbb{R}_+ \to \mathbb{R}^n$ is locally Lipschitz in x and piecewise continuous in t.

Definition C.1. *The origin $x = 0$ is the equilibrium point for (C.1) if*

$$f(0,t) = 0, \qquad \forall t \geq 0. \tag{C.2}$$

Scalar *comparison functions* are important stability tools.

Definition C.2. *A continuous function $\gamma : [0,a) \to \mathbb{R}_+$ is said to belong to class-$\mathcal{K}$ if it is strictly increasing and $\gamma(0) = 0$. It is said to belong to class-$\mathcal{K}_\infty$ if $a = \infty$ and $\gamma(r) \to \infty$ as $r \to \infty$.*

Definition C.3. *A continuous function $\beta : [0,a) \times \mathbb{R}_+ \to \mathbb{R}_+$ is said to belong to class-$\mathcal{K}\mathcal{L}$ if, for each fixed s, the mapping $\beta(r,s)$ belongs to class-$\mathcal{K}$ with respect to r and, for each fixed r, the mapping $\beta(r,s)$ is decreasing with respect to s and $\beta(r,s) \to 0$ as $s \to \infty$. It is said to belong to class-$\mathcal{K}\mathcal{L}_\infty$ if, in addition, for each fixed s, the mapping $\beta(r,s)$ belongs to class-$\mathcal{K}_\infty$ with respect to r.*

Definition C.4. *We say that a continuous function $\rho : \mathbb{R}_+ \times (0,1) \mapsto \mathbb{R}_+$ belongs to class-$\mathcal{K}\mathcal{C}$ if, for each fixed c, the mapping $\rho(s,c)$ belongs to class-$\mathcal{K}$ with respect to s and, for each fixed s, the mapping $\rho(s,c)$ is continuous with respect to c. It belongs to class-$\mathcal{K}\mathcal{C}_\infty$ if, in addition, for each fixed c, the mapping $\rho(s,c)$ belongs to class-$\mathcal{K}_\infty$ with respect to s.*

The main list of stability definitions for ODE systems is given next.

Definition C.5 (Stability). *The equilibrium point $x = 0$ of (C.1) is*

- uniformly stable *if there exist a class-$\mathcal{K}$ function $\gamma(\cdot)$ and a positive constant c, independent of t_0, such that*

$$|x(t)| \leq \gamma(|x(t_0)|), \quad \forall t \geq t_0 \geq 0, \quad \forall x(t_0) \text{ s.t. } |x(t_0)| < c, \tag{C.3}$$

- uniformly asymptotically stable *if there exist a class-$\mathcal{KL}$ function $\beta(\cdot,\cdot)$ and a positive constant c, independent of t_0, such that*

$$|x(t)| \leq \beta(|x(t_0)|, t - t_0), \quad \forall t \geq t_0 \geq 0, \quad \forall x(t_0) \ s.t. \ |x(t_0)| < c, \qquad \text{(C.4)}$$

- exponentially stable *if (C.4) is satisfied with $\beta(r,s) = kre^{-\alpha s}$, $k > 0$, $\alpha > 0$,*
- globally uniformly stable *if (C.3) is satisfied with $\gamma \in \mathcal{K}_\infty$ for any initial state $x(t_0)$,*
- globally uniformly asymptotically stable *if (C.4) is satisfied with $\beta \in \mathcal{KL}_\infty$ for any initial state $x(t_0)$,*
- globally exponentially stable *if (C.4) is satisfied for any initial state $x(t_0)$ and with $\beta(r,s) = kre^{-\alpha s}$, $k > 0$, $\alpha > 0$.*

The following lemma is proved in [55] (Lemma 4.4).

Lemma C.6. *Consider the scalar autonomous differential equation*

$$\dot{y} = -\alpha(y), \quad y(t_0) = y_0, \qquad \text{(C.5)}$$

where α is a locally Lipschitz class-$\mathcal{K}$ function defined on $[0, a]$. For all $0 \leq y_0 < a$, this equation has a unique solution $y(t)$ defined for all $t \geq t_0$. Moreover,

$$y(t) = \sigma(y_0, t - t_0), \qquad \text{(C.6)}$$

where σ is a class-$\mathcal{KL}$ function defined on $[0, a) \times [0, \infty)$.

The main Lyapunov stability theorem is then formulated as follows.

Theorem C.7 (Lyapunov Theorem). *Let $x = 0$ be an equilibrium point of (C.1) and $D = \{x \in \mathbb{R}^n \mid |x| < r\}$. Let $V : D \times \mathbb{R}^n \to \mathbb{R}_+$ be a continuously differentiable function such that $\forall t \geq 0$, $\forall x \in D$,*

$$\gamma_1(|x|) \leq V(x, t) \leq \gamma_2(|x|), \qquad \text{(C.7)}$$

$$\frac{\partial V}{\partial t} + \frac{\partial V}{\partial x} f(x, t) \leq -\gamma_3(|x|). \qquad \text{(C.8)}$$

Then the equilibrium $x = 0$ is

- uniformly stable *if γ_1 and γ_2 are class-$\mathcal{K}$ functions on $[0, r)$ and $\gamma_3(\cdot) \geq 0$ on $[0, r)$,*
- uniformly asymptotically stable *if γ_1, γ_2, and γ_3 are class-$\mathcal{K}$ functions on $[0, r)$,*
- exponentially stable *if $\gamma_i(\rho) = k_i \rho^\alpha$ on $[0, r)$, $k_i > 0$, $\alpha > 0$, $i = 1, 2, 3$,*
- globally uniformly stable *if $D = \mathbb{R}^n$, γ_1 and γ_2 are class-$\mathcal{K}_\infty$ functions, and $\gamma_3(\cdot) \geq 0$ on $\mathbb{R}_+$,*

- globally uniformly asymptotically stable *if* $D = \mathbb{R}^n$, γ_1 *and* γ_2 *are class-$\mathcal{K}_\infty$ functions, and γ_3 is a class-$\mathcal{K}$ function on* $\mathbb{R}_+$,
- globally exponentially stable *if* $D = \mathbb{R}^n$ *and* $\gamma_i(\rho) = k_i \rho^\alpha$ *on* $\mathbb{R}_+$, $k_i > 0$, $\alpha > 0$, $i = 1, 2, 3$.

The following theorem is proved in [55] (Theorem 4.14).

Theorem C.8 (Converse Lyapunov Theorem for Exponential Stability). *Let* $x = 0$ *be an equilibrium point for the nonlinear system*

$$\dot{x} = f(t, x), \tag{C.9}$$

where $f : [0, \infty) \times D \to \mathbb{R}^n$ *is continuously differentiable,* $D = \{x \in \mathbb{R}^n \,|\, \|x\| < r\}$, *and the Jacobian matrix* $\left[\frac{\partial f}{\partial x}\right]$ *is bounded on* D, *uniformly in* t. *Let* κ, λ, *and* r_0 *be positive constants with* $r_0 < \frac{r}{k}$. *Let* $D_0 = \{x \in \mathbb{R}^n \,|\, \|x\| < r_0\}$. *Assume that the trajectories of the system satisfy*

$$\|x(t)\| \leq k \|x(t_0)\| e^{-\lambda(t-t_0)}, \quad \forall x(t_0) \in D_0, \quad \forall t \geq t_0 \geq 0. \tag{C.10}$$

Then, there is a function $V : [0, \infty) \times D_0 \to \mathbb{R}^n$ *that satisfies the inequalities*

$$c_1 \|x\|^2 \leq V(t, x) \leq c_2 \|x\|^2, \tag{C.11}$$

$$\frac{\partial V}{\partial t} + \frac{\partial V}{\partial x} f(t, x) \leq -c_3 \|x\|^2, \tag{C.12}$$

$$\left\| \frac{\partial V}{\partial x} \right\| \leq c_4 \|x\|, \tag{C.13}$$

for some positive constant c_1, c_2, c_3, *and* c_4. *Moreover, if* $r = \infty$ *and the origin is globally exponentially stable, then* $V(t, x)$ *is defined and satisfies the aforementioned inequalities on* $\mathbb{R}^n$. *Furthermore, if the system is autonomous,* V *can be chosen independent of* t.

C.2 Barbalat's Lemma with Its Alternative and LaSalle-Yoshizawa's Theorem

In adaptive control our goal is to achieve convergence to a set. For time-invariant systems, the main convergence tool is LaSalle's invariance theorem. For time-varying systems, a more refined tool is the LaSalle-Yoshizawa theorem. For pedagogical reasons, we introduce it via a technical lemma due to Barbalat. These key results and their proofs are of importance in guaranteeing that an adaptive system will fulfill its tracking task.

Lemma C.9 (Barbalat). *Consider the function* $\phi : \mathbb{R}_+ \to \mathbb{R}$. *If* ϕ *is uniformly continuous and* $\lim_{\tau \to \infty} \int_0^\infty \phi(\tau) d\tau$ *exists and is finite, then*

$$\lim_{t \to \infty} \phi(t) = 0. \tag{C.14}$$

Proof. Suppose that (C.14) does not hold; that is, either the limit does not exist or it is not equal to zero. Then there exists $\varepsilon > 0$ such that for every $T > 0$, one can find $t_1 \geq T$ with $|\phi(t_1)| > \varepsilon$. Since ϕ is uniformly continuous, there is a positive constant $\delta(\varepsilon)$ such that $|\phi(t) - \phi(t_1)| < \varepsilon/2$ for all $t_1 \geq 0$ and all t such that $|t - t_1| \leq \delta(\varepsilon)$. Hence, for all $t \in [t_1, t_1 + \delta(\varepsilon)]$, we have

$$
\begin{aligned}
|\phi(t)| &= |\phi(t) - \phi(t_1) + \phi(t_1)| \\
&\geq |\phi(t_1)| - |\phi(t) - \phi(t_1)| \\
&> \varepsilon - \frac{\varepsilon}{2} = \frac{\varepsilon}{2},
\end{aligned}
\tag{C.15}
$$

which implies that

$$
\left| \int_{t_1}^{t_1 + \delta(\varepsilon)} \phi(\tau) d\tau \right| = \int_{t_1}^{t_1 + \delta(\varepsilon)} |\phi(\tau)| d\tau > \frac{\varepsilon \delta(\varepsilon)}{2},
\tag{C.16}
$$

where the first equality holds since $\phi(t)$ does not change sign on $[t_1, t_1 + \delta(\varepsilon)]$. Noting that $\int_0^{t_1+\delta(\varepsilon)} \phi(\tau) d\tau = \int_0^{t_1} \phi(\tau) d\tau + \int_{t_1}^{t_1+\delta(\varepsilon)} \phi(\tau) d\tau$, we conclude that $\int_0^t \phi(\tau) d\tau$ cannot converge to a finite limit as $t \to \infty$, which contradicts the assumption of the lemma. Thus, $\lim_{t\to\infty} \phi(t) = 0$. $\square$

Lemma C.10 (An Alternative to Barbalat). *Suppose that the function $\phi(t)$ defined on $[0, \infty)$ satisfies the following conditions:*

(i) $\phi(t) \geq 0$ for all $t \in [0, \infty)$,
(ii) $\phi(t)$ is differentiable on $[0, \infty)$ and there exists a constant M such that $\dot{\phi}(t) \leq M, \forall t \geq 0$,
(iii) $\int_0^\infty \phi(t) dt \leq \infty$.

Then we have $\lim_{t\to\infty} \phi(t) = 0$.

Proof. If $M \leq 0$, then $\phi(t)$ is nonincreasing. Hence conditions (i) and (iii) immediately imply $\lim_{t\to\infty} \phi(t) = 0$. We now suppose that $M \geq 0$. We argue by contradiction. If $\lim_{t\to\infty} \phi(t) = 0$ is not true, then there exist a positive constant δ and a sequence $\{t_n\}$ $(n = 1, 2, \cdots)$ with $t_n \to \infty$ as $n \to \infty$ such that $\phi(t_n) \geq \delta$, $n = 1, 2, \cdots$. Let $\Phi(t) = \phi(t) - M(t - t_n) - \phi(t_n)$. By condition (ii), we have $\dot{\Phi}(t) = \dot{\phi}(t) - M \leq 0$. Therefore, we deduce $\Phi(t) \geq \Phi(t_n) = 0$, $\forall 0 \leq t \leq t_n$, that is, $\phi(t) \geq M(t - t_n) + \phi(t_n)$, $\forall 0 \leq t \leq t_n$. Thus we have $\int_{t_n - \frac{\delta}{M}}^{t_n} \phi(t) dt \geq \int_{t_n - \frac{\delta}{M}}^{t_n} [M(t - t_n) + \phi(t_n)] dt = \frac{\delta \phi(t_n)}{M} - \frac{\delta^2}{2M} \geq \frac{\delta^2}{2M}$, $n = 1, 2, \cdots$, which is in contradiction with condition (iii). $\square$

It is important to see that this lemma and the standard Barbalat's lemma do not imply each other. While Barbalat's lemma assumes that $\phi(t)$ is uniformly continuous, the alternative assumes that $\dot{\phi}(t)$ is bounded, but only from above.

Corollary C.11. *Consider the function* $\phi: \mathbb{R}_+ \to \mathbb{R}$. *If* $\phi, \dot{\phi} \in \mathcal{L}_\infty$, *and* $\phi \in \mathcal{L}_p$ *for some* $p \in [1, \infty)$, *then*

$$\lim_{t \to \infty} \phi(t) = 0. \tag{C.17}$$

Theorem C.12 (LaSalle-Yoshizawa). *Let* $x = 0$ *be an equilibrium point of (C.1) and suppose* f *is locally Lipschitz in* x *uniformly in* t. *Let* $V : \mathbb{R}^n \times \mathbb{R}_+ \to \mathbb{R}_+$ *be a continuously differentiable function such that*

$$\gamma_1(|x|) \le V(x, t) \le \gamma_2(|x|), \tag{C.18}$$

$$\dot{V} = \frac{\partial V}{\partial t} + \frac{\partial V}{\partial x} f(x, t) \le -W(x) \le 0, \tag{C.19}$$

$\forall t \ge 0$, $\forall x \in \mathbb{R}^n$, *where* γ_1 *and* γ_2 *are class-*$\mathcal{K}_\infty$ *functions and* W *is a continuous function. Then all solutions of (C.1) are globally uniformly bounded and satisfy*

$$\lim_{t \to \infty} W(x(t)) = 0. \tag{C.20}$$

In addition, if $W(x)$ *is positive definite, then the equilibrium* $x = 0$ *is globally uniformly asymptotically stable.*

Proof. Since $\dot{V} \le 0$, V is nonincreasing. Thus, in view of the first inequality in (C.18), we conclude that x is globally uniformly bounded, that is, $|x(t)| \le B$, $\forall t \ge 0$. Since $V(x(t), t)$ is nonincreasing and bounded from below by zero, we conclude that it has a limit V_∞ as $t \to \infty$. Integrating (C.19), we have

$$\lim_{t \to \infty} \int_{t_0}^t W(x(\tau)) d\tau \le - \lim_{t \to \infty} \int_{t_0}^t \dot{V}(x(\tau), \tau) d\tau$$

$$= \lim_{t \to \infty} \{V(x(t_0), t_0) - V(x(t), t)\}$$

$$= V(x(t_0), t_0) - V_\infty, \tag{C.21}$$

which means that $\int_{t_0}^\infty W(x(\tau)) d\tau$ exists and is finite. Now we show that $W(x(t))$ is also uniformly continuous. Since $|x(t)| \le B$ and f is locally Lipschitz in x uniformly in t, we see that for any $t \ge t_0 \ge 0$,

$$|x(t) - x(t_0)| = \left| \int_{t_0}^t f(x(\tau), \tau) d\tau \right| \le L \int_{t_0}^t |x(\tau)| d\tau$$

$$\le LB|t - t_0|, \tag{C.22}$$

where L is the Lipschitz constant of f on $\{|x| \le B\}$. Choosing $\delta(\varepsilon) = \varepsilon/LB$, we have

$$|x(t) - x(t_0)| < \varepsilon, \qquad \forall |t - t_0| \le \delta(\varepsilon), \tag{C.23}$$

which means that $x(t)$ is uniformly continuous. Since W is continuous, it is uniformly continuous on the compact set $\{|x| \leq B\}$. From the uniform continuity of $W(x)$ and $x(t)$, we conclude that $W(x(t))$ is uniformly continuous. Hence, it satisfies the conditions of Lemma C.9, which then guarantees that $W(x(t)) \to 0$ as $t \to \infty$.

If, in addition, $W(x)$ is positive definite, there exists a class-$\mathcal{K}$ function $\gamma_3(\cdot)$ such that $W(x) \geq \gamma_3(|x|)$. Using Theorem C.7, we conclude that $x = 0$ is globally uniformly asymptotically stable.   □

In applications, we usually have $W(x) = x^T Q x$, where Q is a symmetric positive-semidefinite matrix. For this case, the proof of Theorem C.12 simplifies using Corollary C.11 with $p = 1$.

C.3   Input-to-State Stability

The following definition of input-to-state stability was provided by Sontag [123].

Definition C.13 (Input-to-State Stability). *The system*

$$\dot{x} = f(t, x, u), \qquad (\text{C.24})$$

where f is piecewise continuous in t and locally Lipschitz in x and u, is said to be input-to-state stable *(ISS) if there exist a class-$\mathcal{KL}$ function β and a class-$\mathcal{K}$ function γ such that for any $x(t_0)$ and for any input $u(\cdot)$ continuous and bounded on $[0, \infty)$, the solution exists for all $t \geq 0$ and satisfies*

$$|x(t)| \leq \beta\left(|x(t_0)|, t - t_0\right) + \gamma\left(\sup_{t_0 \leq \tau \leq t} |u(\tau)|\right), \qquad (\text{C.25})$$

for all t_0 and t such that $0 \leq t_0 \leq t$.

The following theorem establishes the connection between the existence of a Lyapunov-like function and the input-to-state stability.

Theorem C.14. *Suppose that for the system (C.24), there exists a C^1 function $V : \mathbb{R}_+ \times \mathbb{R}^n \to \mathbb{R}_+$ such that for all $x \in \mathbb{R}^n$ and $u \in \mathbb{R}^m$,*

$$\gamma_1(|x|) \leq V(t, x) \leq \gamma_2(|x|), \qquad (\text{C.26})$$

$$|x| \geq \rho(|u|) \quad \Rightarrow \quad \frac{\partial V}{\partial t} + \frac{\partial V}{\partial x} f(t, x, u) \leq -\gamma_3(|x|), \qquad (\text{C.27})$$

where γ_1, γ_2, and ρ are class-$\mathcal{K}_\infty$ functions and γ_3 is a class-$\mathcal{K}$ function. Then system (C.24) is ISS with $\gamma = \gamma_1^{-1} \circ \gamma_2 \circ \rho$.

Proof. (Outline) If $x(t_0)$ is in the set

$$R_{t_0} = \left\{ x \in \mathbb{R}^n \;\middle|\; |x| \leq \rho\left(\sup_{\tau \geq t_0} |u(\tau)| \right) \right\}, \tag{C.28}$$

then $x(t)$ remains within the set

$$S_{t_0} = \left\{ x \in \mathbb{R}^n \;\middle|\; |x| \leq \gamma_1^{-1} \circ \gamma_2 \circ \rho\left(\sup_{\tau \geq t_0} |u(\tau)| \right) \right\} \tag{C.29}$$

for all $t \geq t_0$. Define $B = [t_0, T)$ as the time interval before $x(t)$ enters R_{t_0} for the first time. In view of the definition of R_{t_0}, we have

$$\dot{V} \leq -\gamma_3 \circ \gamma_2^{-1}(V), \qquad \forall t \in B. \tag{C.30}$$

Then there exists a class-$\mathcal{KL}$ function β_V such that

$$V(t) \leq \beta_V(V(t_0), t - t_0), \qquad \forall t \in B, \tag{C.31}$$

which implies

$$|x(t)| \leq \gamma_1^{-1}(\beta_V(\gamma_2(|x(t_0)|), t - t_0)) \stackrel{\triangle}{=} \beta(|x(t_0)|, t - t_0), \qquad \forall t \in B. \tag{C.32}$$

On the other hand, by (C.29), we conclude that

$$|x(t)| \leq \gamma_1^{-1} \circ \gamma_2 \circ \rho\left(\sup_{\tau \geq t_0} |u(\tau)| \right) \stackrel{\triangle}{=} \gamma\left(\sup_{\tau \geq t_0} |u(\tau)| \right), \qquad \forall t \in [t_0, \infty] \setminus B. \tag{C.33}$$

Then, by (C.32) and (C.33),

$$|x(t)| \leq \beta(|x(t_0)|, t - t_0) + \gamma\left(\sup_{\tau \geq t_0} |u(\tau)| \right), \qquad \forall t \geq t_0 \geq 0. \tag{C.34}$$

By causality, it follows that

$$|x(t)| \leq \beta(|x(t_0)|, t - t_0) + \gamma\left(\sup_{t_0 \leq \tau \leq t} |u(\tau)| \right), \qquad \forall t \geq t_0 \geq 0. \tag{C.35}$$

$\square$

A function V satisfying the conditions of Theorem C.14 is called an *ISS-Lyapunov function*. The inverse of Theorem C.14 is introduced next, and an equivalent dissipativity-type characterization of ISS is also introduced.

Theorem C.15 (Lyapunov Characterization of ISS). *For the system*

$$\dot{x} = f(t, x, u), \tag{C.36}$$

where f is periodic in t, the following properties are equivalent:

1. *the system is ISS,*
2. *there exists an ISS-Lyapunov function,*
3. *there exist a C^1 function $V : \mathbb{R}_+ \times \mathbb{R}^n \to \mathbb{R}_+$ and class-$\mathcal{K}_\infty$ functions α_1, α_2, α_3, and α_4 such that the following holds for all $t \geq 0$, $x \in \mathbb{R}^n$, $u \in \mathbb{R}^m$:*

$$\alpha_1(|x|) \leq V(t, x) \leq \alpha_2(|x|), \tag{C.37}$$

$$\frac{\partial V(t, x)}{\partial t} + \frac{\partial V(t, x)}{\partial x} f(t, x, u) \leq -\alpha_3(|x|) + \alpha_4(|u|). \tag{C.38}$$

Moreover, if f is autonomous then V can be chosen independent of t and C^∞.

The proof of the time-varying case can be found in [103], whereas the proof for the autonomous case can be found in [124].

The following lemma establishes a useful property that a cascade of two ISS systems is itself ISS.

Lemma C.16. *Suppose that in the system*

$$\dot{x}_1 = f_1(t, x_1, x_2, u), \tag{C.39}$$
$$\dot{x}_2 = f_2(t, x_2, u), \tag{C.40}$$

the x_1-subsystem is ISS with respect to x_2 and u, and the x_2-subsystem is ISS with respect to u; that is,

$$|x_1(t)| \leq \beta_1(|x_1(s)|, t - s) + \gamma_1\left(\sup_{s \leq \tau \leq t} \{|x_2(\tau)| + |u(\tau)|\}\right), \tag{C.41}$$

$$|x_2(t)| \leq \beta_2(|x_2(s)|, t - s) + \gamma_2\left(\sup_{s \leq \tau \leq t} |u(\tau)|\right), \tag{C.42}$$

where β_1 and β_2 are class-$\mathcal{KL}$ functions and γ_1 and γ_2 are class-$\mathcal{K}$ functions. Then the complete $x = (x_1, x_2)$-system is ISS with

$$|x(t)| \leq \beta(|x(s)|, t - s) + \gamma\left(\sup_{s \leq \tau \leq t} |u(\tau)|\right), \tag{C.43}$$

where

$$\beta(r, t) = \beta_1(2\beta_1(r, t/2) + 2\gamma_1(2\beta_2(r, 0)), t/2)$$
$$+ \gamma_1(2\beta_2(r, t/2)) + \beta_2(r, t), \tag{C.44}$$
$$\gamma(r) = \beta_1(2\gamma_1(2\gamma_2(r) + 2r), 0) + \gamma_1(2\gamma_2(r) + 2r) + \gamma_2(r). \tag{C.45}$$

Proof. With $(s,t) = (t/2, t)$, (C.41) is rewritten as

$$|x_1(t)| \leq \beta_1(|x_1(t/2)|, t/2) + \gamma_1 \left(\sup_{t/2 \leq \tau \leq t} \{|x_2(\tau)| + |u(\tau)|\} \right). \tag{C.46}$$

From (C.42), we have

$$\sup_{t/2 \leq \tau \leq t} |x_2(\tau)| \leq \sup_{t/2 \leq \tau \leq t} \left\{ \beta_2(|x_2(0)|, \tau) + \gamma_2 \left(\sup_{0 \leq \sigma \leq \tau} |u(\sigma)| \right) \right\}$$

$$\leq \beta_2(|x_2(0)|, t/2) + \gamma_2 \left(\sup_{0 \leq \tau \leq t} |u(\tau)| \right), \tag{C.47}$$

and from (C.41), we obtain

$$|x_1(t/2)| \leq \beta_1(|x_1(0)|, t/2) + \gamma_1 \left(\sup_{0 \leq \tau \leq t/2} \{|x_2(\tau)| + |u(\tau)|\} \right)$$

$$\leq \beta_1(|x_1(0)|, t/2)$$

$$+ \gamma_1 \left(\sup_{0 \leq \tau \leq t/2} \left\{ \beta_2(|x_2(0)|, \tau) + \gamma_2 \left(\sup_{0 \leq \sigma \leq \tau} |u(\sigma)| \right) + |u(\tau)| \right\} \right)$$

$$\leq \beta_1(|x_1(0)|, t/2)$$

$$+ \gamma_1 \left(\beta_2(|x_2(0)|, 0) + \sup_{0 \leq \tau \leq t/2} \{\gamma_2(|u(\tau)|) + |u(\tau)|\} \right)$$

$$\leq \beta_1(|x_1(0)|, t/2) + \gamma_1 \left(2\beta_2(|x_2(0)|, 0) \right)$$

$$+ \gamma_1 \left(2 \sup_{0 \leq \tau \leq t/2} \{\gamma_2(|u(\tau)|) + |u(\tau)|\} \right), \tag{C.48}$$

where in the last inequality we have used the fact that $\delta(a+b) \leq \delta(2a) + \delta(2b)$ for any class-$\mathcal{K}$ function δ and any nonnegative a and b. Then, substituting (C.47) and (C.48) into (C.46), we get

$$|x_1(t)| \leq \beta_1 \left(\beta_1(|x_1(0)|, t/2) + \gamma_1(2\beta_2(|x_2(0)|, 0)) \right.$$

$$+ \gamma_1 \left(2 \sup_{0 \leq \tau \leq t/2} \{\gamma_2(|u(\tau)|) + |u(\tau)|\} \right)$$

$$+ \gamma_1 \left(\beta_2(|x_2(0)|, t/2) + \gamma_2 \left(\sup_{0 \leq \tau \leq t} |u(\tau)| \right) + \sup_{t/2 \leq \tau \leq t} \{|u(\tau)|\} \right)$$

$$\leq \beta_1(2\beta_1(|x_1(0)|, t/2) + 2\gamma_1(2\beta_2(|x_2(0)|, 0)), t/2)$$

$$+ \gamma_1(2\beta_2(|x_2(0)|, t/2))$$

$$+ \beta_1 \left(2\gamma_1 \left(2 \sup_{0 \leq \tau \leq t} \{ \gamma_2 \left(|u(\tau)| \right) + |u(\tau)| \} \right), 0 \right)$$

$$+ \gamma_1 \left(2 \sup_{0 \leq \tau \leq t} \{ \gamma_2 \left(|u(\tau)| \right) + |u(\tau)| \} \right). \tag{C.49}$$

Combining (C.49) and (C.42), we arrive at (C.43) with (C.44)–(C.45). ☐

Since (C.39) and (C.40) are ISS, then there exist ISS-Lyapunov functions V_1 and V_2 and class-$\mathcal{K}_\infty$ functions $\alpha_1, \rho_1, \alpha_2,$ and ρ_2 such that

$$\frac{\partial V_1}{\partial x_1} f_1(t, x_1, x_2, u) \leq -\alpha_1(|x_1|) + \rho_1(|x_2|) + \rho_1(|u|), \tag{C.50}$$

$$\frac{\partial V_2}{\partial x_2} f_2(t, x_2, u) \leq -\alpha_2(|x_2|) + \rho_2(|u|). \tag{C.51}$$

The functions $V_1, V_2, \alpha_1, \rho_1, \alpha_2,$ and ρ_2 can *always* be found such that

$$\rho_1 = \alpha_2/2. \tag{C.52}$$

Then the ISS-Lyapunov function for the complete system (C.39)–(C.40) can be defined as

$$V(x) = V_1(x_1) + V_2(x_2), \tag{C.53}$$

and its derivative

$$\dot{V} \leq -\alpha_1(|x_1|) - \frac{1}{2}\alpha_2(|x_2|) + \rho_1(|u|) + \rho_2(|u|) \tag{C.54}$$

establishes the ISS property of (C.39)–(C.40) by part 3 of Theorem C.15.

In some applications of input-to-state stability, the following lemma is useful, as it is much simpler than Theorem C.14.

Lemma C.17. *Let v and ρ be real-valued functions defined on $\mathbb{R}_+$, and let b and c be positive constants. If they satisfy the differential inequality*

$$\dot{v} \leq -cv + b\rho(t)^2, \qquad v(0) \geq 0, \tag{C.55}$$

then the following hold:

(i) If $\rho \in \mathcal{L}_\infty$, then $v \in \mathcal{L}_\infty$ and

$$v(t) \leq v(0)e^{-ct} + \frac{b}{c}\|\rho\|_\infty^2. \tag{C.56}$$

(ii) If $\rho \in \mathcal{L}_2$, then $v \in \mathcal{L}_\infty$ and

$$v(t) \leq v(0)e^{-ct} + b\|\rho\|_2^2. \tag{C.57}$$

Proof. (i) From Lemma B.3, we have

$$v(t) \leq v(0)e^{-ct} + b \int_0^t e^{-c(t-\tau)}\rho(\tau)^2 d\tau$$

$$\leq v(0)e^{-ct} + b \sup_{\tau \in [0,t]} \{\rho(\tau)^2\} \int_0^t e^{-c(t-\tau)} d\tau$$

$$\leq v(0)e^{-ct} + b\|\rho\|_\infty^2 \frac{1}{c}\left(1 - e^{-ct}\right)$$

$$\leq v(0)e^{-ct} + \frac{b}{c}\|\rho\|_\infty^2. \tag{C.58}$$

(ii) From (B.9), we have

$$v(t) \leq v(0)e^{-ct} + b \sup_{\tau \in [0,t]} \left\{e^{-c(t-\tau)}\right\} \int_0^t \rho(\tau)^2 d\tau$$

$$= v(0)e^{-ct} + b\|\rho\|_2^2. \tag{C.59}$$

$\square$

Remark C.18. From Lemma C.17, it follows that if

$$\dot{v} \leq -cv + b_1\rho_1(t)^2 + b_2\rho_2(t)^2, \qquad v(0) \geq 0, \tag{C.60}$$

and $\rho_1 \in \mathcal{L}_\infty$ and $\rho_2 \in \mathcal{L}_2$, then $v \in \mathcal{L}_\infty$ and

$$v(t) \leq v(0)e^{-ct} + \frac{b_1}{c}\|\rho_1\|_\infty^2 + b_2\|\rho_2\|_2^2. \tag{C.61}$$

This, in particular, implies the input-to-state stability with respect to two inputs: ρ_1 and $\|\rho_2\|_2$. ◇

Appendix D

D.1 Parameter Projection

Our adaptive designs rely on the use of parameter projection in our identifiers. We provide a treatment of projection for a general convex parameter set. The treatment for some of our designs where projection is used for only a scalar estimate $\hat{D}$ is easily deduced from the general case.

Let us define the following convex set:

$$\Pi = \left\{ \hat{\theta} \in \mathbb{R}^p \,\middle|\, \mathcal{P}(\hat{\theta}) \leq 0 \right\}, \tag{D.1}$$

where by assuming that the convex function $\mathcal{P} : \mathbb{R}^p \to \mathbb{R}$ is smooth, we ensure that the boundary $\partial\Pi$ of Π is smooth. Let us denote the interior of Π by $\overset{\circ}{\Pi}$ and observe that $\nabla_{\hat{\theta}}\mathcal{P}$ represents an outward normal vector at $\hat{\theta} \in \partial\Pi$. The standard projection operator is

$$\mathrm{Proj}\{\tau\} = \begin{cases} \tau, & \hat{\theta} \in \overset{\circ}{\Pi} \text{ or } \nabla_{\hat{\theta}}\mathcal{P}^T\tau \leq 0, \\[2mm] \left(I - \Gamma \dfrac{\nabla_{\hat{\theta}}\mathcal{P}\,\nabla_{\hat{\theta}}\mathcal{P}^{rmT}}{\nabla_{\hat{\theta}}\mathcal{P}^T\Gamma\nabla_{\hat{\theta}}\mathcal{P}}\right)\tau, & \hat{\theta} \in \partial\Pi \text{ and } \nabla_{\hat{\theta}}\mathcal{P}^T\tau > 0, \end{cases} \tag{D.2}$$

where Γ belongs to the set $\mathcal{G}$ of all positive-definite symmetric $p \times p$ matrices. Although Proj is a function of three arguments, τ, $\hat{\theta}$, and Γ, for compactness of notation, we write only $\mathrm{Proj}\{\tau\}$.

The meaning of (D.2) is that when $\hat{\theta}$ is in the interior of Π or at the boundary with τ pointing inward, then $\mathrm{Proj}\{\tau\} = \tau$. When $\hat{\theta}$ is at the boundary with τ pointing outward, then Proj projects τ on the hyperplane tangent to $\partial\Pi$ at $\hat{\theta}$.

In general, the mapping (D.2) is discontinuous. This is undesirable for two reasons. First, the discontinuity represents a difficulty for implementation in continuous time. Second, since the Lipschitz continuity is violated, we cannot use standard theorems for the existence of solutions. Therefore, we sometimes want to smooth the projection operator. Let us consider the following convex set:

$$\Pi_\varepsilon = \left\{ \hat{\theta} \in \mathbb{R}^p \,\middle|\, \mathcal{P}(\hat{\theta}) \leq \varepsilon \right\}, \tag{D.3}$$

which is a union of the set Π and an $O(\varepsilon)$-boundary layer around it. We now modify (D.2) to achieve continuity of the transition from the vector field τ on the boundary of Π to the vector field $\left(I - \Gamma \frac{\nabla_{\hat{\theta}}\mathcal{P}\,\nabla_{\hat{\theta}}\mathcal{P}^T}{\nabla_{\hat{\theta}}\mathcal{P}^T\Gamma\nabla_{\hat{\theta}}\mathcal{P}}\right)\tau$ on the boundary of Π_ε:

$$\text{Proj}\{\tau\} = \begin{cases} \tau, & \hat{\theta}\in\overset{\circ}{\Pi}\ \text{or}\ \nabla_{\hat{\theta}}\mathcal{P}^T\tau\leq 0, \\ \left(I - c(\hat{\theta})\Gamma\dfrac{\nabla_{\hat{\theta}}\mathcal{P}\,\nabla_{\hat{\theta}}\mathcal{P}^T}{\nabla_{\hat{\theta}}\mathcal{P}^T\Gamma\nabla_{\hat{\theta}}\mathcal{P}}\right)\tau, & \hat{\theta}\in\Pi_\varepsilon\backslash\overset{\circ}{\Pi}\ \text{and}\ \nabla_{\hat{\theta}}\mathcal{P}^T\tau > 0, \end{cases} \tag{D.4}$$

$$c(\hat{\theta}) = \min\left\{1, \frac{\mathcal{P}(\hat{\theta})}{\varepsilon}\right\}. \tag{D.5}$$

It is helpful to note that $c(\partial\Pi) = 0$ and $c(\partial\Pi_\varepsilon) = 1$.

In our proofs of stability of adaptive systems, we use the following technical properties of the projection operator (D.4).

Lemma D.1 (Projection Operator). *The following are the properties of the projection operator (D.4):*

(i) *The mapping* $\text{Proj}: \mathbb{R}^p \times \Pi_\varepsilon \times \mathcal{G} \to \mathbb{R}^p$ *is locally Lipschitz in its arguments* $\tau, \hat{\theta},$ *and* Γ.

(ii) $\text{Proj}\{\tau\}^T\Gamma^{-1}\text{Proj}\{\tau\} \leq \tau^T\Gamma^{-1}\tau$, $\forall\hat{\theta}\in\Pi_\varepsilon$.

(iii) *Let* $\Gamma(t), \tau(t)$ *be continuously differentiable and*

$$\dot{\hat{\theta}} = \text{Proj}\{\tau\}, \qquad \hat{\theta}(0)\in\Pi_\varepsilon.$$

Then, on its domain of definition, the solution $\hat{\theta}(t)$ *remains in* Π_ε.

(iv) $-\tilde{\theta}^T\Gamma^{-1}\text{Proj}\{\tau\} \leq -\tilde{\theta}^T\Gamma^{-1}\tau$, $\forall\hat{\theta}\in\Pi_\varepsilon, \theta\in\Pi$.

Proof. (i) The proof of this point is lengthy but straightforward and is omitted here.

(ii) For $\hat{\theta}\in\overset{\circ}{\Pi}$ or $\nabla_{\hat{\theta}}\mathcal{P}^T\tau\leq 0$, we have $\text{Proj}\{\tau\} = \tau$ and (ii) trivially holds with equality. Otherwise, a direct computation gives

$$\text{Proj}\{\tau\}^T\Gamma^{-1}\text{Proj}\{\tau\} = \tau^T\Gamma^{-1}\tau - 2c(\hat{\theta})\frac{\left(\nabla_{\hat{\theta}}\mathcal{P}^T\tau\right)^2}{\nabla_{\hat{\theta}}\mathcal{P}^T\Gamma\nabla_{\hat{\theta}}\mathcal{P}} + c(\hat{\theta})^2\frac{\left|\nabla_{\hat{\theta}}\mathcal{P}\nabla_{\hat{\theta}}\mathcal{P}^T\tau\right|_\Gamma^2}{\left(\nabla_{\hat{\theta}}\mathcal{P}^T\Gamma\nabla_{\hat{\theta}}\mathcal{P}\right)^2}$$

$$= \tau^T\Gamma^{-1}\tau - c(\hat{\theta})\left(2 - c(\hat{\theta})\right)\frac{\left(\nabla_{\hat{\theta}}\mathcal{P}^T\tau\right)^2}{\nabla_{\hat{\theta}}\mathcal{P}^T\Gamma\nabla_{\hat{\theta}}\mathcal{P}}$$

$$\leq \tau^T\Gamma^{-1}\tau, \tag{D.6}$$

where the last inequality follows by noting that $c(\hat{\theta})\in[0,1]$ for $\hat{\theta}\in\Pi_\varepsilon\backslash\overset{\circ}{\Pi}$.

(iii) Using the definition of the Proj operator, we get

$$
\nabla_{\hat{\theta}} \mathcal{P}^T \text{Proj}\{\tau\} =
\begin{cases}
\nabla_{\hat{\theta}} \mathcal{P}^T \tau, & \hat{\theta} \in \overset{\circ}{\Pi} \text{ or } \nabla_{\hat{\theta}} \mathcal{P}^T \tau \leq 0, \\
\left(1 - c(\hat{\theta})\right) \nabla_{\hat{\theta}} \mathcal{P}^T \tau, & \hat{\theta} \in \Pi_\varepsilon \backslash \overset{\circ}{\Pi} \text{ and } \nabla_{\hat{\theta}} \mathcal{P}^T \tau > 0,
\end{cases}
\tag{D.7}
$$

which, in view of the fact that $c(\hat{\theta}) \in [0, 1]$ for $\hat{\theta} \in \Pi_\varepsilon \backslash \overset{\circ}{\Pi}$, implies that

$$
\nabla_{\hat{\theta}} \mathcal{P}^T \text{Proj}\{\tau\} \leq 0 \text{ whenever } \hat{\theta} \in \partial \Pi_\varepsilon ;
\tag{D.8}
$$

that is, the vector $\text{Proj}\{\tau\}$ either points inside Π_ε or is tangential to the hyperplane of $\partial \Pi_\varepsilon$ at $\hat{\theta}$. Since $\hat{\theta}(0) \in \Pi_\varepsilon$, it follows that $\hat{\theta}(t) \in \Pi_\varepsilon$ as long as the solution exists.

(iv) For $\hat{\theta} \in \overset{\circ}{\Pi}$, (iv) trivially holds with equality. For $\hat{\theta} \in \Pi_\varepsilon \backslash \overset{\circ}{\Pi}$, since $\theta \in \Pi$ and $\mathcal{P}$ is a convex function, we have

$$
(\theta - \hat{\theta})^T \nabla_{\hat{\theta}} \mathcal{P} \leq 0 \text{ whenever } \hat{\theta} \in \Pi_\varepsilon \backslash \overset{\circ}{\Pi}.
\tag{D.9}
$$

With (D.9), we now calculate

$$
-\tilde{\theta}^T \Gamma^{-1} \text{Proj}\{\tau\} = -\tilde{\theta}^T \Gamma^{-1} \tau
$$

$$
+
\begin{cases}
0, & \hat{\theta} \in \overset{\circ}{\Pi} \text{ or } \nabla_{\hat{\theta}} \mathcal{P}^T \tau \leq 0 \\
c(\hat{\theta}) \frac{(\tilde{\theta}^T \nabla_{\hat{\theta}} \mathcal{P})(\nabla_{\hat{\theta}} \mathcal{P}^T \tau)}{\nabla_{\hat{\theta}} \mathcal{P}^T \Gamma \nabla_{\hat{\theta}} \mathcal{P}}, & \hat{\theta} \in \Pi_\varepsilon \backslash \overset{\circ}{\Pi} \text{ and} \\
& \nabla_{\hat{\theta}} \mathcal{P}^T \tau > 0
\end{cases}
$$

$$
\leq -\tilde{\theta}^T \Gamma^{-1} \tau ,
\tag{D.10}
$$

which completes the proof.

$\square$

Bibliography

[1] Anfinsen, H., and Aamo, O.-M. (2019). *Adaptive Control of Hyperbolic PDEs*, Springer.

[2] Artstein, Z. (1982). Linear systems with delayed controls: A reduction, *IEEE Transactions on Automatic Control*, vol. 27, no. 4, pp. 869–879.

[3] Astrom, K. J., and Wittenmark, B. (2013). *Adaptive Control*, Courier Corporation.

[4] Bekiaris-Liberis, N. (2014). Simultaneous compensation of input and state delays for nonlinear systems, *Systems and Control Letters*, vol. 73, pp. 96–102.

[5] Bekiaris-Liberis, N., Jankovic, M., and Krstic, M. (2012). Compensation of state-dependent state delay for nonlinear systems, *Systems and Control Letters*, vol. 61, pp. 849–856.

[6] Bekiaris-Liberis, N., Jankovic, M., and Krstic, M. (2013). Adaptive stabilization of LTI systems with distributed input delay, *International Journal of Adaptive Control and Signal Processing*, vol. 27, pp. 47–65.

[7] Bekiaris-Liberis, N., and Krstic, M. (2010). Compensating the distributed effect of a wave PDE in the actuation or sensing path of MIMO LTI systems, *Systems and Control Letters*, vol. 59, pp. 713–719.

[8] Bekiaris-Liberis, N., and Krstic, M. (2010). Delay-adaptive feedback for linear feedforward systems, *Systems and Control Letters*, vol. 59, pp. 277–283.

[9] Bekiaris-Liberis, N., and Krstic, M. (2010). Stabilization of linear strict-feedback systems with delayed integrators, *Automatica*, vol. 56, pp. 1902–1910.

[10] Bekiaris-Liberis, N., and Krstic, M. (2011). Compensating the distributed effect of diffusion and counter-convection in multi-input and multi-output LTI systems, *IEEE Transactions on Automatic Control*, vol. 56, pp. 637–642.

[11] Bekiaris-Liberis, N., and Krstic, M. (2011). Lyapunov stability of linear
      predictor feedback for distributed input delays, *IEEE Transactions on
      Automatic Control*, vol. 56, pp. 655–660.

[12] Bekiaris-Liberis, N., and Krstic, M. (2012). Compensation of time-varying
      input and state delays for nonlinear systems, *Journal of Dynamic Sys-
      tems, Measurement, and Control*, vol. 134, paper 011009.

[13] Bekiaris-Liberis, N., and Krstic, M. (2013). Compensation of state-
      dependent input delay for nonlinear systems, *IEEE Transactions on Auto-
      matic Control*, vol. 58, pp. 275–289.

[14] Bekiaris-Liberis, N., and Krstic, M. (2013). Nonlinear control under de-
      lays that depend on delayed states, *European Journal of Control*, vol. 19,
      pp. 389–398.

[15] Bekiaris-Liberis, N., and Krstic, M. (2013). *Nonlinear Control under Non-
      constant Delays*, Society for Industrial and Applied Mathematics.

[16] Bekiaris-Liberis, N., and Krstic, M. (2013). Robustness of nonlinear pre-
      dictor feedback laws to time- and state-dependent delay perturbations,
      Automatica, vol. 49, pp. 1576–1590.

[17] Bekiaris-Liberis, N., and Krstic, M. (2014). Compensation of wave actu-
      ator dynamics for nonlinear systems, *IEEE Transactions on Automatic
      Control*, vol. 59, pp. 1555–1570.

[18] Bekiaris-Liberis, N., and Krstic, M. (2015). Predictor-feedback stabiliza-
      tion of multi-input nonlinear systems, *Proceedings of IEEE Conference
      on Decision and Control*, pp. 7078–7083, Osaka, Japan, December
      15–18.

[19] Bekiaris-Liberis, N., and Krstic, M. (2016). Stability of predictor-based
      feedback for nonlinear systems with distributed input delay, *Automatica*,
      vol. 70, pp. 195–203.

[20] Bekiaris-Liberis, N., and Krstic, M. (2017). Predictor-feedback stabiliza-
      tion of multi-input nonlinear systems, *IEEE Transactions on Automatic
      Control*, vol. 62, no. 2, pp. 516–531.

[21] Boyd, S., Ghaoui, L.-E., Feron, and E., Balakrishnan, V. (1994). *Linear
      Matrix Inequality in Systems and Control Theory*, vol. 15, SIAM, Studies
      in Applied Mathematics.

[22] Bresch-Pietri, D., Chauvin, J., and Petit, N. (2010). Adaptive backstep-
      ping controller for uncertain systems with unknown input time-delay.
      Application to SI engines, *49th IEEE Conference on Decision and Con-
      trol*, Atlanta, GA, December 15–17.

[23] Bresch-Pietri, D., Chauvin, J., and Petit, N. (2011). Adaptive backstep-
      ping for uncertain systems with time-delay on-line update laws, *Proceed-
      ings of 2011 American Control Conference*, San Francisco, CA, June 29–
      July 1.

[24] Bresch-Pietri, D., Chauvin, J., and Petit, N. (2011). Output feedback con-
      trol of time delay systems with adaptation of delay estimate, *Proceedings
      of the 18th IFAC World Congress*, Milano, Italy, August 28–September 2.

[25] Bresch-Pietri, D., Chauvin, J., and Petit, N. (2012). Adaptive control
      scheme for uncertain time-delay systems, *Automatica*, vol. 48, pp. 1536–
      1552.

[26] Bresch-Pietri, D., Chauvin, J., and Petit, N. (2014). Prediction-based stabi-
      lization of linear systems subject to input-dependent input delay of integral
      type, *IEEE Transactions on Automatic Control*, vol. 59, pp. 2385–2399.

[27] Bresch-Pietri, D., and Krstic, M. (2009). Adaptive trajectory tracking
      despite unknown input delay and plant parameters, *Automatica*, vol. 45,
      pp. 2074–2081.

[28] Bresch-Pietri, D., and Krstic, M. (2010). Delay-adaptive predictor feed-
      back for systems with unknown long actuator delay, *IEEE Transactions
      on Automatic Control*, vol. 55, no. 9, pp. 2106–2112.

[29] Bresch-Pietri, D., and Krstic, M. (2014). Delay-adaptive control for non-
      linear systems, *IEEE Transactions on Automatic Control*, vol. 59, no. 5,
      pp. 1203–1218.

[30] Bresch-Pietri, D., and Krstic, M. (2014). Output-feedback adaptive con-
      trol of a wave PDE with boundary anti-damping, *Automatica*, vol. 50,
      pp. 1407–1415.

[31] Bresch-Pietri, D., Mazenc, F., and Petit, N. (2018). Robust compensation
      of a chattering time-varying input delay with jumps, *Automatica*, vol. 92,
      pp. 225–234.

[32] Bresch-Pietri, D., Prieur, C., and Trelat, E. (2018). Alternative formula-
      tion of predictors for finite-dimensional linear control systems with input
      delay, *Systems and Control Letters*, vol. 113, pp. 9–16.

[33] Cai, X., Bekiaris-Liberis, N., and Krstic, M. (2018). Input-to-state sta-
      bility and inverse optimality of linear time-varying-delay predictor feed-
      backs, *IEEE Transactions on Automatic Control*, vol. 63, pp. 233–240.

[34] Diagne, M., Bekiaris-Liberis, N., and Krstic, M. (2017). Compensation
      of input delay that depends on delayed input, *Automatica*, vol. 85,
      pp. 362–373.

[35] Diagne, M., Bekiaris-Liberis, N., and Krstic, M. (2017). Time- and state-dependent input delay-compensated Bang-Bang control of a screw extruder for 3D printing, *International Journal of Robust and Nonlinear Control*, vol. 27, pp. 3727–3757.

[36] Diop, S., Kolmanovsky, I., Moraal, P. E., and Van Nieuwstadt, M. (2001). Preserving stability/performance when facing an unknown time-delay, *Control Engineering Practice*, vol. 9, pp. 1319–1325.

[37] Evesque, S., Annaswamy, A. M., Niculescu, S., and Dowling, A. P. (2003). Adaptive control of a class of time-delay systems, *ASME Transactions on Dynamics, Systems, Measurement, and Control*, vol. 125, pp. 186–193.

[38] Fridman, E. (2001). New Lyapunov-Krasovskii functionals for stability of linear retarded and neutral type systems, *Systems and Control Letters*, vol. 43, no. 4, pp. 309–319.

[39] Fridman, E. (2002). Stability of linear descriptor systems with delay: A Lyapunov-based approach, *Journal of Mathematical Analysis and Applications*, vol. 273, no. 1, pp. 24–44.

[40] Fridman, E. (2010). A refined input delay approach to sampled-data control, *Automatica*, vol. 46, pp. 421–427.

[41] Fridman, E. (2014). *Introduction to Time-Delay Systems: Analysis and Control*, Birkhauser.

[42] Fridman, E. (2014). Tutorial on Lyapunov-based methods for time-delay systems, *European Journal of Control*, vol. 20, pp. 271–283.

[43] Fridman, E., Seuret, A., and Richard, J.-P. (2004). Robust sampled-data stabilization of linear systems: An input delay approach, *Automatica*, vol. 40, no. 8, pp. 1441–1446.

[44] Fridman, E., and Shaked, U. (2001). A new bounded real lemma representation for time-delay systems and its applications, *IEEE Transactions on Automatic Control*, vol. 46, no. 12, pp. 1973–1979.

[45] Fridman, E., and Shaked, U. (2002). A descriptor system approach to H_∞ control of linear time-delay systems, *IEEE Transactions on Automatic Control*, vol. 47, no. 2, pp. 253–270.

[46] Fridman, E., and Shaked, U. (2002). An improved stabilization method for linear time-delay systems, *IEEE Transactions on Automatic Control*, vol. 47, no. 11, pp. 1931–1937.

[47] Fridman, E., and Shaked, U. (2003). Delay-dependent stability and H_∞ control: Constant and time-varying delays, *International Journal of Control*, vol. 76, no. 1, pp. 48–60.

[48] Goodwin, G. C., and Sin, K. S. (2014). *Adaptive Filtering, Prediction and Control*, Courier Corporation.

[49] Gu, K. (1997). Discretized LMI set in the stability problem of linear time-delay systems, *International Journal of Control*, vol. 68, pp. 923–934.

[50] Gu, K. (2010). Stability problem of systems with multiple delay channels, *Automatica*, vol. 46, pp. 743–751.

[51] Gu, K., Kharitonov, V., and Chen, J. (2003). *Stability of Time-Delay Systems*, Birkhauser.

[52] Gu, K., and Niculescu, S.-I. (2001). Further remarks on additional dynamics in various model transformations of linear delay systems, *IEEE Transactions on Automatic Control*, vol. 46, pp. 497–500.

[53] Gu, K., and Niculescu, S.-I. (2003). Survey on recent results in the stability and control of time-delay systems, *Transactions of ASME*, vol. 125, pp. 158–165.

[54] Gu, K., Niculescu, S.-I., and Chen, J. (2005). On stability of crossing curves for general systems with two delays, *Journal of Mathematical Analysis and Applications*, vol. 311, pp. 231–253.

[55] Halil, H. (2002). *Nonlinear Systems*, Prentice Hall, 3rd ed.

[56] Hardy, G. H., Littlewood, J. E., and Polya, G. (1959). *Inequalities*, Cambridge University Press, 2nd ed.

[57] Ioannou, P., and Sun, J. (1996). *Robust Adaptive Control*, PTR Prentice-Hall.

[58] Karafyllis, I. (2004). The non-uniform in time small-gain theorem for a wide class of control systems with outputs, *European Journal on Control*, vol. 10, pp. 307–323.

[59] Karafyllis, I. (2006). Finite-time global stabilization by means of time-varying distributed delay feedback, *SIAM Journal on Control and Optimization*, vol. 45, pp. 320–342.

[60] Karafyllis, I. (2006). Lyapunov theorems for systems described by retarded functional differential equations, *Nonlinear Analysis*, vol. 64, pp. 590–617.

[61] Karafyllis, I. (2009). Stability results for systems described by coupled retarded functional differential equations and functional difference equations, *Nonlinear Analysis*, vol. 71, pp. 3339–3362.

[62] Karafyllis, I. (2010). Necessary and sufficient Lyapunov-like conditions for robust nonlinear stabilization, *ESAIM Control, Optimization and Calculus of Variations*, vol. 16, pp. 887–928.

[63] Karafyllis, I. (2011). Stabilization by means of approximate predictors for systems with delayed input, *SIAM Journal on Control and Optimization*, vol. 49, pp. 1100–1123.

[64] Karafyllis, I., and Jiang, Z.-P. (2007). A small-gain theorem for a wide class of feedback systems with control applications, *SIAM Journal on Control and Optimization*, vol. 46, pp. 1483–1517.

[65] Karafyllis, I., and Jiang, Z.-P. (2011). *Stability and Stabilization of Nonlinear Systems*, Springer.

[66] Karafyllis, I., and Krstic, M. (2012). Nonlinear stabilization under sampled and delayed measurements, and with inputs subject to delay and zero-order hold, *IEEE Transactions on Automatic Control*, vol. 57, pp. 1141–1154.

[67] Karafyllis, I., and Krstic, M. (2013). Delay-robustness of linear predictor feedback without restriction on delay rate, *Automatica*, vol. 49, pp. 1761–1767.

[68] Karafyllis, I., and Krstic, M. (2013). *Predictor Feedback for Delay Systems: Implementations and Approximations*, Birkhauser.

[69] Karafyllis, I., and Krstic, M. (2019). *Input-to-State Stability for PDEs*, Springer.

[70] Karafyllis, I., Pepe, P., and Jiang, Z.-P. (2008). Input-to-output stability for systems described by retarded functional differential equations, *European Journal of Control*, vol. 6, pp. 539–555.

[71] Karafyllis, I., and Tsinias, J. (2004). Nonuniform in time input-to-state stability and the small gain theorem, *IEEE Transactions on Automatic Control*, vol. 49, pp. 196–216.

[72] Kharitonov, V. L. (2014). An extension of the prediction scheme to the case of systems with both input and state delay, *Automatica*, vol. 50, pp. 211–217.

[73] Kharitonov, V. L. (2017). Prediction-based control for systems with state and several input delays, *Automatica*, vol. 79, pp. 11–16.

[74] Krasovskii, N. (1962). On the analytic construction of an optimal control in a system with time lags, *Journal of Applied Mathematics and Mechanics*, vol. 26, pp. 50–67.

[75] Krasovskii, N. (1963). *Stability of Motion* (in Russian), Stanford University Press.

[76] Krstic, M. (2004). Feedback linearizability and explicit integrator forwarding controllers for classes of feedforward systems, *IEEE Transactions on Automatic Control*, vol. 49, pp. 1668–1682.

[77] Krstic, M. (2008). Lyapunov tools for predictor feedbacks for delay sys-
      tems: Inverse optimality and robustness to delay mismatch, *Automatica*,
      vol. 44, pp. 2930–2935.

[78] Krstic, M. (2008). On compensating long actuator delays in nonlinear
      control, *IEEE Transactions on Automatic Control*, vol. 53, pp. 1684–1688.

[79] Krstic, M. (2009). Adaptive control of an anti-stable wave PDE, 2009
      American Control Conference, pp. 1505–1510, St. Louis, MO, June 10–12.

[80] Krstic, M. (2009). *Delay Compensation for Nonlinear, Adaptive, and PDE
      Systems*. Berlin.

[81] Krstic, M. (2010). Adaptive control of an anti-stable wave PDE, *Dynam-
      ics of Continuous, Discrete, and Impulsive Systems*, vol. 17, pp. 853–
      882, Invited paper in the special issue in the honor of Professor Hassan
      Khalil.

[82] Krstic, M. (2010). Compensation of infinite-dimensional actuator and sen-
      sor dynamics: Nonlinear and delay–adaptive systems, *IEEE Control Sys-
      tems Magazine*, vol. 30, pp. 22–41.

[83] Krstic, M. (2010). Input delay compensation for forward complete and
      feedforward nonlinear systems, *IEEE Transactions on Automatic Control*,
      vol. 55, pp. 287–303.

[84] Krstic, M. (2010). Lyapunov stability of linear predictor feedback for
      time-varying input delay, *IEEE Transactions on Automatic Control*,
      vol. 55, pp. 554–559.

[85] Krstic, M., and Banaszuk, A. (2006). Multivariable adaptive control of
      instabilities arising in jet engines, *Control Engineering Practice*, vol. 14,
      pp. 833–842.

[86] Krstic, M., and Bekiaris-Liberis, N. (2010). Compensation of infinite-
      dimensional input dynamics, *IFAC Annual Reviews*, vol. 34, pp. 233–244.

[87] Krstic, M., and Bekiaris-Liberis, N. (2013). Nonlinear stabilization in infi-
      nite dimension, *IFAC Annual Reviews*, vol. 37, pp. 220–231.

[88] Krstic, M., and Bresch-Pietri, D. (2009). Delay-adaptive full-state predic-
      tor feedback for systems with unknown long actuator delay, *Proceedings
      of 2009 American Control Conference*, St. Louis, MO, June 10–12.

[89] Krstic, M., Kanellakopoulos, I., and Kokotovic, P. V. (1995). *Nonlinear
      and Adaptive Control Design*, Wiley.

[90] Krstic, M., and Smyshlyaev, A. (2008). Adaptive boundary control for
      unstable parabolic PDEs—Part I: Lyapunov design, *IEEE Transactions
      on Automatic Control*, vol. 53, pp. 1575–1591.

[91] Krstic, M., and Smyshlyaev, A. (2008). Adaptive control of PDEs, *Annual Reviews in Control*, vol. 32, pp. 149–160.

[92] Krstic, M., and Smyshlyaev, A. (2008). Backstepping boundary control for first-order hyperbolic PDEs and application to systems with actuator and sensor delays, *Systems and Control Letters*, vol. 57, no. 9, pp. 750–758.

[93] Krstic, M., and Smyshlyaev, A. (2008). *Boundary Control of PDEs: A Course on Backstepping Designs*, SIAM.

[94] Kwon, W. H., and Pearson, A. E. (1980). Feedback stabilization of linear systems with delayed control, *IEEE Transactions on Automatic Control*, vol. 25, no. 2, pp. 266–269.

[95] Liu, K., and Fridman, E. (2012). Wirtinger's inequality and Lyapunov-based sampled-data stabilization, *Automatica*, vol. 48, pp. 102–108.

[96] Liu, K., and Fridman, E. (2014). Delay-dependent methods and the first delay interval, *Systems and Control Letters*, vol. 64, pp. 57–63.

[97] Liu, K., and Fridman, E. (2015). Discrete-time network-based control under scheduling and actuator constraints, *International Journal of Robust and Nonlinear Control*, vol. 25, pp. 1816–1830.

[98] Liu, K., Fridman, E., and Hetel, L. (2012). Network-based control via a novel analysis of hybrid systems with time-varying delays. *Proceedings of the 51st IEEE Conference on Decision and Control*, Hawaii.

[99] Liu, K., Fridman, E., and Hetel, L. (2012). Stability and L2-gain analysis of networked control systems under Round-Robin scheduling: A time-delay approach, *Systems and Control Letters*, vol. 61, pp. 666–675.

[100] Liu, K., Suplin, V., and Fridman, E. (2010). Stability of linear systems with a general sawtooth delay, *IMA Journal of Mathematical Control and Information*, vol. 27, pp. 419–436, special issue on time-delay systems.

[101] Liu, W.-J., and Krstic, M. (2001). Adaptive control of Burgers' equation with unknown viscosity, *International Journal of Adaptive Control and Signal Processing*, vol. 15, pp. 745–766.

[102] Liu, Y., Pan, C., Gao, H., and Guo, G. (2017). Cooperative spacing control for interconnected vehicle systems with input delays, *IEEE Transactions on Vehicular Technology*, vol. 66, pp. 10692–10704.

[103] Malisoff, M., and Mazenc, F. (2005). Further remarks on strict input-to-state stable Lyapunov functions for time-varying systems, *Automatica*, vol. 41, pp. 1973–1978.

[104] Manitius, A. Z., and Olbrot, A. W. (1979). Finite spectrum assignment problem for systems with delays, *IEEE Transactions on Automatic Control*, vol. 24, no. 4, pp. 541–552.

[105] Mayne, D. Q. (1968). Control of linear systems with time delay, *Electronics Letters*, vol. 4, pp. 439–440.

[106] Michiels, W., and Niculescu, S.-I. (2003). On the delay sensitivity of Smith predictors, *International Journal of Systems Science*, vol. 34, pp. 543–551.

[107] Mondie, S., and Michiels, W. (2003). Finite spectrum assignment of unstable time-delay systems with a safe implementation, *IEEE Transactions on Automatic Control*, vol. 48, pp. 2207–2212.

[108] Narendra, K.-S., and Annaswamy, A.-M. (2012). *Stable Adaptive Systems*, Courier Corporation.

[109] Niculescu, S.-I. (2001). Delay effects on stability: A robust control approach, *Lecture Notes in Control and Information Sciences*, vol. 269, Springer.

[110] Niculescu, S.-I., and Annaswamy, A. M. (2003). An adaptive Smith-controller for time-delay systems with relative degree $n^* \leq 2$, *Systems and Control Letters*, vol. 49, pp. 347–358.

[111] Niculescu, S.-I., and Lozano, R. (2001). On the passivity of linear delay systems, *IEEE Transactions on Automatic Control*, vol. 46, pp. 460–464.

[112] Ortega, R., and Lozano, R. (1988). Globally stable adaptive controller for systems with delay, *International Journal of Control*, vol. 47, pp. 17–23.

[113] Ploeg, J., Van De Wouw, N., and Nijmeijer, H. (2014). Lp string stability of cascaded systems: Application to vehicle platooning, *IEEE Transactions on Control Systems Technology*, vol. 22, no. 2, pp. 786–793.

[114] Pyrkin, A., Bobtsov, A., Aranovskiy, S., Kolyubin, S., and Gromov, V. (2014). Adaptive controller for linear plant with parametric uncertainties, input delay and unknown disturbance dynamics, *The 19th World Congress of the International Federation of Automatic Control*, Cape Town, South Africa, pp. 11294–11298.

[115] Razumikhin, B. (1956). On the stability of systems with a delay (in Russian), *Prikladnaya Matematika i Mekhanika*, vol. 20, pp. 500–512.

[116] Richard, J.-P. (2003). Time-delay systems: An overview of some recent advances and open problems, *Automatica*, vol. 39, pp. 1667–1694.

[117] Sastry, S., and Bodson, M. (2011). *Adaptive Control: Stability, Convergence and Robustness*, Courier Corporation.

[118] Seuret, A., and Gouaisbaut, F. (2013). Wirtinger-based integral inequality: Application to time-delay systems, *Automatica*, vol. 49, pp. 2860–2866.

[119] Smith, O.J.M. (1959). A controller to overcome dead time, *ISA*, vol. 6, pp. 28–33.

[120] Smyshlyaev, A., and Krstic, M. (2007). Adaptive boundary control for unstable parabolic PDEs—Part II: Estimation-based designs, *Automatica*, vol. 43, pp. 1543–1556.

[121] Smyshlyaev, A., and Krstic, M. (2007). Adaptive boundary control for unstable parabolic PDEs—Part III: Output-feedback examples with swapping identifiers, *Automatica*, vol. 43, pp. 1557–1564.

[122] Smyshlyaev, A., and Krstic, M. (2010). *Adaptive Control of Parabolic PDEs*, Princeton University Press.

[123] Sontag, E. D. (1989). Smooth stabilization implies coprime factorization, *IEEE Transactions on Automatic Control*, vol. 34, pp. 435–443.

[124] Sontag, E. D., and Wang, Y. (1995). On characterization of the input-to-state stability property, *Systems and Control Letters*, vol. 24, pp. 351–359.

[125] Suplin, V., Fridman, E., and Shaked, U. (2006). H_∞ control of linear uncertain time-delay systems—A projection approach, *IEEE Transactions on Automatic control*, vol. 51, no. 4.

[126] Tao, G. (2003). *Adaptive Control Design and Analysis*, Wiley.

[127] Tsubakino, D., Oliveira, T. R., and Krstic, M. (2015). Predictor-feedback for multi-input LTI systems with distinct delays, *Proceedings of 2015 American Control Conference*, pp. 571–576, Chicago, IL, July 1–3.

[128] Tsubakino, D., Krstic, M., and Oliveira, T. R. (2016). Exact predictor feedbacks for multi-input LTI systems with distinct input delays, *Automatica*, vol. 71, pp. 143–150.

[129] Tsypkin, Y. Z. (1946). The systems with delayed feedback, *Avtomatika i Telemekhnika*, vol. 7, pp. 107–129.

[130] Vazquez, R., and Krstic, M. (2008). Control of 1-D parabolic PDEs with Volterra nonlinearities—Part I: Design, *Automatica*, vol. 44, pp. 2778–2790.

[131] Vazquez, R., and Krstic, M. (2008). Control of 1-D parabolic PDEs with Volterra nonlinearities—Part II: Analysis, *Automatica*, vol. 44, pp. 2791–2803.

[132] Wang, W., and Wen, C. (2011). Adaptive compensation for infinite number of actuator faults and failures, *Automatica*, vol. 47, pp. 2197–2210.

[133] Watanabe, K., and Ito, M. (1981). An observer for linear feedback control laws of multivariable systems with multiple delays in controls and outputs, *Systems and Control Letters*, vol. 1, pp. 54–59.

[134] Witrant, E., Fridman, E., Sename, O., and Dugard, L. (2016). *Recent Results on Time-Delay Systems: Analysis and Control.* Advances in Delays and Dynamics, Springer.

[135] Xiao, L., and Gao, F. (2011). Practical string stability of platoon of adaptive cruise control vehicles, *IEEE Transactions on Intelligent Transportation Systems*, vol. 12, no. 4, pp. 1184–1194.

[136] Zhong, Q.-C. (2006). *Robust Control of Time-Delay Systems*, Springer.

[137] Zhou, B. (2014). Input delay compensation of linear systems with both state and input delays by nested prediction, *Automatica*, vol. 50, pp. 1434–1443.

[138] Zhou, B. (2014). *Truncated Predictor Feedback for Time-Delay Systems*, Springer-Verlag.

[139] Zhou, B., Duan, G.-R., and Lin, Z. (2010). Global stabilization of the double integrator system with saturation and delay in the input, *IEEE Transactions on Circuits and Systems I: Regular Papers*, vol. 57, pp. 1371–1383.

[140] Zhou, B., and Egorov, A. V., (2016). Razumikhin and Krasovskii stability theorems for time-varying time-delay systems, *Automatica*, vol. 71, pp. 281–291.

[141] Zhou, B., Gao, H., Lin, Z., and Duan, G.-R. (2012). Stabilization of linear systems with distributed input delay and input saturation, *Automatica*, vol. 48, pp. 712–724.

[142] Zhou, B., Li, Z.-Y., and Lin, Z. (2013). Observer based output feedback control of linear systems with input and output delays, *Automatica*, vol. 49, pp. 2039–2052.

[143] Zhou, B., Li, Z.-Y., Zheng, W.-X., and Duan, G.-R. (2012). Stabilization of some linear systems with both state and input delays, *Systems and Control Letters*, vol. 61, pp. 989–998.

[144] Zhou, B., and Lin, Z. (2014). Consensus of high-order multi-agent systems with large input and communication delays, *Automatica*, vol. 50, pp. 452–464.

[145] Zhou, B., Lin, Z., and Duan, G.-R. (2009). Properties of the parametric Lyapunov equation-based low-gain design with applications in stabilization of time-delay systems, *IEEE Transactions on Automatic Control*, vol. 54, pp. 1698–1704.

[146] Zhou, B., Lin, Z., and Duan, G.-R. (2010). Global and semi-global sta-
       bilization of linear systems with multiple delays and saturations in the
       input, *SIAM Journal of Control and Optimization*, vol. 48, no. 8, pp. 5294–
       5332.

[147] Zhou, B., Lin, Z., and Duan, G.-R. (2010). Stabilization of linear sys-
       tems with input delay and saturation-a parametric Lyapunov equation
       approach, *International Journal of Robust and Nonlinear Control*, vol. 20,
       pp. 1502–1519.

[148] Zhou, B., Lin, Z., and Duan, G.-R. (2012). Truncated predictor feed-
       back for linear systems with long time-varying input delays, *Automatica*,
       vol. 48, pp. 2387–2399.

[149] Zhou, J., and Krstic, M. (2017). Adaptive predictor control for stabilizing
       pressure in a managed pressure drilling system under time-delay, *Journal
       of Process Control*, vol. 40, pp. 106–118.

[150] Zhou, J., Wang, W., and Wen, C. (2008). Adaptive backstepping control
       of uncertain systems with unknown input time delay, *Proceedings of the
       17th IFAC World Congress*, pp. 13361–13366, Seoul, Korea.

[151] Zhou, J., Wen, C., and Wang, W. (2009). Adaptive backstepping control
       of uncertain systems with unknown input time-delay. *Automatica*, vol. 45,
       no. 6, pp. 1415–1422.

[152] Zhu, Y., Krstic, M., and Su, H. (2015). Lyapunov-based backstepping
       control of a class of linear systems without overparametrization, tuning
       functions or nonlinear damping, *Proceedings of 2015 European Control
       Conference*, pp. 3609–3616, Linz, Austria, July 15–17.

[153] Zhu, Y., Krstic, M., and Su, H. (2015). Prediction-based boundary control
       of linear delayed systems without restriction on relative degree, *Proceed-
       ings of 2015 IEEE Conference on Decision and Control*, pp. 7728–7735,
       Osaka, Japan, December 15–18.

[154] Zhu, Y., Krstic, M., and Su, H. (2017). Adaptive output feedback control
       for uncertain linear time-delay systems, *IEEE Transactions on Automatic
       Control*, vol. 62, pp. 545–560.

[155] Zhu, Y., Krstic, M., and Su, H. (2018). Adaptive global stabilization of
       uncertain multi-input linear time-delay systems by PDE full-state feed-
       back, *Automatica*, vol. 96, pp. 270–279.

[156] Zhu, Y., Krstic, M., and Su, H. (2018). PDE boundary control of multi-
       input LTI systems with distinct and uncertain input delays, *IEEE Trans-
       actions on Automatic Control*, DOI: 10.1109/TAC.2018.2810038.

[157] Zhu, Y., Krstic, M., and Su, H. (2019). PDE output feedback control of LTI systems with uncertain multi-input delays, plant parameters and ODE state, *Systems and Control Letters*, vol. 123, pp. 1–7.

[158] Zhu, Y., Krstic, M., Su, H., and Xu, C. (2016). Linear backstepping output feedback control for uncertain linear systems, *International Journal of Adaptive Control and Signal Processing*, vol. 30, pp. 1080–1098.

[159] Zhu, Y., Su, H., and Krstic, M. (2015). Adaptive backstepping control of uncertain linear systems under unknown actuator delay, *Automatica*, vol. 54, pp. 256–265.

[160] Zhu, Y., Wen, C., Su, H., Xu, W., and Wang, L. (2013). Adaptive modular control for a class of nonlinear systems with unknown time-varying parameters, *Proceedings of 2013 American Control Conference*, pp. 2631–2636, Washington, DC, June 17–19.

[161] Zhu, Y., Wen, C., Su, H., and Liu, X. (2014). Modular-based adaptive control of uncertain nonlinear time-varying systems with piecewise jumping parameters, *International Journal of Adaptive Control and Signal Processing*, vol. 28, pp. 1266–1289.

Index

actuator delays, 1
adaptive control, 11, 36, 134, 237
Agmon's inequality, 110, 180, 294

backstepping, 22
backstepping-forwarding transformation, 221, 226
backstepping transformation, 13, 26, 31, 123, 126
Barbalat's lemma, 78, 101, 114, 305
boundary control, 12

Cauchy-Schwartz inequality, 71, 95, 293
certainty equivalence, 38, 134, 237
class-$\mathcal{K}$ function, 303
class-$\mathcal{KL}$ function, 303
comparison principle, 34, 224, 302
converse Lyapunov theorem, 305

delay compensation, 5, 7
delay kernel, 235, 267
delay mismatch, 44
delay systems, 1
discrete delays, 4, 19, 122
distributed delays, 5, 219

exponential stability, 24, 28, 33, 128, 132, 223, 229, 304

finite spectrum assignment, 3
finite-dimensional, 12
first-order hyperbolic PDEs, 6

Gronwall's lemma, 302

Hölder's inequality, 296
hyperbolic PDEs, xiii

infinite-dimensional, 12
input delays, 4
input-to-state stability, 308

Kreisselmeier-filters, 61, 86, 102

linear matrix inequality, 21
linear parametrization, 37, 136
linear time-invariant systems, 20, 122, 219
Lyapunov-Krasovskii functional, 24
Lyapunov stability, 304

multi-input, 135, 199

networked control systems, 6

observer canonical form, 59, 84
ODE-PDE cascade, 12
ordinary differential equation, 37
output feedback, 264, 282, 287

partial differential equation, 38
Poincaré inequality, 110, 180, 293
predictor feedback, 7, 21
projection operator, 71, 314

reduction approach, 5, 220, 225
reference trajectory, 49
relative degree, 60, 84
robust, 264, 283
robustness, 44

sampled-data, 6
sensor delays, 1
set-point regulation, 52
single-input, 35, 59, 84
Smith predictor, 3

stability, 303
state feedback, 243, 251, 268, 276

target system, 22, 23, 27, 31
trajectory tracking, 49
truncated L_p norm, 296

uniform asymptotic stability, 304
update laws, 70, 71, 95

Young's convolution theorem, 296
Young's inequality, 71, 95, 293

www.ingramcontent.com/pod-product-compliance
Ingram Content Group UK Ltd.
Pitfield, Milton Keynes, MK11 3LW, UK
UKHW031839161224
452263UK00003B/105